AF686370

Sustainability of Agriculture: The Impact of Climate Change on Crops

Sustainability of Agriculture: The Impact of Climate Change on Crops

Guest Editors

Fokion Papathanasiou
Elissavet Ninou
Ioannis Mylonas

MDPI

Basel • Beijing • Wuhan • Barcelona • Belgrade • Novi Sad • Cluj • Manchester

Guest Editors
Fokion Papathanasiou
Department of Agriculture
University of Western
Macedonia
Florina
Greece

Elissavet Ninou
Department of Agriculture
International Hellenic
University
Thessaloniki
Greece

Ioannis Mylonas
Institute of Plant Breeding and
Genetic Resources
Hellenic Agricultural
Organization–"Demeter"
Thessaloniki
Greece

Editorial Office
MDPI AG
Grosspeteranlage 5
4052 Basel, Switzerland

This is a reprint of the Special Issue, published open access by the journal *Sustainability* (ISSN 2071-1050), freely accessible at: https://www.mdpi.com/journal/sustainability/special_issues/96CO415B08.

For citation purposes, cite each article independently as indicated on the article page online and as indicated below:

Lastname, A.A.; Lastname, B.B. Article Title. *Journal Name* **Year**, *Volume Number*, Page Range.

ISBN 978-3-7258-7593-1 (Hbk)
ISBN 978-3-7258-7594-8 (PDF)
https://doi.org/10.3390/books978-3-7258-7594-8

© 2026 by the authors. Articles in this reprint are Open Access and distributed under the Creative Commons Attribution (CC BY) license. The reprint as a whole is distributed by MDPI under the terms and conditions of the Creative Commons Attribution-NonCommercial-NoDerivs (CC BY-NC-ND) license (https://creativecommons.org/licenses/by-nc-nd/4.0/).

Contents

About the Editors

Fokion Papathanasiou

Fokion Papathanasiou is a Professor of Botany and Plant Physiology at the School of Agricultural Sciences, University of Western Macedonia, Greece, where he also serves as Dean of the School. His research interests focus on the effect of environmental conditions on secondary plant metabolites, the impact of abiotic stresses on plant growth and product quality, the application of new technologies in agriculture, and the study of biodiversity in plant genetic resources. He has participated in many research projects, including HORIZON, ERASMUS+, and LIFE programmes, as well as 14 European educational programmes and one INTERREG project. He has published over 152 research papers in international journals and conference proceedings and serves as Invited Editor and Reviewer for numerous international scientific journals.

Elissavet Ninou

Elissavet Ninou is an Assistant Professor of Agronomy at the Department of Agriculture, International Hellenic University, Sindos, Thessaloniki, Greece. Her research interests include the evaluation and characterisation of plant genetic resources, genotype-by-environment interaction, digital phenotyping, and agricultural biodiversity, with particular focus on field crops under abiotic stress conditions. She has participated in approximately 20 competitive research programmes, including FP7, HORIZON, and ERASMUS+ projects. She serves as principal investigator in two EIP-AGRI projects related to the conservation and valorization of Greek local varieties and coordinates a Work Package in the HORIZON project Legumes Generation. She has published over 30 peer-reviewed papers and 75 contributions in national and international conference proceedings.

Ioannis Mylonas

Ioannis Mylonas is a Principal Researcher in Cereal Genetics and Breeding at the Institute of Plant Breeding and Genetic Resources, Hellenic Agricultural Organisation ELGO-DIMITRA, Thermi, Thessaloniki, Greece, and has served as Associate Manager of the Greek Gene Bank since 2018. His research interests encompass the selection of superior genotypes, evaluation of varieties for resistance to biotic and abiotic stresses, genotype discrimination using molecular techniques, adaptation of varieties to climate change and low-input systems, and the study of genotype-by-environment interaction in multi-location field trials. He has participated in 30 European and national research programmes. He has published 40 papers in peer-reviewed international journals and 75 contributions in national and international conference proceedings.

Preface

Agriculture stands at the crossroads of two defining challenges of our era: the urgent need to feed a growing global population and the accelerating pressure of climate change on the natural systems that make food production possible. Rising temperatures, erratic rainfall, prolonged droughts, and shifting growing seasons are no longer distant projections — they are present realities that farmers, agronomists, and policymakers confront every season. It is within this context that the Special Issue *Sustainability of Agriculture: The Impact of Climate Change on Crops*, published in the MDPI journal *Sustainability*, was conceived and developed.

This Reprint gathers twelve original research articles that address, from multiple scientific angles, the complex relationships between climatic variables and crop performance. The studies range from crop yield modelling under future climate scenarios and agronomic adaptation strategies to circular resource management, policy frameworks, and socioeconomic dimensions of agricultural resilience. Geographically, the contributions span Europe, the Middle East, Africa, and South America, reflecting the global urgency and universality of the subject. Crops examined include barley, maize, potato, strawberry, subterranean clover, date palm, and tea, among others, while analytical approaches embrace simulation modelling, field experimentation, remote sensing, and social surveys.

The aim of this Reprint is to provide the scientific community, agricultural practitioners, students, and policymakers with a rigorous and multidisciplinary reference that consolidates current knowledge on climate-smart agriculture. It is our hope that the findings and methodologies presented here will inform the design of more resilient cropping systems and contribute to evidence-based agricultural policy at national and international levels.

We sincerely thank all contributing authors for sharing their research, and the reviewers for their invaluable expertise and dedication.

Fokion Papathanasiou, Elissavet Ninou, and Ioannis Mylonas
Guest Editors

sustainability

MDPI

Article

Circular Mining Wastes Management for Sustainable Production of *Camellia sinensis* (L.) O. Kuntze

Lyudmila S. Malyukova [1], Nikita V. Martyushev [2,*], Valeriya Valerievna Tynchenko [3,4], Viktor V. Kondratiev [5], Vladimir V. Bukhtoyarov [6,7], Vladimir Yu. Konyukhov [8], Kirill Aleksandrovich Bashmur [6], Tatyana Aleksandrovna Panfilova [6] and Vladimir Brigida [1,9]

1 Federal Research Centre the Subtropical Scientific Centre of the Russian Academy of Sciences, 354002 Sochi, Russia; malukovals@mail.ru (L.S.M.); 1z011@inbox.ru (V.B.)
2 Department of Advanced Technologies, Tomsk Polytechnic University, 634050 Tomsk, Russia
3 Department of Computer Science, Institute of Space and Information Technologies, Siberian Federal University, 660041 Krasnoyarsk, Russia; vvtynchenko@sfu-kras.ru
4 Department of Computer Science and Computer Engineering, Institute of Computer Science and Telecommunications, Reshetnev Siberian State University of Science and Technology, 660037 Krasnoyarsk, Russia
5 Laboratory of Geochemistry of Ore Formation and Geochemical Methods of Prospecting, A. P. Vinogradov Institute of Geochemistry of the Siberian Branch of the Russian Academy of Sciences, 664033 Irkutsk, Russia; v.kondratiev@igc.irk.ru
6 Department of Technological Machines and Equipment of Oil and Gas Complex, School of Petroleum and Natural Gas Engineering, Siberian Federal University, 660041 Krasnoyarsk, Russia; vladber@list.ru (V.V.B.); bashmur@bk.ru (K.A.B.); tpanfilova@sfu-kras.ru (T.A.P.)
7 Artificial Intelligence Technology Scientific and Education Center, Bauman Moscow State Technical University, 105005 Moscow, Russia
8 Department of Automation and Control, Irkutsk National Research Technical University, 664074 Irkutsk, Russia; konyukhov_vyu@mail.ru
9 Department of Biomedical, Veterinary and Ecological Directions, RUDN University, 117198 Moscow, Russia
* Correspondence: martjushev@tpu.ru

Abstract: Mining operations have a significant negative impact on the surrounding ecosystems. The operation of mines and quarries creates a large amount of waste that accumulate and are practically unrecyclable in the environment. The involvement of these wastes in economic activity is an extremely urgent task. This can make the economy more sustainable and reduce its influence on ecosystems. This work presents the attempts of using quarry wastes as a fertilizer applied for growing tea crops. The novelty of this research involves revealing the quarry wastes as a fertilizer when growing *Camellia sinensis* (L.) O. Kuntze and assessing changes in the productivity of this plant when applying these calcium wastes. The waste of a quarry intended for extracting crushed stone was studied in this article. The composition of the waste was analyzed. Fertilizers used for manuring the soil were prepared based on the waste. Two experimental sites were selected. One of them was a control, where *Camellia sinensis* (L.) O. Kuntze was grown without using fertilizers. Fertilizers obtained from the waste were applied on the second site. The experimental work proceeded for 10 years. When discussing the results, special attention was paid to climatic conditions. This was caused by the need to show that it was the use of the fertilizer that influenced the change in the yield, not the climatic conditions. As a result of using calcium fertilizers based on the waste, the productivity of *Camellia sinensis* (L.) O. Kuntze was increased. The application of the fertilizers based on the quarry wastes was shown to provide an increase in the yield. The possibility of using calcium fertilizers to overcome unfavorable agroclimatic conditions during the tea cultivation was also demonstrated. To assess the climatic impact of applying new fertilizers, three-dimensional modeling in the "gnuplot v.5.4" software was used. As a result, an increase in the average annual precipitation, from 1000 to 1980 mm/year, in the range of the average annual air temperature, from 14 to 16 °C, was found to lead to an increase (when using a new fertilizer) in the yield of *Camellia sinensis* (L.) O. Kuntze up to 4.8 times (from 20 to 95 centner/ha). The results have shown that applying fertilizers based on the quarry wastes is also possible in unfavorable climatic conditions.

Keywords: circular economy; sustainable agriculture; mining waste; *Camellia sinensis* (L.) O. Kuntze

1. Introduction

According to the latest data, CO_2 emissions from rivers in the permafrost zone will increase three times over the next 10 years due to climate change [1,2], which can significantly affect life expectancy in certain regions of the world [3]. Differences in the regional policy when constructing the carbon trading market do not always provide a tangible effect of industrial sector decarbonization [4]. Along with the increase in the average annual air temperature, the negative consequences of global climate changes are also manifested in an increased risk of water scarcity, which may exceed USD 2.7 trillion [5]. In subtropical zone conditions of the Caucasus in the last 5 years, a change in the trend of the average annual air temperature towards an increase was traced [6], which deteriorates the growing conditions for crops. In addition, in the spring period (from March to May), climatic factors have been proven to have a dominating effect, causing soil organic carbon growth in the layers of 0–10 cm and 10–30 cm [7]. The temperature of the late rainy season can significantly limit the radial growth of S. superba plantings in the southern rims of the subtropical forests of Southeastern China [8].

The influence of industrial wastes on agriculture. The mining that exploits "brown mining technologies" aggravates the anthropogenic component effect (for example, the influence of landfills of household wastes [9]) on the existing biocenoses [10–12]. The mining of rich ores at shallow depths is replaced by complicated mining conditions, and the transportation of interspersed ores to the extraction is accompanied by an increased consumption of high-tech goods. For example, in the Mufulira River, the presence of the slurry pulp obtained from the Konkola Copper Mines Plc mine tailings led to an increased SO_4^{2}—concentration (1396 mg/L) and dissolved Al (2115 μg/L), Co (909 μg/L), Cu (7405 μg/L), Ni (51.5 μg/L), Pb (161 μg/L), and Zn (346 μg/L), including a decreased pH (2.04). This fact poses a serious threat to agriculture [11]. In the southwest of Spain, a rather long-term extraction of sulfide deposits led to the migration of dissolved and suspended metals and arsenic compounds into surface waters [13]. The authors identified the arsenic concentrations of 80 μg/L As^{3+} and 5 mg/L As^{5+}. At the same time, the concentration of arsenic compounds increases, and the As^{3+}/As^{5+} ratio decreases downstream of the Agua Agria brook (in the Tarsis Mine area). Many authors point to the presence of uneven contamination of the soil and surface waters. For example, in Zambia, the soils in the zone of influence of the Koperbelt mining area (4700 km^2) reflect a significant variability in the geochemical properties of the soils, as well as different contamination degrees (a high proportion of copper in soils equal to 85.5 mg/kg is found even in seemingly uncontaminated areas) [14]. The levels of heavy metals in various fractions of soil particles (>250, 63–250, and <63 microns) of the landscapes in suburban areas of Saudi Arabia are determined by mining (Cd, Cu, Zn, and Pb) [15].

Open-pit mining causes serious damage to land resources and completely transforms the land relief [16]. Land amelioration is mainly aimed at restoring the ecosystem and the value of the land exposed to mining, which is still a difficult task [17]. The results of the high-performance sequencing (using the PIECRUST method) of microbial communities of aluminum quarry soils have shown the efficiency of using yellow rose, lespedeza, and sweet wormwood herb to improve the soil quality [18]. When compared to this fact, spontaneous succession at an abandoned dolomite limestone quarry in Missouri can lead to the emergence of plant communities, including 74 native species (and 21 non-native species), which is close to the species diversity of natural communities [19].

The hydroseeding of *Leguminosae*, *Poaceae*, and *Cyperaceae* is effectively used on the landscapes to reclaim sites and reduce soil erosion in the post-mining period [20]. The increase in the related environmental problems [21–23] actualizes not only the need to develop a green mining strategy [24], green technologies for mineral enrichment [25], involving other sustainable technologies [26] but also a transition to a sustainable use of mineral resources based on circular economy models [27–29]. A circular economy implies the closure of resource flows without forming associated wastes [30–32]. This direction is the dominant tool in developing national strategies used for sustainable development all over the world [33]. The process of zero fluid discharge used for extracting struvite and regenerating the drinking water obtained from wastewater is quite a promising direction when introducing a closed-cycle economy [34]. When forming sustainable agricultural processes, a circular economy is most often used in the context of the life cycle assessment (LCA) theory [35], while waste management issues remain insufficiently studied. Circular waste management is the most efficient method of soil reclamation. In this regard, a rather promising direction in solving the problem of waste is developing the directions for their use in agriculture or closure in the in-house technological cycle. For example, using composted sewage sludge of an organic fertilizer in sugar cane fertilizers increases soil fertility and reduces the need for mineral fertilizers [36]. The wastes obtained from a highly acidic dump in the form of soil substitutes obtain a nutrient medium with the following parameters: (N (0.37–0.51%), P (0.23–0.47%), K (1.78–3.17%), Ca (4.93–8.39%), and Mg (1.16–1.71%)). This allows their use for agricultural purposes [37]. In some cases, for example, using waste drilling mud ($\geq$6%) has a negative effect on the germination of Maize seeds (*Zea mays* L.), which does not yet allow it to be used for sustainable agriculture purposes [38,39]. In another study, the influence of using organic wastes on the physical properties of the soil in arid regions affected by the negative consequences of coal mining was established [40].

One example of involving the calcium wastes obtained from quarries in the production process is adding them to cement [41,42] or the thermochemical modification of granular copper slag [43] for using it as a binder substitute. At the same time, in the regions of the agricultural or recreational-tourist specialization, their use as fertilizers is more productive. Investing in innovations of this kind [44,45] will allow the achievement of the carbon neutrality of individual industries [46], in addition to minimizing the anthropogenic impact.

Meteorological conditions for growing the plants. Tea cultivation in the subtropics of different regions of the world, as a rule, is associated with the risks of yield losses in dry years. In the end, the yield losses in adverse meteorological conditions can be 50–60% or more. According to a number of researchers [47,48], hydrothermal stress leads to the loss of more than 50% of the tea harvest during rain-fed cultivation. For example, in the case of chickpeas or soybeans, drought can pose a serious threat [49–52], causing the need to ensure sustainable agribusiness [53–55]. In this regard, it is interesting to develop the possibility of cleaning ponds for sedimentation tanks followed by their subsequent use in agriculture [56] or to use the synergy of agricultural and energy production [57]. The progressive aridization of the climate actualizes the task of increasing the plants' resistances to drought. In this regard, much attention in the research devoted to the tea culture is paid to studying the mechanisms of its drought resistance, searching for the most informative markers of drought resistance, as well as creating new resistant varieties and the influence of exogenous inducers, including mineral fertilizers. Therefore, the influence of applying the calcined fertilizers on the tea yield is an urgent specific research issue in circular waste management when ensuring sustainable agriculture. This study hypothesis consisted in the fact that the introduction of calcium-containing quarry wastes would increase the tea yield despite the insufficient rainfall.

In this connection, the aim of this study is to assess the productivity of growing *Camellia sinensis* (L.) O. Kuntze in conditions of the cyclic management of calcium wastes.

2. Materials and Methods

2.1. Properties of Calcium-Containing Wastes

The object of research and the source of the calcifying fertilizer is the "Kamensky" quarry, which is located at 43°32′04.5″ N 39°58′59.4″ E (Figure 1). To characterize the chemical composition of the sample material, the gross content of the chemical elements was determined with quantitative methods. They are mass spectrometric and atomic emission analyzers using inductively coupled plasma (ICP-MS and ICP-AES) according to the certified NSAM methodology. The 499-NPP/MS method, "Determination of the elemental composition of rocks, soils and bottom sediments using the atomic emission method, having inductively coupled plasma, and the mass spectral method involving inductively coupled plasma", was used to conduct these studies [58]. The equipment included the inductively coupled plasma mass spectrometer Elan-6100 ("Perkin Elmer", Waltham, MA, USA) and the atomic emission inductively coupled plasma spectrometer Optima-4300 DV ("PerkinElmer", Waltham, MA, USA).

Figure 1. Location of the waste collection (43.534586, 39.983163 Copyright © Google Earth).

Limestone ($CaCO_3$) and dolomite ($CaMg(CO_3)_2$) crushed stone are mined at the facility, whose sludge waters flow along the streams into the Mzymta river (Figure 2). The sampling was carried out as follows. During the license revocation period, the facility was examined to determine the oversized tailing at the industrial site where the rock was crushed (Figure 2a). To ensure the sample representativeness, the material was selected once (in three stages). Approximately 2000 kg of the small rocks was filled into bags (45–50 kg each). As a result, 142 bags were filled for the sampling and then transported to the warehouse of the scientific center (FRC SSC RAS). Later, before each application, the packed material was additionally prepared (ground in a laboratory mill) and sieved up to a grinding thickness of 0.25 mm. The material is characterized by a slightly alkaline reaction in water (pH_{H_2O} = 8.2).

Figure 2. Functioning "Kamensky" quarry: (**a**)—view of the quarry; (**b**)—its sludge water drains flowing into the Mzymta river (**b**).

The obtained results of the elemental composition of the bulk wastes are presented in Table 1. The data presented in Table 1 are the average values for 10 measurements. The maximum deviation for the content of the elements over 400 μg/g was 6.3%, for the content of the elements within 1–400 μg/g it was 8.7%%, for the content of the elements within 0.1–1 μg/g it was 0.05 μg/g, and for the values less than 0.1 μg/g it was 0.02 μg/g. This substance is characterized by a slightly alkaline reaction of an aqueous suspension (pH is 8.20). It contains calcium in an exchange form (10.9 mmol (eq)/100 g) and a smaller amount of magnesium of 2.2 mmol (eq/100 g). The calcium-containing natural material is the clay-lime waste formed when crushing the limestones into small fractions of the crushed stone.

Table 1. Chemical composition of the Ca-containing mining waste.

Elements	Concentration μg/g	Elements	Concentration μg/g	Elements	Concentration μg/g	Elements	Concentration μg/g
Li	5.76	Sr	709.80	Tb	0.15	Th	0.84
Be	0.15	Nb	1.23	Dy	0.58	U	0.19
Sc	1.43	Mo	0.07	Ho	0.28	Na	783
V	3.71	Sn	1.84	Er	0.56	Mg	3848
Cr	2.22	Sb	0.02	Tm	0.04	Al	3870
Co	1.21	Ba	19.50	Yb	0.41	K	2086
Ni	12.60	Pr	0.62	Lu	0.12	Ti	114.80
Cu	1.920	Nd	2.52	Hf	0.27	Mn	421
Zn	11.13	Sm	0.89	Ta	0.72	Fe	1485
Ga	0.54	Eu	0.17	Pb	4.40	Cd	0.36
Rb	7.50	Gd	0.89	Bi	0.04	I	0.72
As	0.59						

After checking the fact that individual chemical elements did not exceed the specified critical concentrations, a standard biotesting method was applied to them. The studied quarry wastes turned out to be chemically neutral, as a result of biotesting its extracts [59] using the single-celled algae *Chlorella Vulgaris*. The test results showed no significant deviations from the control in either acute or chronic experiments. This proves that they do not have reliable toxic properties and can be used in agriculture.

2.2. Experiment Parameters and Possible Limitations

The basis for choosing the plant type was the fact that tea is an industrially significant crop for this region, which has been grown in the Black Sea coastal area of the Krasnodar

Territory for more than 100 years. Due to the favorable conditions of tea cultivation in the subtropics, as well as the presence of experimental plantations, the existing full-aged tea plantation *Camellia sinensis* (L.) O. Kuntze (Colchis variety) was chosen to be used for further experiments (Figure 3).

a b

Figure 3. Influence of calcium wastes on the yield of tea leaves in different meteorological conditions (ZAO "Dagomyschai" Copyright © Google Earth): (**a**)—a variant of using calcium wastes obtained from a quarry; (**b**)—a basic variant of cultivation (N250P70K90).

The studies were conducted for nine years, from 2011 to 2020. The object was 2 variants of the field experiment: (1) control (background) was N240P70K90 (in kg d.v./ha) as a traditional fertilizer system; (2) background + natural calcium-containing material, whose description is provided in [60]. Tea grew on chernozem (black soil) soil with a humus content of 6.5%. The granulometric composition of the soil was generally uniform along the profile. There was no noticeable movement of silt particles. The predominance of silty–silty fractions in the soil caused a high absorption capacity and moisture capacity.

The material was sieved to obtain a fractional composition of less than 0.25 mm in fineness. The area of the experimental plots was 100 m^2. The tea plantation was founded in 1983 with a high-yielding variety of Colchis in subtropical conditions (Dagomys, Russia); the experiment scheme was imposed in 2003. The tea plantation plots are located at 43°41′04.8″ N 39°38′14.0″ E (43.684655, 39.637222). Annual fertilization in the amount of 250 kg/ha of the wastes in the form of a dry white-grayish powder was conducted in the spring period. The fertilizer was evenly applied to the root zone (manually). The crop yield was recorded (manual harvesting) during the periods approaching the tea leaf harvesting, according to agricultural regulations [61]. When designing the experiment plan, the following problems were taken into account, which could affect the obtained results. In this case, the factors of "uncontrolled environmental parameters" and insignificant climatic anomalies were not observed, and the intra-annual variability was conditionally stable for the entire observation period. The variability factor of the calcium-containing waste products, due to the uneven distribution of the conditions of the processed rock formations on the ledges, could significantly affect the results. To neutralize this factor, the material was taken once from one source, while the composition for the chemical analysis was taken from the random 10 bags. The "experiment duration" factor was undoubtedly of significant interest in the context of assessing the duration of a positive effect (which was one of the main research issues this time). In this regard, the observation period was approaching ten years.

During the study period, the soil was analyzed in a standard way. The exchange acidity of the soils was determined with the potentiometric method using the ionometer "pH-121 EKOSTAB" (OOO "EKOINSTRUMENT", Moscow, Russia). Easily hydrolyzable nitrogen was determined with the Tyurin–Kononova method (colorimetric method involving the Nessler reagent), including mobile phosphorus (with Oniani having a colorimetric ending according to Deniges) using the "USF-01" spectrophotometer (FSUE All-Russian Research Institute of Optical and Physical Measurements, Moscow, Russia). Mobile potassium was determined with Oniani using an atomic-absorption spectrometer "KVANT-AFA" (OOO "KORTEK", Moscow, Russia). The content of the exchange forms of Ca^{2+} and Mg^{2+} was determined with the trigonometric method.

2.3. Mathematical Processing of the Results

When choosing the methods for processing arrays of three-dimensional data, we proceeded from the experience of solving such types of problems in the relevant fields of knowledge [62–65]. The closest analog is the spatial–temporal problems of geoecology: the assessment of the aerogas regime of the surface layer (earth breathing) [66–69], forecasting the long-term time series of data [70], geostatistical modeling using GIS technologies [71–73], applying fractal models [74], deep learning [75,76], or perceptrons [77,78]. In contrast to using splines [79,80] or the local polynomial regression (LOESS) method [81,82], the primary data were not smoothed in this work due to the small sample size. The primary data were processed using Microsoft Excel 2010 software; three-dimensional models (using R. Renka algorithm implemented as a Python script) were formed based on a standard sequence of built-in commands: "gnuplot v.5.4" software and "Scilab 6.1.1" software for searching the regression equations.

3. Results

The essence of the experiment was a long (more than 5–7 years) introduction of a constant volume of the technogenic fertilizer (whose 40% of the weight was Ca) to confirm the effectiveness of its application in situ. The main effect of its application in the subtropical zone was to restore soil fertility and to assist in overcoming unfavorable climatic conditions of growing *Camellia sinensis* (L.) O. Kuntze. For this purpose, up to 2000 kg of the primary fine-dispersed material was selected at the quarry, which after storage, was periodically sifted and, in a spring period, was manually introduced into the root zone of tea rows. The material consumption in each period was 250 kg per hectare of the tea plantation. The three-leave flush was harvested manually; the average yield of tea is provided in Table 2. The experimental field was divided into nine land plots from which the tea was harvested. The table provides the average values based on the measurement results. The maximum deviation from the average value was 3.2%. In our work, we carried out mathematical processing of the obtained results and for each of the years determined the values of Student's *t*-test—t_S. Based on the obtained values of the criterion for each of the years, we accepted the null hypothesis at a significance level of 0.001. The only exception was in 2012. This year the average temperature was minimal, and there was little rainfall. This year, the Student's *t*-test passed with a significance level of 0.05. By a null hypothesis, we mean that the samples are not equivalent, which in turn, indicates a statistically proven effect of fertilizers on productivity. In most cases, the significance level was less than 0.01. This indicates that there was no relationship between the results.

Table 2. Results of tea yield when Ca was applied to the soil as a fertilizer (ZAO "Dagomyschai").

Time Period, (Year)	Control, Kg/ha (Centner/ha)	Control, Standard Deviation	Yield When Ca Is Introduced in the Soil, Kg/ha (Centner/ha)	With Ca, Standard Deviation	t_S	Percentage Increase, %s	Average Annual Air Temperature, (°C/Year)	Total Precipitation, (mm/Year)
2011	5860 (58.6)	81	6373 (63.73)	93	12.5	9	14.6	1825.5
2012	2610 (26.1)	39	2563 (25.63)	41	2.5	3	14.1	1396.8
2013	9150 (91.5)	98	11,656 (116.56)	150	42.0	27	15.5	1927.8
2014	6510 (65.1)	74	8241 (82.41)	81	47.3	27	15.2	1433.9
2015	2900 (29.0)	35	4040 (40.40)	46	59.2	39	15.3	1325.6
2017	4470 (44.7)	51	5683 (56.83)	56	48.0	27	14.3	1439.3
2018	4560 (45.6)	58	6210 (62.10)	68	55.4	36	16.0	1671.1
2019	4930 (49.3)	67	6110 (61.10)	78	34.4	24	15.3	1653.2
2020	3520 (35.2)	38	3730 (37.30)	45	10.7	6	15.9	1029.8

In the period of study (2011–2020), there were years when the meteorological conditions were relatively favorable, as when the yield varied from 6500 to 11,700 kg/ha (65 to 117 centner/ha). And it was commensurate with the potentially achievable yield for this variety (9000–11,000 kg/ha or 90–110 centner/ha) and relatively unfavorable conditions, in which the yield decreased by an average of 50% and was 26–40 kg/ha (26–40 centner/ha).

The application of modern methods of three-dimensional interpolation has revealed the nature of the relationship between the culture cultivation parameters and the culture yield level (Figure 4).

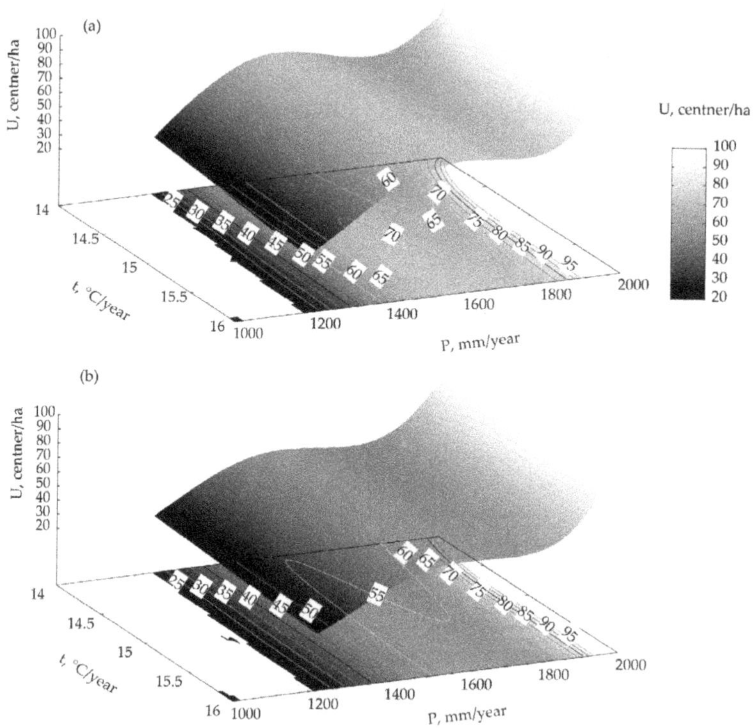

Figure 4. Influence of calcium wastes on the yield of tea leaves in various meteorological conditions: (**a**)—variant of using calcium wastes obtained from the quarry; (**b**)—basic variant of cultivation (N250P70K90).

The analysis of Figure 4 implies the fact that the surface of the distribution of crop yield parameters has pronounced nonlinearities, which determines the presence of three pronounced extremes (two maxima and one minimum). At the same, the overall appearance of the response surface is the same in both cases; in the case of waste use, the overall trend is clearly higher relative to the Z-axis.

As to the base case, a logarithmic-type dependence has been revealed (the solution is shown in Figure 5), whose formula has the following view ($R^2 = 0.84$):

$$U = -646558.11(\ln t)^2 + 79331.38(\ln t)^3 - 0.23P - 0.006P^2, \quad (1)$$

where *U*—tea yield in the control version of the experiment using NPK, centner/ha; *t*—average annual air temperature (from October of the previous year to September of the current year), C/year; *P*—annual precipitation (for the same period), mm/year.

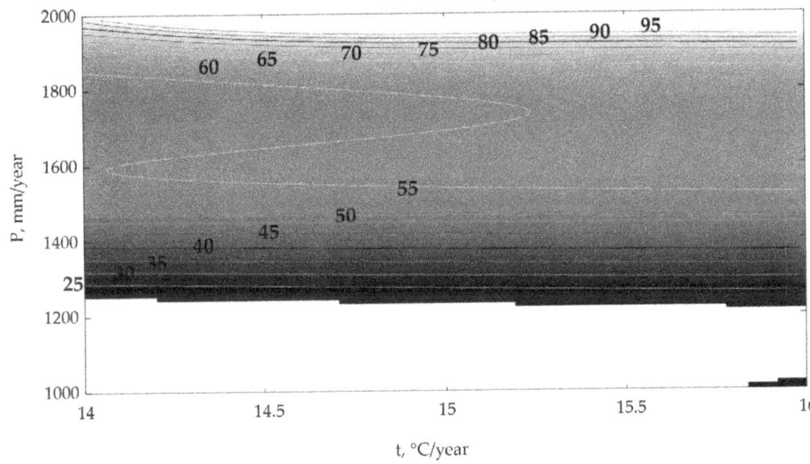

Figure 5. Projection of the logarithmic dependence of the tea yield (1) onto the t—P plane in the case of the basic variant of growing over a period of 2011–2020.

Figure 5 shows that a precipitation (P) increase from 1230 to 2000 mm/year at an average annual temperature of T = 14 °C increases the yield of green tea leaves from 20 to 90 centner/ha (4.5 times along a nonlinear trajectory). Simultaneously, a logarithmic increase from 20 centner/ha is traced up to forming a local maximum (54 centner/ha) within 1580 and 1600 mm/year, followed by a slight decline (at the level of 46–48 centner/ha) in the range of 1670–1800 mm/year, and subsequent growth to maximum values (90 centner/ha) when P = 2000 mm/year.

In turn, a temperature increase of up to 16 °C against the same background of precipitation leads to a similar pattern of yield changes from 20 centner/ha (starting from P = 1200 mm/year) to a maximum of $\geq$100 centner/ha (at P = 2000 mm/year), providing a five-time increase in the yield. A change in the air temperature from 14 to 16 °C results in a minimum rate of crop growth from 20 centner/ha (14 °C) up to 23 centner/ha (16 °C). In the case of a high amount of precipitation (2000 mm/year), in the same range of changes in the average annual air temperature, the tea yield gradually increases by 25%.

In general, a favorable combination of the average annual temperature and the sum of the precipitations intended for tea cultivation (control, traditional fertilizer system) based on the data analysis is the ranges of T = 14–16 °C and P = 1800–2000 mm/year, when the yield is approximately more than 80 c/ha. The green tea leaf increase in optimal weather conditions when using the traditional fertilizer system is 17.0%.

The introduction of calcium wastes obtained from the quarry into the soil can significantly improve the dynamics of its yield due to the growth in the resistance of the tea plants (*Camellia sinensis* (L.) O. Kuntze) to adverse meteorological conditions. As a result of approximating the experimental data, a logarithmic-type dependence has been established (its solution is shown in Figure 6), whose formula has the following view (R^2 = 0.83):

$$U = 144095.12(\ln t)^3 + 0.90P - 0.009P^2, \tag{2}$$

where U—tea yield when using a calcium-containing material, centner/ha; the other designations are the same as they are in Formula (1).

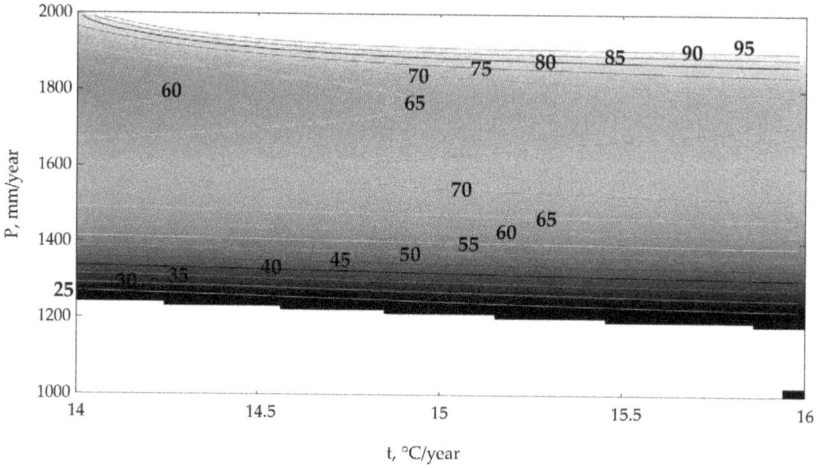

Figure 6. Projection of the logarithmic dependence of the tea yield (2) onto the t—P plane in the case of using the calcium wastes obtained from the quarry.

The nonlinearities typical of Figure 5 are also inherent in the studied Figure 6 (when using quarry wastes), but their parameters have specific features, namely the stable optimal productivity zone is located higher (along the Z-axis) than that in the base case.

Figure 6 shows that during the same research period, an increase in the amount of precipitation from 1200 to 2000 mm/year at an average annual temperature above 14 °C ensures a yield increase in the tea culture from 20 to more than 100 centner/ha. At the same time, the logarithmic increase from 20 centner/ha is traced to the formation of a local maximum (68 centner/ha) when P = 1600 mm/year. This is then replaced by a slight decline (at the level of 53 centner/ha) in the range of 1800–1823 mm/year and a subsequent increase up to the maximum values (78 centner/ha) when P = 2000 mm/year.

In turn, a temperature increase of up to 16 °C against the same background of precipitations leads to a similar pattern of yield changes from 20 centner/ha (starting from P = 1180 mm/year) to a maximum of $\geq$100 centner/ha (when P $\geq$ 1900 mm/year), providing a yield increase of more than five times. An air temperature change from 14 to 16 °C results in a minimum rate of crop growth ranging from 20 centner/ha (14 °C) to 28 centner/ha (16 °C). In general, a combination of the average annual temperature and precipitations that are favorable for the tea cultivation (control, traditional fertilizer system) proceeding from the data analysis is the ranges of T = 14–16 °C and P = 1800–2000 mm/year, when the yield is about more than 95 centner/ha. The crop increase in the green tea leaves in optimal weather conditions and using the calcium wastes obtained from the quarry increase the maximum productivity area by 50% (when T = 16 °C and P = 2000 mm/year) compared to the base case.

The combined profiles of the two variants are shown in Figure 7 to visually assess the additional effects that the wastes have on tea cultivation efficiency.

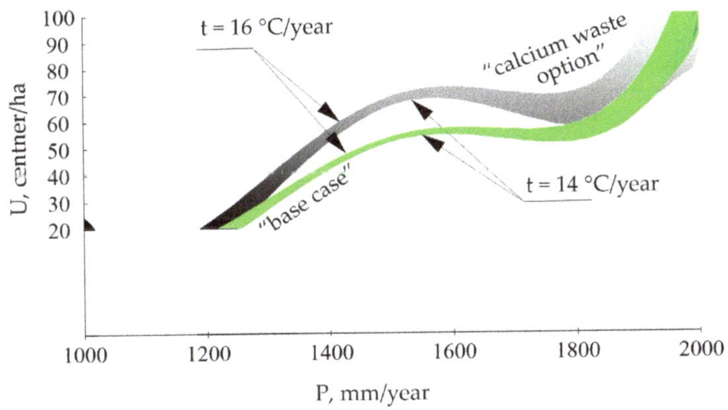

Figure 7. Profile surfaces of the productivity of growing *Camellia sinensis* (L.) O. Kuntze when adding the calcium wastes obtained from the "Kamensky" quarry over the period of 2011–2020.

The analysis of the productivity dynamics of the tea cultivation shown in Figure 7 implies the fact that when using a differentiated calculation of the areas under the average profiles of the two surfaces, the difference is about 22%, which determines a direct effect gained from using the wastes in these microclimatic conditions. The initial agrochemical characteristic of pH_{KCl} was 3.75 ± 0.14. On average, over the period, the variation range for the base case was 2.95–3.3, and in the case of the experimental site, it was in the range of 3.2–3.4. The value of mobile phosphorus (P_2O_5) was 294 ± 37; in the case of the base case, it equaled 780–1100, and it was 560–890 mg/kg when applying the fertilizers. The value of exchangeable potassium (K_2O) was 273 ± 25 and for the base case was 300–450, and in the case of the fertilization, it was 400–660 mg/kg. The value of easily hydrolyzable nitrogen (N_{lg}) was 48 ± 6; in the case of the base case, it was equal to 55–270, and in the case of the fertilization, it was 110–450 mg/kg. The value of the fractions of exchangeable calcium and magnesium ($Ca^{2+} + Mg^{2+}$) was 8.40 ± 1.40; in the case of the base case, it was 2.9–4.5, and when applying the fertilizers, it amounted to 3.9–5.3 mmol (eq)/100 g.

4. Discussion

The main factor studied in this work is undoubtedly the drilled fertilizers based on the waste obtained from the quarries where crushed stone was extracted. Conducting a study on the influence of the fertilizers on productivity requires the consideration of climatic conditions. The climatic conditions are the most important factor influencing both the productivity of plants and the work of the fertilizer itself. The absence of precipitation entails the absence of the fertilizers' influence since it will not be dissolved in water, being in the soil. In this regard, considerable attention is to be paid to climatic conditions. Firstly, this demonstrates the fact that it was the fertilizer that influenced the productivity increase of the tea culture under study. Secondly, this fact clarified the difference in the productivity of the tea culture when applying the fertilizers in different years.

The influence of meteorological conditions on the yield of tea plantations was analyzed throughout the entire period of tea cultivation in the subtropics [83–85]. P.M. Bushin [84] established the temperature factor influence on the 1 May harvesting of tea leaves. The subsequent harvestings (6–8th) depended on the precipitation amount. L.S. Malyukova had previously found that the unfavorable thermal regime of the spring period (in particular, frosts in March–April) leads to a significant loss in the first and, accordingly, the gross yield of tea leaves [86]. Insufficient precipitation (less than 700 mm over the period of April–October) during the growing season additionally inhibits the shoot formation process, reducing shoot activity. An average daily temperature increase in the leaf-harvesting period up to 23–25 °C—conditioned, as a rule, by a significant increase in daytime temperatures up to the values that reduce the photosynthesis process intensity—led to a drop in the tea plantation yield. At the same time, the precipitation of 390 mm over the leaf-harvesting period was of critical value for the shoot-forming activity of the tea plant. The regularities obtained in this study allowed for the detailing of the optimal ranges of the average annual temperature and annual precipitation specified for tea cultivation: in the case of the traditional fertilizer system, T = 14.76–16 °C and P = 1575–1780 mm/year; in the case of using the calcium wastes obtained from the "Kamensky" quarry, T = 14–16 °C and P = 1500–1830 mm/year. The revealed increase in the tea yield against the background of applying the calcium-containing material in the case of different combinations of climatic factors was 22%. Similar to the study of conditions in Southwestern Ethiopia [87], fluctuations in yield in specific years seem to be associated with seasonal climate fluctuations (their influence on the resulting yield values is the topic of further research). In the case of the rather arid tea-growing conditions in the subtropical zone of the Caucasus (compared to the optimal range of 2500–3000 mm/year [88]), a slight change in precipitation leads to significant deviations in the tea yield.

The obtained results do not contradict the studies conducted for the Indian tea variety, where the presence of the seasonal variability of biochemical compounds in the samples of *Camellia sinensis* (L.) O. Kuntze has been convincingly proven [89,90]. In such climatic growing conditions, the increase in applying the fertilizer from 0 to 300 kg/ha can significantly improve the final product quality due to the improved ability of the tea leaf to inhibit antioxidants [91]. According to the literature data, the material concentration change in the *cytosol* represents the first stage in the cell's recognition of external influences and launches a *signal transduction* system for causing a feedback response [92–94]. The reason may be the influence of abiotic stresses on the growth of *metabolites* in tea [95]. An important group of sensors involved in the cascade of calcium ion signals in the cells of higher plants is *Ca-dependent protein kinases* [96,97]. Under the action of calcium, there are changes in the growth, photosynthesis, and water–air regime of the plants, the work of stomata, as well as the accumulation of stress proteins [98,99]. The influence of calcium-entailed changes was noted in the growth, photosynthesis, and water–air regime of the plants, stomata functioning, as well as the antioxidant system induction [100,101]. This contributed to an increase in the tea plant's resistance to the combined effects of stress factors by maintaining the enzymatic activity of the leaves at a higher level, *induction of secondary metabolites synthesis*—the adaptive restructuring of the pigmentary fund, which on the whole, provided a higher functional activity of the plants during subsequent *rehydration*.

When cultivating tea on brown acidic soils, the use of the calcium-containing material was determined by the need to verify the hypothesis of a possible replenishment of the calcium content (when leaching it in acidic conditions) and, accordingly, to improve the resistance of *Camellia sinensis* (L.) O. Kuntze to abiotic stresses. But at the same time, it became possible to identify the existence of the pH shift towards the acidic interval accompanied by a simultaneous increase in the content of exchangeable calcium and magnesium (see the results of Section 3). This conditions the hypothesis, in which the mechanism explaining these changes in improving the nutritional regime of the tea plants in conditions of insufficient water supply is the intensification of the respiratory and enzymatic activity of soils. Verifying this hypothesis is a prospect for further research.

5. Conclusions

In conditions of humid subtropics and high variability of microclimatic cultivation parameters, the calcium wastes obtained from quarries can be successfully applied by agricultural enterprises cyclically when growing tea plants. At the same time, the effect of using the wastes in these microclimatic conditions when growing *Camellia sinensis* (L.) O. Kuntze is about 22%. Finally, the conducted studies have established the fact that an increase in the average annual precipitation from 1000 to 1980 mm/year, in the range of the average annual air temperature from 14 to 16 °C, leads to the 4.8-time yield growth of *Camellia sinensis* (L.) O. Kuntze with the logarithmic dependence (from 20 to 95 centner/ha). The work made it possible to reveal for the first time that the crop yield of *Camellia sinensis*

(L.) O. Kuntze increases more than five times with the logarithmic dependence (from 20 to more than 100 centner/ha) when adding the wastes obtained from the "Kamensky" quarry. At the same time, the zone and absolute values of the region of productive tea cultivation increase, and the maximum productivity is achieved in the case of less precipitation.

The practical significance of applying mining wastes as a non-traditional type of fertilizer for forming sustainable agriculture may consist in the subsequent development of the technological regulations of their use, including transportation, storage, and introduction into the soil.

The obtained data have confirmed the positive role, previously revealed by a set of biophysiochemical indicators, of the calcium-containing natural material in adapting tea plants to drought and the efficiency of its use in tea plantations. The above-mentioned conclusions and recommendations are valid only for the range of conditions of the subtropical zone of the Caucasus (acidic soils, P ranging from 1000 to 2000 mm/year, and t varying between 14 and 16 °C). Further research should be focused on identifying within-year climate variability during the growing season and the productivity of the fertilizer application when cultivating tea.

Author Contributions: Conceptualization, L.S.M., N.V.M. and V.B.; methodology, V.V.T. and V.V.K.; software, K.A.B. and T.A.P.; validation, K.A.B. and T.A.P.; formal analysis, V.V.T. and V.V.K.; investigation, V.V.B. and V.Y.K.; resources, V.V.T. and V.V.K.; data curation, V.V.B. and V.Y.K.; writing—original draft preparation, L.S.M., N.V.M. and V.B.; writing—review and editing, L.S.M., N.V.M. and V.B.; visualization, K.A.B. and T.A.P.; funding acquisition, V.V.B. and V.Y.K. All authors have read and agreed to the published version of the manuscript.

Funding: This research received no external funding.

Institutional Review Board Statement: Not applicable.

Informed Consent Statement: Not applicable.

Data Availability Statement: The data of this work are given in the text of the article.

Conflicts of Interest: The authors declare no conflict of interest.

References

1. Vorobyev, S.N.; Pokrovsky, O.S.; Korets, M.; Shirokova, L.S. A snap-shot assessment of carbon emission and export in a pristine river draining permafrost peatlands (Taz River, Western Siberia). *Front. Environ. Sci.* **2022**, *10*, 987596. [CrossRef]
2. Lyubas, A.A.; Tomilova, A.A.; Kondakov, A.V.; Konopleva, E.S.; Vikhrev, I.V.; Gofarov, M.Y.; Eliseeva, T.A.; Aksenova, O.V.; Bovykina, G.V.; Kryuk, D.V.; et al. Phylogeography and Genetic Diversity of Duck Mussel Anodonta anatina (Bivalvia: Unionidae) in Eurasia. *Diversity* **2023**, *15*, 260. [CrossRef]
3. Polcyn, J.; Voumik, L.C.; Ridwan, M.; Ray, S.; Vovk, V. Evaluating the Influences of Health Expenditure, Energy Consumption, and Environmental Pollution on Life Expectancy in Asia. *Int. J. Environ. Res. Public Health* **2023**, *20*, 4000. [CrossRef] [PubMed]
4. Wu, R.; Tan, Z.; Lin, B. Does carbon emission trading scheme really improve the CO_2 emission efficiency? Evidence from China's iron and steel industry. *Energy* **2023**, *277*, 127743. [CrossRef]
5. Chen, X.; Zhao, B.; Shuai, C.; Qu, S.; Xu, M. Global spread of water scarcity risk through trade. *Resour. Conserv. Recycl.* **2022**, *187*, 106643. [CrossRef]
6. Yaitskaya, N.A.; Dzaganiia, L.M.; Brigida, V.S. Geoecological hazards in context of climate change of territories of Caucasus subtropical zone. *Geol. I Geofiz. Yuga Ross. Geol. Geophys. Russ. South* **2023**, *13*, 118–132. (In Russian) [CrossRef]
7. Cheng, X.; Zhou, T.; Liu, S.; Sun, X.; Zhou, Y.; Xu, L.; Xie, B.; Ying, J.; Shi, Y. Effects of Climate on Variation of Soil Organic Carbon and Alkali-Hydrolyzed Nitrogen in Subtropical Forests: A Case Study of Zhejiang Province, China. *Forests* **2023**, *14*, 914. [CrossRef]
8. Jiang, S.; Guo, X.; Zhao, P.; Liang, H. Radial Growth–Climate Relationship Varies with Spatial Distribution of *Schima superba* Stands in Southeast China's Subtropical Forests. *Forests* **2023**, *14*, 1291. [CrossRef]
9. Godio, A.; Chiampo, F. Geophysical Monitoring of Leachate Injection in Pretreated Waste Landfill. *Appl. Sci.* **2023**, *13*, 5661. [CrossRef]
10. Golik, V.I.; Alekseev, I.A. Metal leaching from low-grade raw materials in North Ossetia–Alania. *MIAB. Mining Inf. Anal. Bull.* **2023**, *5-1*, 135–144. (In Russian)
11. Křibek, B.; Nyambe, I.; Sracek, O.; Mihaljevič, M.; Knésl, I. Impact of Mining and Ore Processing on Soil, Drainage and Vegetation in the Zambian Copperbelt Mining Districts: A Review. *Minerals* **2023**, *13*, 384. [CrossRef]

12. Kumar, R.; Kumar, K.; Singh, R.; Sá, J.C.; Carvalho, S.; Santos, G. Modeling Environmentally Conscious Purchase Behavior: Examining the Role of Ethical Obligation and Green Self-Identity. *Sustainability* **2023**, *15*, 6426. [CrossRef]
13. Sarmiento, A.M.; Caraballo, M.A.; Sanchez-Rodas, D.; Nieto, J.M.; Parviainen, A. Dissolved and particulate metals and arsenic species mobility along a stream affected by Acid Mine Drainage in the Iberian Pyrite Belt (SW Spain). *Appl. Geochem.* **2012**, *27*, 1944–1952. [CrossRef]
14. Křibek, B.; Majer, V.; Knésl, I.; Nyambe, I.; Mihaljevič, M.; Ettler, V.; Sracek, O. Concentrations of arsenic, copper, cobalt, lead and zinc in cassava (*Manihot esculenta* Crantz) growing on uncontaminated and contaminated soils of the Zambian Copperbelt. *J. Afr. Earth Sci.* **2014**, *99*, 713–723. [CrossRef]
15. Al-Swadi, H.A.; Usman, A.R.A.; Al-Farraj, A.S.; Al-Wabel, M.I.; Ahmad, M.; Al-Faraj, A. Sources, toxicity potential, and human health risk assessment of heavy metals-laden soil and dust of urban and suburban areas as affected by industrial and mining activities. *Sci. Rep.* **2022**, *12*, 8972. [CrossRef]
16. Guan, Y.; Zhou, W.; Bai, Z.; Cao, Y.; Wang, J. Delimitation of supervision zones based on the soil property characteristics in a reclaimed opencast coal mine dump on the loess plateau, China. *Sci. Total Environ.* **2021**, *772*, 145006. [CrossRef]
17. Pratiwi; Narendra, B.H.; Siregar, C.A.; Turjaman, M.; Hidayat, A.; Rachmat, H.H.; Mulyanto, B.; Suwardi; Iskandar; Maharani, R.; et al. Managing and Reforesting Degraded Post-Mining Landscape in Indonesia: A Review. *Land* **2021**, *10*, 658. [CrossRef]
18. Zhao, D.; Hou, H.; Liu, H.; Wang, C.; Ding, Z.; Xiong, J. Microbial Community Structure and Predictive Functional Analysis in Reclaimed Soil with Different Vegetation Types: The Example of the Xiaoyi Mine Waste Dump in Shanxi. *Land* **2023**, *12*, 456. [CrossRef]
19. Stephan, K.; Hubbart, J.A. Plant Community, Soil and Microclimate Attributes after 70 Years of Natural Recovery of an Abandoned Limestone Quarry. *Land* **2023**, *12*, 117. [CrossRef]
20. Narendra, B.H.; Pratiwi, P. Adaptability of some legume trees on quartz tailings of a former tin mining area in Bangka Island, Indonesia. *J. Degrad. Min. Lands Manag.* **2016**, *4*, 671–674. [CrossRef]
21. Oggeri, C.; Vinai, R.; Fenoglio, T.M.; Godio, A. Large Scale Trials of Waste Mine Burden Backfilling in Pit Lakes: Impact on Sulphate Content and Suspended Solids in Water. *Sustainability* **2023**, *15*, 7387. [CrossRef]
22. Radzhabov, G.K.; Musaev, A.M.; Islamova, F.I.; Aliev, A.M. Analysis of accumulation of volatile organic compounds in plants of Hyssopus officinalis L., introduced in mountain conditions. *Sustain. Dev. Mt. Territ.* **2023**, *15*, 174–181. (In Russian) [CrossRef]
23. Korchak, P.A.; Karasev, M.A. Geomechanical prediction of the brittle fracture zones in rocks in the vicinity of the excavation junction of Ltd. "Apatit" mines. *Sustain. Dev. Mt. Territ.* **2023**, *15*, 67–80. (In Russian) [CrossRef]
24. Doulati Ardejani, F.; Maghsoudy, S.; Shahhosseini, M.; Jodeiri Shokri, B.; Doulati Ardejani, S.; Shafaei, F.; Amirkhani Shiraz, F.; Rajaee, A. Developing a Conceptual Framework of Green Mining Strategy in Coal Mines: Integrating So-cio-economic, Health, and Environmental Factors. *J. Min. Environ.* **2022**, *13*, 101–115. [CrossRef]
25. Chipakwe, V.; Karlkvist, T.; Rosenkranz, J.; Chelgani, S.C. Beneficial effects of a polysaccharide-based grinding aid on magnetite flotation: A green approach. *Sci. Rep.* **2022**, *12*, 6502. [CrossRef]
26. Chiampo, F.; Shanthakumar, S.; Ricky, R.; Pattukandan Ganapathy, G. Tannery: Environmental impacts and sustainable technologies. *Mater. Today Proc.* **2023**, *in press*. [CrossRef]
27. Marinina, O.; Kirsanova, N.; Nevskaya, M. Circular Economy Models in Industry: Developing a Conceptual Framework. *Energies* **2022**, *15*, 9376. [CrossRef]
28. Shutaleva, A.; Martyushev, N.; Nikonova, Z.; Savchenko, I.; Abramova, S.; Lubimova, V.; Novgorodtseva, A. Environmental Behavior of Youth and Sustainable Development. *Sustainability* **2022**, *14*, 250. [CrossRef]
29. Singh, R.; Khan, S.; Centobelli, P. Investigating the Interplay between Social Performance and Organisational Factors Supporting Circular Economy Practices. *Sustainability* **2022**, *14*, 16781. [CrossRef]
30. Xu, M.; Yan, J. Energy and circular economy in sustainability transitions. *Resour. Conserv. Recycl.* **2021**, *169*, 105471. [CrossRef]
31. Zhanbayev, R.A.; Yerkin, A.Y.; Shutaleva, A.V.; Irfan, M.; Gabelashvili, K.; Temirbaeva, G.R.; Chazova, I.Y.; Abdykadyrkyzy, R. State asset management paradigm in the quasi-public sector and environmental sustainability: Insights from the Republic of Kazakhstan. *Front. Environ. Sci.* **2023**, *10*, 1037023. [CrossRef]
32. Singh, R.; Khan, S.; Dsilva, J. A framework for assessing critical factor for circular economy practice implementation. *J. Model. Manag.* **2022**, *2*, 1746–5664. [CrossRef]
33. Almulhim, A.I.; Abubakar, I.R. Understanding Public Environmental Awareness and Attitudes toward Circular Economy Transition in Saudi Arabia. *Sustainability* **2021**, *13*, 10157. [CrossRef]
34. Mavhungu, A.; Masindi, V.; Foteinis, S.; Mbaya, R.; Tekere, M.; Kortidis, I.; Chatzisymeon, E. Advocating circular economy in wastewater treatment: Struvite formation and drinking water reclamation from real municipal effluents. *J. Environ. Chem. Eng.* **2020**, *8*, 103957. [CrossRef]
35. Stillitano, T.; Spada, E.; Iofrida, N.; Falcone, G.; De Luca, A.I. Sustainable Agri-Food Processes and Circular Economy Pathways in a Life Cycle Perspective: State of the Art of Applicative Research. *Sustainability* **2021**, *13*, 2472. [CrossRef]
36. Silva, R.D.S.; Jalal, A.; Nascimento, R.E.N.D.; Elias, N.C.; Kawakami, K.C.; Abreu-Junior, C.H.; Oliveira, F.C.; Jani, A.D.; He, Z.; Zhao, F.; et al. Composted Sewage Sludge Application Reduces Mineral Fertilization Requirements and Improves Soil Fertility in Sugarcane Seedling Nurseries. *Sustainability* **2022**, *14*, 4684. [CrossRef]

37. Bauerek, A.; Diatta, J.; Pierzchała, Ł.; Więckol-Ryk, A.; Krzemień, A. Development of Soil Substitutes for the Sustainable Land Reclamation of Coal Mine-Affected Areas. *Sustainability* **2022**, *14*, 4604. [CrossRef]
38. Wang, Z.; Hao, M. Effects of Spent Drilling Fluids from Natural Gas Fields on Seed Germination and Root Development of Maize (*Zea mays* L.). *Sustainability* **2021**, *13*, 1510. [CrossRef]
39. Brigida, V.S.; Zinchenko, N.N. Methane Release in Drainage Holes Ahead of Coal Face. *J. Min. Sci.* **2014**, *50*, 60–64. [CrossRef]
40. Abdelrhman, A.A.; Gao, L.; Li, S.; Lu, J.; Song, X.; Zhang, M.; Zheng, F.; Wu, H.; Wu, X. Long-Term Application of Organic Wastes Improves Soil Carbon and Structural Properties in Dryland Affected by Coal Mining Activity. *Sustainability* **2021**, *13*, 5686. [CrossRef]
41. Adesanya, E.; Ezu, A.; Nguyen, H.; Kinnunen, P.; Illikainen, M. Hydration of blended ladle slag and calcium aluminate cement. *J. Build. Eng.* **2023**, *66*, 105855. [CrossRef]
42. Gujar, P.; Murali, N.; Ilango, N.K.; Santhanam, M.; Ghosh, P. Engineering interfacial strength of polymer coated hydrating cement paste by tuning calcium characteristics. *Mater. Struct.* **2023**, *56*, 65. [CrossRef]
43. Sun, L.; Feng, Y.; Wang, D.; Qi, C.; Zeng, X. Influence of CaO on Physical and Environmental Properties of Granulated Copper Slag: Melting Behavior, Grindability and Leaching Behavior. *Int. J. Environ. Res. Public Health* **2022**, *19*, 13543. [CrossRef] [PubMed]
44. Inthavong, P.; Rehman, K.U.; Masood, K.; Shaukat, Z.; Hnydiuk-Stefan, A.; Ray, S. Impact of organizational learning on sustainable firm performance: Intervening effect of organizational networking and innovation. *Heliyon* **2023**, *9*, e16177. [CrossRef] [PubMed]
45. Lin, B.; Xie, Y. Positive or negative? R&D subsidies and green technology innovation: Evidence from China's renewable ener-gy industry. *Renew. Energy* **2023**, *213*, 148–156. [CrossRef]
46. Huang, S.; Du, C.; Jin, X.; Zhang, D.; Wen, S.; Wang, Y.; Cheng, Z.; Jia, Z. The Boundary of Porter Hypothesis: The Energy and Economic Impact of China's Carbon Neutrality Target in 2060. *Energies* **2022**, *15*, 9162. [CrossRef]
47. Bhagat, R.M.; Baruah, R.D.; Cacigue, S. Climate tea [*Camellia sinensis* (L.), O. Kuntze] production with special reference to north eastern India: A review. *J. Environ. Res. Dev.* **2010**, *4*, 1017–1028.
48. Maritim, T.K.; Kamunya, S.M.; Mireji, P.; Wendia, C.M.; Muoki, R.C.; Cheruiyot, E.K.; Wachira, F.N. Physiological biochemical response of tea (*Camellia sinensis* (L.), O. Kuntze) to water-deficit stress. *J. Hortic. Sci. Biotechnol.* **2015**, *90*, 395–400. [CrossRef]
49. Mishra, N.; Tripathi, M.K.; Tripathi, N.; Sharma, A.; Shrivastav, M.K. Changes in biochemical and antioxidant enzymes activities play significant role in drought tolerance in soybean. *Int. J. Agric. Technol.* **2021**, *17*, 1425–1446.
50. Tiwari, P.N.; Tiwari, S.; Sapre, S.; Babbar, A.; Tripathi, N.; Tiwari, S.; Tripathi, M.K. Screening and Selection of Drought-Tolerant High-Yielding Chickpea Genotypes Based on Physio-Biochemical Selection Indices and Yield Trials. *Life* **2023**, *13*, 1405. [CrossRef]
51. Tripathi, N.; Tripathi, M.K.; Tiwari, S.; Payasi, D.K. Molecular Breeding to Overcome Biotic Stresses in Soybean: Update. *Plants* **2022**, *11*, 1967. [CrossRef] [PubMed]
52. Mishra, N.; Tripathi, M.K.; Tiwari, S.; Tripathi, N.; Sapre, S.; Ahuja, A.; Tiwari, S. Cell Suspension Culture and In Vitro Screening for Drought Tolerance in Soybean Using Poly-Ethylene Glycol. *Plants* **2021**, *10*, 517. [CrossRef]
53. Akhmetov, B.Z.; Kasenova, A.Z.; Merzlyakov, O.E.; Tatarintsev, L.M. Modeling of sustainable agricultural land use on the basis of agroecological assessment of the territory. *Sustain. Dev. Mt. Territ.* **2023**, *15*, 17–26. (In Russian) [CrossRef]
54. Ali, E.B.; Gyamfi, B.A.; Bekun, F.V.; Ozturk, I.; Nketiah, P. An empirical assessment of the tripartite nexus between environmental pollution, economic growth, and agricultural production in Sub-Saharan African countries. *Environ. Sci. Pollut. Res.* **2023**, *30*, 71007–71024. [CrossRef] [PubMed]
55. Garima; Dhingra, A.; Centobelli, P.; Cerchione, R. Factors and Activities Considered by First Generation Agripreneurs for Agri-Business Sustainable Development: A Study of Haryana, India. *Sustainability* **2023**, *15*, 7109. [CrossRef]
56. Yu, H.; Zahidi, I.; Fai, C.M. Reclaiming abandoned mine tailings ponds for agricultural use: Opportunities and challenges. *Environ. Res.* **2023**, *232*, 116336. [CrossRef]
57. Sarr, A.; Soro, Y.M.; Tossa, A.K.; Diop, L. Agrivoltaic, a Synergistic Co-Location of Agricultural and Energy Production in Perpetual Mutation: A Comprehensive Review. *Processes* **2023**, *11*, 948. [CrossRef]
58. Federal Scientific and Methodological Center for Laboratory Research and Certification of Mineral Raw Materials "VIMS" MPR of Russia 499-NPP/MS "Determination of the Elemental Composition of Rocks, Soils, Soils and Bottom Sediments Using the Atomic Emission Method Having Inductively Coupled Plasma and Mass Spectral Method Involving Inductively Coupled Plasma". Available online: https://www.russiangost.com/p-274885-fr131201211927.aspx (accessed on 1 May 2023).
59. Hydrochemical Institute—Roshydromet R 52.24.690-2006—"Assessment of Toxic Pollution of Watercourses and Reservoirs of Different Salinity and Areas of Mixing of River and Sea Waters by Biotesting Methods", Rostov-on-Don. 2006. Available online: https://docs.cntd.ru/document/1200070999 (accessed on 1 May 2023).
60. Malyukova, L.S.; Pritula, Z.V.; Kozlova, N.V.; Velikiy, A.V.; Rogozhina, E.V.; Kerimzade, V.V.; Samarina, L. Effects if calcium-containing natural fertilizer on *Camellia sinensis* (L.) kunize Bangladesh. *J. Bot.* **2021**, *50*, 179–187. [CrossRef]
61. Alekseeva, T.P. *Guidelines for the technology of tea cultivation in the subtropical zone of the Krasnodar Territory, Russia, Ministry of Agriculture*; Russian Research Institute of mountain gardening and floriculture; Firm "Krasnodar tea": Sochi, Russia, 1977.
62. Bosikov, I.I.; Martyushev, N.V.; Klyuev, R.V.; Tynchenko, V.S.; Kukartsev, V.A.; Eremeeva, S.V.; Karlina, A.I. Complex Assessment of X-ray Diffraction in Crystals with Face-Centered Silicon Carbide Lattice. *Crystals* **2023**, *13*, 528. [CrossRef]
63. Colatosti, M.; Carboni, B.; Fantuzzi, N.; Trovalusci, P. Composite material identification as micropolar continua via an optimization approach. *Compos. Part C Open Access* **2023**, *11*, 100362. [CrossRef]

64. Ji, Z.; Shi, Y.; Da, X.; Cao, J.; Gong, Q.; Wang, Z.; Huang, X. Influence of Installation Deviation of Thrust Bearing on Oil Film Flow of 1000 MW Hydraulic Turbine Unit. *Water* **2023**, *15*, 1649. [CrossRef]
65. Matveeva, V.A.; Bech, J.; Danilov, A.S. Prospects of nitrogen removal from mine drainage by micro-algae in northern environments. *Sustain. Dev. Mt. Territ.* **2023**, *15*, 134–142. [CrossRef]
66. Guo, Q.; Peng, H.; Hong, B.; Yao, H.; Zhu, Y.; Ding, H.; An, N.; Hong, Y. Variations of methane stable isotopic values from an Alpine peatland on the eastern Qing-hai-Tibetan Plateau. *Acta Geochim.* **2021**, *40*, 473–483. [CrossRef]
67. Yao, H.; Peng, H.; Hong, B.; Guo, Q.; Ding, H.; Hong, Y.; Zhu, Y.; Cai, C.; Chi, J. Environmental Controls on Multi-Scale Dy-namics of Net Carbon Dioxide Exchange From an Alpine Peatland on the Eastern Qinghai-Tibet Plateau. *Front. Plant Sci.* **2022**, *12*, 791343. [CrossRef]
68. Dastpak, P.; Sousa, R.L.; Dias, D. Soil Erosion Due to Defective Pipes: A Hidden Hazard Beneath Our Feet. *Sustainability* **2023**, *15*, 8931. [CrossRef]
69. Yao, H.; Peng, H.; Hong, B.; Ding, H.; Hong, Y.; Zhu, Y.; Wang, J.; Cai, C. Seasonal and diurnal variation in ecosystem respiration and environmental controls from an alpine wetland in arid northwest China. *J. Plant Ecol.* **2022**, *15*, 933–946. [CrossRef]
70. Klyuev, R.V.; Morgoev, I.D.; Morgoeva, A.D.; Gavrina, O.A.; Martyushev, N.V.; Efremenkov, E.A.; Mengxu, Q. Methods of Forecasting Electric Energy Consumption: A Literature Review. *Energies* **2022**, *15*, 8919. [CrossRef]
71. Bocharnikov, M.V. Bioclimatic variants of the Northeastern Transbaikalia orobiome. *Sustain. Dev. Mt. Territ.* **2023**, *15*, 143–159. (In Russian) [CrossRef]
72. Yang, P.; Esmaeili, K.; Goodfellow, S.; Ordóñez Calderón, J.C. Mine Pit Wall Geological Mapping Using UAV-Based RGB Imaging and Unsupervised Learning. *Remote Sens.* **2023**, *15*, 1641. [CrossRef]
73. Zaalishvili, V.; Burdzieva, O.; Kanukov, A.; Zaks, T. Eco-Geophysical and Geoecological Factors in Assessing the State of the Geological Environment Based on the Analysis of Spatial Databases of the Territory of the Republic of North Ossetia–Alania. *Appl. Sci.* **2022**, *12*, 2644. [CrossRef]
74. Khan, N.; Ahmad, Z.; Shah, J.; Murtaza, S.; Albalwi, M.D.; Ahmad, H.; Baili, J.; Yao, S.-W. Dynamics of chaotic system based on circuit design with Ulam stability through frac-tal-fractional derivative with power law kernel. *Sci. Rep.* **2023**, *13*, 5043. [CrossRef] [PubMed]
75. Zhao, B.; Shuai, C.; Qu, S.; Xu, M. Using Deep Learning to Fill Data Gaps in Environmental Footprint Accounting. *Environ. Sci. Technol.* **2022**, *56*, 11897–11906. [CrossRef]
76. Abdolmaleki, M.; Consens, M.; Esmaeili, K. Ore-Waste Discrimination Using Supervised and Unsupervised Classification of Hyperspectral Images. *Remote Sens.* **2022**, *14*, 6386. [CrossRef]
77. Naveen, N.S.; Kishore, P.S.; Pujari, S.; Silas Kumar, M.D.; Jogi, K. Optimization through Taguchi and artificial neural networks on thermal performance of a radiator using graphene-based coolant. *Proc. Inst. Mech. Eng. A J. Power Energy* **2022**, *236*, 1680–1693. [CrossRef]
78. Szwajka, K.; Zielińska-Szwajka, J.; Trzepieciński, T. Improving the Surface Integrity of 316L Steel in the Context of Bioimplant Applications. *Materials* **2023**, *16*, 3460. [CrossRef] [PubMed]
79. Bosikov, I.I.; Martyushev, N.V.; Klyuev, R.V.; Savchenko, I.A.; Kukartsev, V.V.; Kukartsev, V.A.; Tynchenko, Y.A. Modeling and Complex Analysis of the Topology Parameters of Ventilation Networks When Ensuring Fire Safety While Developing Coal and Gas Deposits. *Fire* **2023**, *6*, 95. [CrossRef]
80. Golik, V.I.; Razorenov, Y.U.I.; Brigida, V.S.; Burdzieva, O.G. Mechanochemical technology of metal mining from enriching tails. *Bull. Tomsk. Polytech. Univ. Geo Assets Eng.* **2020**, *331*, 175–183. (In Russian) [CrossRef]
81. Malozyomov, B.V.; Golik, V.I.; Brigida, V.; Kukartsev, V.V.; Tynchenko, Y.A.; Boyko, A.A.; Tynchenko, S.V. Substantiation of Drilling Parameters for Undermined Drainage Boreholes for Increasing Methane Production from Unconventional Coal-Gas Collectors. *Energies* **2023**, *16*, 4276. [CrossRef]
82. Golik, V.I.; Klyuev, R.V.; Martyushev, N.V.; Brigida, V.; Efremenkov, E.A.; Sorokova, S.N.; Mengxu, Q. Tailings Utilization and Zinc Extraction Based on Mechanochemical Activation. *Materials* **2023**, *16*, 726. [CrossRef]
83. Samarina, L.; Wang, S.; Malyukova, L.; Bobrovskikh, A.; Doroshkov, A.; Koninskaya, N.; Shkhalakhova, R.; Matskiv, A.; Fedorina, J.; Fizikova, A.; et al. Long-term cold, freezing and drought: Overlapping and specific regulatory mechanisms and signal transduction in tea plant (*Camellia sinensis* (L.) Kuntze). *Front. Plant Sci.* **2023**, *14*, 1145793. [CrossRef]
84. Bushin, P.M. On the influence of air temperature and humidity on the yield of tea leaves in the subtropical zone of the Krasnodar Territory. *Meteorology and Hydrology* **1975**, *3 C*, 93–100.
85. Wang, Y.; Samarina, L.; Mallano, A.I.; Tong, W.; Xia, E. Recent progress and perspectives on physiological and molecular mechanisms underlying cold tolerance of tea plants. *Front. Plant Sci.* **2023**, *14*, 1145609. [CrossRef]
86. Malyukova, L.S.; Koninskaya, N.G.; Orlov, Y.L.; Samarina, L.S. Effects of exogenous calcium on the drought response of the tea plant (*Camellia sinensis* (L.) Kuntze). *PeerJ* **2022**, *10*, e13997. [CrossRef]
87. Benti, T.; Debela, A.; Bekele, Y.; Suleman, S. Effect of seasonal variation on yield and leaf quality of tea clone (*Camellia sinensis* (L.) O. Kuntze) in South West Ethiopia. *Heliyon* **2023**, *9*, e14051. [CrossRef]
88. Jayasinghe, S.L.; Kumar, L.; Hasan, M.K. Relationship between Environmental Covariates and Ceylon Tea Cultivation in Sri Lanka. *Agronomy* **2020**, *10*, 476. [CrossRef]

89. Kottawa-Arachchi, J.D.; Ranatunga, M.A.B.; Amarakoon, A.M.T.; Gunasekare, M.T.K.; Attanayake, R.N.; Sharma, R.K.; Chaudhary, H.K.; Sood, V.K.; Katoch, R.; Banyal, D.K.; et al. Variation of catechin and caffeine content in exotic collection of tea [*Camellia sinensis* (L.) O. Kuntze] in Sri Lanka and potential implication in breeding cultivars with enhanced quality and medicinal properties. *Food Chem. Adv.* **2022**, *1*, 100108. [CrossRef]
90. Jayasinghe, S.L.; Kumar, L. Climate Change May Imperil Tea Production in the Four Major Tea Producers According to Climate Prediction Models. *Agronomy* **2020**, *10*, 1536. [CrossRef]
91. Benti, T.; Debela, A.; Bekele, Y.; Suleman, S. Influence of clone and nitrogen application level on quality of green tea in some selected tea (*Camellia sinensis* (L.) O. Kuntze) in Southwest Ethiopia. *Heliyon* **2022**, *8*, e10179. [CrossRef]
92. Li, C.-L.; Xu, J.; Xu, H.-M.; Liu, J.; Zhang, L.-X.; Wang, Z.-K. Analysis of growth resistance mechanisms and causes in tea plants (*Camellia sinensis*) in high-pH regions of Northern China. *Front. Nutr.* **2023**, *10*, 1131380. [CrossRef]
93. Song, W.Y.; Zhang, Z.B.; Shao, H.B.; Guo, X.L.; Cao, H.X.; Zhao, H.B.; Fu, Z.Y.; Hu, X.J. Relationship between calcium decoding elements and plant abiotic-stress resistance. *Int. J. Biol. Sci.* **2008**, *4*, 116–125. [CrossRef]
94. Hetherington, A.M.; Brownlee, C. The generation of Ca^{2+} signals in plants. *Annu. Rev. Plant Biol.* **2004**, *55*, 401–427. [CrossRef]
95. Samynathan, R.; Shanmugam, K.; Nagarajan, C.; Murugasamy, H.; Ilango, R.; Victor, J.; Shanmugam, A.; Venkidasamy, B.; Thiruvengadam, M. The effect of abiotic and biotic stresses on the production of bioactive compounds in tea (*Camellia sinensis* (L.) O. Kuntze). *Plant Gene* **2021**, *27*, 100316. [CrossRef]
96. Kim, M.C. Calcium and calmodulin-mediated regulation of gene expression in plant. *Mol. Plant* **2009**, *2*, 13–21. [CrossRef] [PubMed]
97. Saidi, Y.; Finka, A.; Muriset, M.; Bromberg, Z.; Weiss, Y.G.; Maathuis, F.J.; Goloubinoff, P. The heat shock response in moss plants is regulated by specific calcium-permeable channels in the plasma membrane. *Plant Cell* **2009**, *21*, 2829–2843. [CrossRef] [PubMed]
98. Rikhvanov, E.G.; Fedoseeva, I.V.; Pyatrikas, D.V.; Borovsky, G.B.; Voinikov, V.K. The mechanism of functioning of the calcium signaling system in plants under the action of heat stress. The role of mitochondria in this process. *Plant Physiol.* **2014**, *61*, 155–169. [CrossRef]
99. Samarina, L.S.; Malyarovskaya, V.I.; Reim, S.; Koninskaya, N.G.; Matskiv, A.O.; Tsaturyan, G.A.; Rakhmangulov, R.S.; Shkhalakhova, R.M.; Shurkina, E.S.; Kulyan, R.V.; et al. Genetic Diversity in Diospyros Germplasm in the Western Caucasus Based on SSR and ISSR Polymorphism. *Biology* **2021**, *10*, 341. [CrossRef] [PubMed]
100. Shu, M.Y.; Fan, M.Q. Effect of osmotic stress and calcium on membrane-lipid peroxidation and the activity of defense enzymes in fir seedling. *For. Res.* **2000**, *4*, 391–396.
101. Upadhyaya, H.; Panda, S.K.; Dutta, B.K. $CaCl_2$ improves post-drought recovery potential in *Camellia sinensis* (L.), O. Kuntze. *Plant Cell Rep.* **2011**, *30*, 495–503. [CrossRef]

Disclaimer/Publisher's Note: The statements, opinions and data contained in all publications are solely those of the individual author(s) and contributor(s) and not of MDPI and/or the editor(s). MDPI and/or the editor(s) disclaim responsibility for any injury to people or property resulting from any ideas, methods, instructions or products referred to in the content.

sustainability

MDPI

Article

Influence of Regional Temperature Anomalies on Strawberry Yield: A Study Using Multivariate Copula Analysis

Poornima Unnikrishnan [1,*], Kumaraswamy Ponnambalam [1,*] and Fakhri Karray [2,3]

[1] Department of Systems Design Engineering, University of Waterloo, Waterloo, ON N2L 3G1, Canada
[2] Department of Electrical & Computer Engineering, University of Waterloo, Waterloo, ON N2L 3G1, Canada; karray@uwaterloo.ca
[3] Department of Machine Learning, Mohamed bin Zayed University of Artificial Intelligence, Masdar City, Abu Dhabi 50819, United Arab Emirates
* Correspondence: poornima.unnikrishnan@uwaterloo.ca (P.U.); ponnu@uwaterloo.ca (K.P.)

Abstract: A thorough understanding of the impact of climatic factors on agricultural production is crucial for improving crop models and enhancing predictability of crop prices and yields. Fluctuations in crop yield and price can have significant implications for the market sector and farming community. Given the projected increase in frequency and intensity of extreme events, reliable modelling of cropping patterns becomes essential. Temperature anomalies are expected to play a prominent role in future extreme events, emphasizing the need to comprehend their influence on crop yield. Forecasting extreme yield, which encompasses both the highest and lowest levels of agricultural production within a given time period, along with peak crop prices representing the highest market values, poses greater challenges in forecasting compared to other values. Probability-based predictions, accounting for uncertainty and variability, offer a more accurate approach for extreme value estimation and risk assessment. In this study, we employ a multivariate analysis based on vine copula to explore the interdependencies between temperature anomalies and daily strawberry yield in Santa Maria, California. By considering the maximum and minimum daily yields each month, we observe an increased probability of yield loss with rising temperature anomalies. While we do not explicitly consider the specific impacts of temperature anomalies under individual Representative Concentration Pathway (RCP) scenarios, our analysis is conducted within the broader context of the current global warming scenario. This allows us to capture the overall anticipated effects of regional temperature anomalies on agriculture. The findings of this study have potential impacts and consequences for understanding the vulnerability of agricultural systems and improving crop model predictions. By enhancing our understanding of the relationships between temperature anomalies and crop yield, we can inform decision-making processes related to the impact of climate change on agriculture. This research contributes to the ongoing efforts in improving agricultural sustainability and resilience in the face of changing climatic conditions.

Keywords: extreme yield; temperature anomaly; vine copula; conditional probability

1. Introduction

Agricultural production is highly sensitive to climatic variabilities [1–3]. The recurrence of extreme events such as heat waves, droughts, flood, etc., are pushing the global crop productivity to an extremely vulnerable state and this will have a profound impact on food security [2,4]. Among the climatic factors influencing crop yield, temperature anomalies are most likely to influence the crop yields negatively [4–8] and can ultimately impact a crop's market values. Considering the current climate change crisis, quantifying the impact of temperature anomalies on crop yields will be of significant value in planning adaptive strategies to assist agriculture and to avoid risk of food insecurity. The response of crops to high temperature stress has been studied in detail for different crops such as groundnut, wheat, maize, and rice, most of them involving experimental studies [9–14].

Temperatures that lie outside the range of optimum temperature levels for crops can have severe repercussions on the irrigated crop yields [15]. Both high and low temperatures can create a distress in the rate of agricultural production [4,16]. Understanding crop productivity responses to the regional weather and their explicit inclusion in crop models will aid in improving the planning of agricultural production.

Over the past few years, significant progress has been made in understanding the climate–yield relationships by many researchers. For instance, Goulart et al., 2021 [17] used a random forest model and global warming scenarios to predict an increase in soybean crop failures in the Midwestern United States due to warmer summer conditions. They introduced the concept of impact-analogues, which are expected to increase with warmer climates, providing a comprehensive risk estimation of future scenarios. Zhang et al., 2023 [18] examined the impact of climate extremes on vegetation growth on the Tibetan Plateau, utilizing event coincidence analysis to assess the susceptibility of different vegetation types to drought, extreme wet, hot, and cold events. They highlighted the importance of understanding vegetation responses to climate variability and identified ecologically sensitive regions that may be at risk due to changing climate conditions. Feng et al., 2021 [19] examined the evolving relationship between climate variables and crop yield, highlighting the impact of concurrent changes in precipitation–yield and temperature–yield relationships on the risk of crop yield reduction. They proposed a statistical approach to assess these risks, especially under compound dry–hot conditions, and underscored the importance of adapting agricultural planning to these changing climate–yield dynamics. Powell et al., 2015 [20] examined the impact of extreme weather events on wheat yields in the Netherlands, utilizing econometric techniques on farm-level panel data and established a correlation between increased frequency of such events and reduced yields, highlighting the importance of local-level analysis for accurate measurement and forecasting. Shayanmehr et al., 2020 [21] examined the impact of climate change on potato yield and variability in Iran's Agro-Ecological Zones (AEZs) for the 2050s, using models to project local climate scenarios and assess future agricultural outcomes. They highlighted the need for region-specific strategies to mitigate yield reduction and ensure food security in the face of changing climatic conditions.

Some of the previous similar studies focus on the influence of isolated extreme events such as heat waves and droughts on crop yields [22,23]. Recent studies have shown that the concurrent weather extremes can create larger impacts to the crop yield than isolated extreme events [19,24].

While different crop models and AI models were able to predict the non-extreme yield and price of strawberry and other crops with reasonable accuracy, the accurate prediction of extreme (both maximum and minimum) yield and price is found to be difficult [25,26]. However, these extreme yields are highly important in quantifying the profits for different stakeholders such as farmers and the market sector. Identifying the interdependence of extreme values in the crop sector can play a crucial role in improving the extreme value prediction performance of models. An integrated, probabilistic model by incorporating the uncertainties and interdependencies between the crop yield and weather anomalies can improve the prediction of crop yield, especially the extreme values [27].

This study focuses on analyzing the variations in strawberry yield in relation to the anomalies in regional air temperature within the study area in California, United States. Strawberry is a lucrative farm produce and has a critically low shelf life. Currently, the market for strawberries alone is worth USD 3.02 billion and they are one of the top ten most valued commodities for California and the United States' economy [28].

Strawberries reach maturity and bear fruit relatively quickly due to their shallow root systems, necessitating optimal light conditions for high yield and fruit quality. Additionally, proper water management is essential to maximize both yield and fruit quality in strawberries [29]. With an aim of explaining the influence of regional weather conditions on strawberries, several studies have been carried out [29–31]. Some of these studies have shown a critical correlation between strawberry yield and temperature changes [29,30,32].

Temperature affects nutrient uptake and photosynthesis in strawberry plants [33]. High temperature can also inhibit developmental stages during flower formation and can eventually bog down the fruit quality [34].

In our study, we made a deliberate choice to primarily focus on temperature anomalies as the key climatic factor influencing crop yield. This decision is supported by compelling justifications based on the unique characteristics of the study area and the current irrigation practices used to grow this commercial crop. Firstly, the area is characterized by strict irrigation practices, where crop growth and yield are predominantly reliant on controlled water supply rather than natural precipitation. This reduces the impact of variations in precipitation on crop yield [35], making temperature anomalies one of the primary climatic factors of interest. While soil moisture is undoubtedly crucial for crop growth and development, strict irrigation practices in the study area ensure that soil moisture levels are maintained at optimal levels, minimizing their variability and impact on crop yield. Additionally, extensive analysis of historical data revealed a strong and consistent correlation between temperature anomalies and crop yield in the study area. In contrast, the correlation between precipitation and yield was found to be comparatively weaker and hence is not reported here. This finding highlights the dominant role of temperature anomalies in driving yield variations, underscoring the need to investigate and understand their impact for improved agricultural predictions.

Furthermore, our choice to focus on temperature anomalies aligns with previous studies conducted in similar agricultural contexts. Existing research consistently highlights the significant influence of temperature fluctuations on crop growth, development, and yield potential [36–38]. By emphasizing temperature anomalies, we build upon this knowledge and contribute to the growing understanding of their impact on agricultural systems. Considering the expected temperature rise and its potential consequences for agricultural production [39], understanding the influence of temperature anomalies becomes crucial. Higher temperatures can have profound effects on critical physiological processes in crops, such as photosynthesis, respiration [39], and water use efficiency [36]. By focusing on temperature anomalies, we aim to uncover the specific interdependencies between temperature variations and crop performance, contributing to the development of more robust and accurate agricultural models and predictions. Here, we present a data-based analysis of the influence of temperature anomalies on the extreme strawberry yield over the Southwestern United States. This study here is trying to analyze and figure out the importance of extreme value analyses. A multivariate copula analysis has been used to understand the conditional probability of decreasing yield given the temperature anomalies. Understanding of the influence of temperature anomalies on crop yield will help in the development of appropriate adaptation measures for agriculture planning and management. The findings of this work help in understanding responses of strawberry extreme yields to the regional temperature anomalies, which is critical to address the needs of the farming communities and to help the agricultural marketing industry in making appropriate price predictions.

2. Materials and Methods

2.1. Data

In this study, the strawberry yield from Santa Maria County in California, United States, was analyzed with respect to the regional air temperature anomalies. California holds significant importance in the global strawberry industry, being responsible for producing over 80% of the fresh market and processed strawberries grown in the United States. Moreover, it accounts for approximately 20% of the world's total strawberry production [40]. The Central Coast and Santa Maria Valley of California collectively host two-thirds of the total acreage dedicated to strawberry production [28,40].

Growers in the Santa Maria region have historically employed drip irrigation systems. In the Santa Maria Valley, strawberry fields are typically divided into drip-irrigated sections ranging from 1 to 5 acres. The drip lines used have lengths varying between 200 and

325 feet, with high-flow, 4 mil drip tape featuring emitters spaced every 8 inches and installed at depths of 1 to 3 inches. Additionally, drip tapes are replaced annually to ensure optimal irrigation efficiency [41]. Furthermore, to address concerns regarding the representativeness of monitored meteorological conditions, it is important to note that the cultivation practices described herein are widely adopted across the entire region [41]. This ensures that temperature data collected accurately reflect the prevailing environmental conditions experienced by strawberry crops in the Santa Maria Valley.

The study period spans from 2011 to 2019, capturing a comprehensive dataset of strawberry yield variations in the Santa Maria Valley. It is pertinent to note that throughout the study period, consistent irrigation practices were maintained, with minimal deviations observed. This ensures the reliability of yield data and allows for a more accurate assessment of the impact of temperature anomalies on strawberry yield dynamics.

The daily data of strawberry yield and temperature pertaining to Santa Maria County in California for a time period from 2011 to 2019 were collected from two publicly available websites [42,43]. In this study, the influence of temperature anomalies on the strawberry yields was analyzed. For this purpose, the monthly maximum temperature anomalies were estimated as the deviations from the average monthly temperature from 1991 to 2021 that is publicly available on the website [44].

In order to define the yield loss with respect to the daily strawberry yield extremes, the following guidelines were proposed in this study.

The standardised yield was estimated by standardizing the maximum and minimum daily yield over each month by means of z score transformation as given below.

$$\text{Std. } Y_{max} = \frac{Y_{max} - \mu_{max}}{\sigma_{max}} \quad (1)$$

$$\text{Std. } Y_{min} = \frac{Y_{min} - \mu_{min}}{\sigma_{min}} \quad (2)$$

where Std. Y_{max} and Std. Y_{min} are the standardised daily maximum and minimum yield over each month, Y_{max} and Y_{min} are the maximum and minimum daily yield over each month, μ_{max} and μ_{min} are the mean of the maximum and minimum yield time series data (for every month in the time period), and σ_{max} and σ_{min} are the standard deviation of the maximum and minimum yield time series data.

In this study, for better understanding, the yield loss is defined as three types of losses (low, moderate, and high). The three categories of yield loss are defined in this study as follows:

$$\begin{array}{lcc} -1.0 < Std.Y \leq -0.5 & : & \textit{Low Yield loss} \\ -1.5 < Std.Y \leq -1.0 & : & \textit{Moderate Yield loss} \\ Std.Y \leq -1.5 & : & \textit{High Yield loss} \end{array} \quad (3)$$

The monthly temperature anomalies were analyzed with the standardised maximum and minimum strawberry yield at Santa Maria for the time period from 2011 to 2019.

2.2. Copula Analysis

Sklar [45] first introduced copula in 1959. Copulas are largely useful in describing the dependencies between random variables [46]. They are mathematical functions by which we can obtain a joint distribution of random variables using their univariate distributions. The major utility of copulas comes where the random variables under consideration do not follow the same marginal distributions in which case it is difficult to find joint distribution without the use of a copula.

Sklar's theorem states that for a given joint multivariate distribution function and relevant marginal distributions for the corresponding random variables, there always exists a copula function that relates the marginal distributions of the variables [45] which can be mathematically derived as follows:

Consider the bivariate case with two random variables, X_1 and X_2, with distribution functions $F_1(X_1)$ and $F_2(X_2)$, respectively. As per Sklar's theorem there always exists a copula function (C) such that,

$$F(X_1 = x_1, X_2 = x_2) = C(F_1(x_1), F_2(x_2)) \tag{4}$$

where $C(u_1,u_2)$ is itself a distribution function where u_1 and u_2 are $F_1(x_1)$, $F_2(x_2)$), respectively.

A copula can be described as a cumulative distribution function with a uniform marginal. Each of the marginal distributions produces a probability of events. The copula function maps these probabilities to a joint probability, by giving a relationship on the probabilities. Hence, copulas build multivariate distributions by merely determining the dependence between the random variables without having any influence on the marginals. The detailed mathematical formulations and derivations of copulas can be seen in many scientific articles [47–49]

In contrast to bivariate copulas, which are very popular in various fields such as finance, marketing, hydrology [50–54], etc., applicability of multivariate copulas was seldom explored until recently due to the complexity in the construction of multivariate copulas. Vine copula is a flexible multi-variate copula construction method [49,55,56] which has a large potential for many multi-variate data analysis applications. The underlying theory for the vine copula is from Joe [57]. The use of vine copulas is gaining popularity in various fields [58–63].

2.3. Construction of a 3-Dimensional Vine Copula

The selection of the copula function that best describes the given data is a crucial step in the analysis of data using copulas. In the current study, the focus will be on a 3-dimensional vine copula coupling the three marginal distributions of three random variables, X, Y and Z (monthly maximum yield, monthly minimum yield, and monthly temperature anomaly). We assume that the samples of all three variables were each transformed (U, V, and W) so that all three variables are now approximately uniformly distributed on [0, 1]. The fundamental idea of vine copulas is to construct multivariate copulas based on a hierarchical mixing of bivariate copulas [64]. All pairwise dependencies are modelled with bivariate copulas. If all mutual dependencies defined are with respect to the same variable, the construction is called a canonical vine (C-vine). If all mutual dependences are considered one after the other variable, this is called a D-vine [65]. C- and D-vines are special cases of regular vines.

Fitting a C-vine might be advantageous when a particular variable is known to be a key variable that governs interactions in the data set. In such a case, this variable can be placed at the root of the canonical vine structure [66]. Different vine copula decompositions produce different multivariate distributions [65]. The selection of the vine copula structure is an important stage in the modelling process, which can be performed by any of the available methods in the literature such as based on the absolute empirical Kendall tau [67], and maximum likelihood method [66]. Here, the maximum likelihood method is used, by considering different copula families and the best fit is determined by the highest log-likelihood value. The estimations in this paper were performed using R 4.2.2, which is an open-source software environment for statistical computing, and the package "VineCopula" [68].

2.4. Calculation of Joint Probability of Occurrence of Events Using Copula

The major focus of this study is to understand the dependency of standardised strawberry yields (both maximum and minimum) on temperature anomaly. A 3D copula is used in this study and the joint probabilities of occurrence of strawberry yield and temperature anomalies for various scenarios are estimated. Different combinations of relations with the fitted copula distribution and the marginal distributions can be used to estimate the joint probabilities of occurrence of events. Some of which, those that were used in this study, are given below:

2.4.1. Univariate Probability

Univariate probability denotes the probability of an event by considering the marginal distribution of the random variable alone. The mutual dependencies of the variable with other random variables are not considered here. This type of probability will be referred to as UP in this paper from here onwards.

$$\mathrm{UP}(X \leq x) = F(x) \tag{5}$$

$$\mathrm{UP}(X > x) = 1 - F(x) \tag{6}$$

where F is the cumulative distribution function of the random variable (X).

2.4.2. Joint Probability for Tri-Variate Events

In multivariate frequency analysis involving three variables using copula, the following probability distributions can be used where C is the copula function.

$$P(X_1 \leq x_1) = F_1(x_1) = u; P(X_2 \leq x_2) = F_2(x_2) = v; P(X_3 \leq x_3) = F_3(x_3) = w; \tag{7}$$

$$C_{12}(u, v) = F_{12}(x_1, x_2) = P_{12}(X_1 \leq x_1, X_2 \leq x_2); \tag{8}$$

$$C_{13}(u, w) = F_{13}(x_1, x_3) = P_{13}(X_1 \leq x_1, X_3 \leq x_3) \tag{9}$$

$$C_{23}(v, w) = F_{23}(x_2, x_3) = P_{23}(X_2 \leq x_2, X_3 \leq x_3) \tag{10}$$

$$C(u, v, w) = F(x_1, x_2, x_3) = P(X_1 \leq x_1, X_2 \leq x_2, X_3 \leq x_3) \tag{11}$$

For tri-variate random variables, X_1, X_2, and X_3, some of the formulae of tri-variate probability distributions is given as follows [49]:

$$\begin{aligned} P(X_1 \leq x_1, X_2 \leq x_2, X_3 > x_3) &= P_{12}(X_1 \leq x_1, X_2 \leq x_2) - P(X_1 \leq x_1, X_2 \leq x_2, X_3 \leq x_3) \\ &= C_{12}(u, v) - C(u, v, w) \end{aligned} \tag{12}$$

$$\begin{aligned} & P(X_1 > x_1, X_2 > x_2, X_3 \leq x_3) \\ = \; & P(X_3 \leq x_3) - P_{13}(X_1 \leq x_1, X_3 \leq x_3) - P_{23}(X_2 \leq x_2, X_3 \leq x_3) + P(X_1 \leq x_1, X_2 \leq x_2, X_3 \leq x_3) \\ = \; & w - C_{13}(u, w) - C_{23}(v, w) + C(u, v, w) \end{aligned} \tag{13}$$

2.4.3. Conditional Probability of Tri-Variate Events

For tri-variate random variables, X_1, X_2, and X_3, the conditional probability of occurrence of events can be estimated with the help of the underlying copulas. An example case is defined below [49]:

Case 1:

$$X_1 > x_1 \cap X_2 > x_2 \mid X_3 = x_3 \text{ (or } X_1 > x_1 \cap X_3 > x_3 \mid X_2 = x_2, X_2 > x_2 \cap X_3 > x_3 \mid X_1 = x_1)$$

In this case, under the condition $X_3 = x_3$, both values of X_1 and X_2 are exceeded. In this case:

$$(P(X_1 > x_1 \cap X_2 > x_2 \mid X_3 = x_3) = 1 - P(X_1 \leq x_1 \mid X_3 = x_3) - P(X_2 \leq x_2 \mid X_3 = x_3) + P(X_1 \leq x_1, X_2 \leq x_2 \mid X_3 = x_3) \tag{14}$$

where

$$P(X_1 \leq x_1 \mid X_3 = x_3) = \left.\frac{\partial C_{13}(u, w)}{\partial w}\right|_{W=w}, P(X_2 \leq x_2 \mid X_3 = x_3) = \left.\frac{\partial C_{23}(v, w)}{\partial w}\right|_{W=w} \tag{15}$$

More detailed mathematical expressions can be found in Zhang and Singh [49].

3. Preliminary Data Analysis

Before starting any modelling process, it is always advisable to carry out a preliminary data analysis. The primary focus of this study is on strawberry yield affected by temperature anomalies. Hence, the maximum and minimum daily yield over each month have been estimated and are given in Figure 1a. It can be observed that the yield data are an approximately stationary process with a predominant seasonal pattern with peak yield during June. Similarly, the maximum monthly temperature observed at Santa Maria from 2011 to 2019 and the average maximum temperature over months based on temperature data from 1990 to 2021 [44] are given in Figure 1b. The monthly maximum temperature anomalies are estimated by calculating the deviation of the observed maximum temperature from the average maximum temperature over months and are shown in Figure 1b,c. As the data length is comparatively short to estimate long-term trends in the data, the temperature anomaly does not show any predominant trend (Figure 1c). It can be inferred that the yield and the temperature anomaly data are approximate stationary processes during the time period of 2011–2019. The basic statistics of the data are given in Table 1.

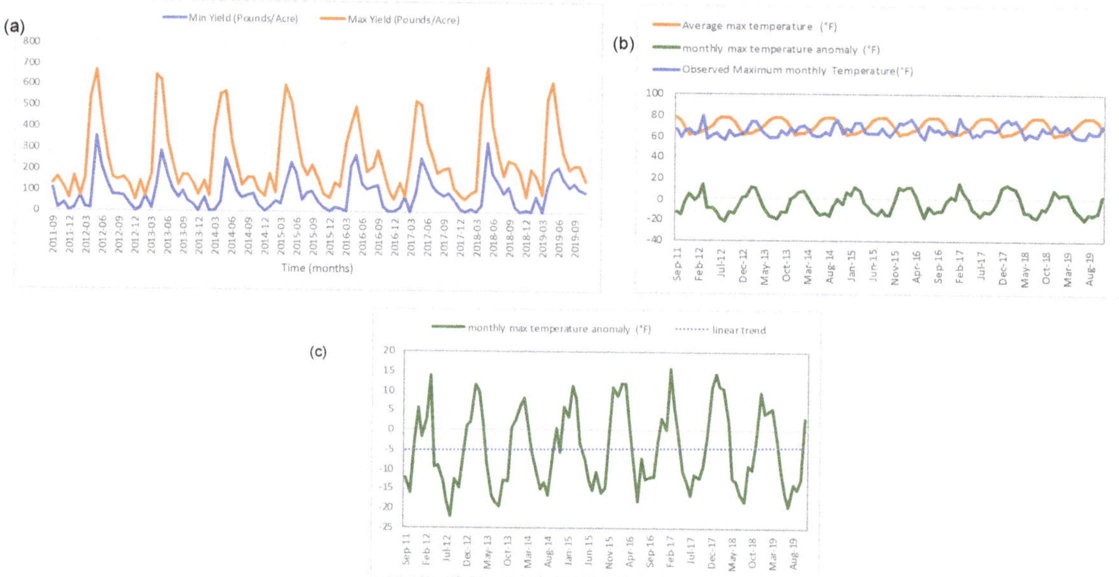

Figure 1. (**a**) Maximum and minimum monthly strawberry yield at Santa Maria from 2011 to 2019; (**b**) the observed temperature at Santa Maria from 2011 to 2019 and the temperature anomalies estimated; (**c**) trend of monthly maximum temperature anomalies observed from 2011 to 2019.

Table 1. Basic statistics of the yield and temperature anomaly data.

Statistics	Max Yield (Pounds/Acre)	Min Yield (Pounds/Acre)	Max Temp Anomaly (°F)
Min.	57.94	1	−11.9
1st Qu.	128.5	17.72	−2.6
Median	185	73.74	4.5
Mean	245.66	86.21	5.839
3rd Qu	327.67	123.15	13.85
Max.	686	356.32	25.9

4. Results and Discussion

In order to identify the correlation of strawberry yield and the temperature anomaly, a 3D copula analysis was carried out with standardised maximum and minimum daily strawberry yield over each month and monthly maximum temperature anomaly as the three random variables. To conduct multi-variate copula analysis, a vine copula structure was selected for this study. Apart from a multivariate analysis, individual univariate frequency analyses of the random variables were also conducted to identify the relevance of using copula analysis in conducting analysis of dependent random variables.

4.1. Univariate Analysis

As the first case, univariate frequency analysis of the random variables is conducted without considering the interdependence between the random variables. The random variables (standardised maximum yield, standardised minimum yield, and the temperature anomaly) were analyzed individually. The histograms of the variables are given in Figure 2. It can be observed from the histograms that the standardised maximum and minimum yields are positively skewed whereas the temperature anomalies show both skewness and heavy tails.

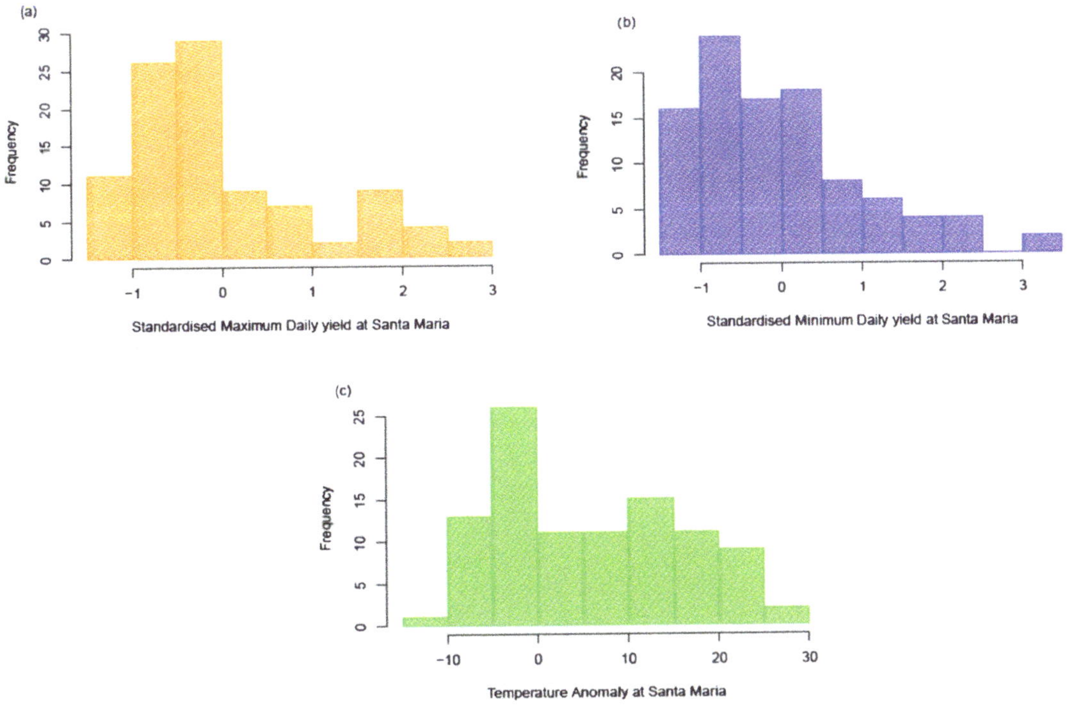

Figure 2. Histogram of (**a**) standardised maximum strawberry yield at Santa Maria; (**b**) standardised minimum strawberry yield at Santa Maria; (**c**) temperature anomaly at Santa Maria for each month.

Different probability distributions were tested and the marginal probability distribution that describes the process best was selected for each of the random variables. The "GAMLSS" [69,70] package in R was used for the purpose. Skew exponential power type III (SEP3) distribution was found to be the best fit for standardised maximum and minimum yield data whereas skewed exponential power type II (SEP2) distribution was found to best describe the temperature anomaly process.

Hence, the identified univariate probability distributions for the three variables were fitted to the data. From the marginal distribution, the probability of different cases (events) as described in Equation (3) were calculated using Equations (5) and (6) and are tabulated, which is given in Table 2.

Table 2. Probability of the yield loss events and temperature anomaly (T_A) to be more than the given thresholds using the univariate probability distribution of the random variables.

Events	Probability of Occurrence	Remarks
Std. $Y_{max} \leq -0.5$, Std. $Y_{min} \leq -0.5$	0.004	Yield loss
Std. $Y_{max} \leq -1.0$, Std. $Y_{min} \leq -1.0$	0.0001	Moderate or high yield loss
Std. $Y_{max} \leq -1.5$, Std. $Y_{min} \leq -1.5$	0.25×10^{-6}	High yield loss
$T_A > 1$, $T_A \leq 2$	0.036	Temperature anomaly is between the range of 1 °F and 2 °F
$T_A > 2$ $T_A \leq 3$	0.042	Temperature anomaly is between the range of 2 °F and 3 °F
$T_A > 3$	0.61	Temperature anomaly greater than 3 °F

As the standardised maximum and minimum yields follow the same marginal distribution (SEP3), their joint probability can be found by directly applying the product rule for independent events as follows:

$$P(\text{Std. } Y_{max} \leq \text{a, Std. } Y_{min} \leq \text{b}) = P(\text{Std. } Y_{max} \leq \text{a}) \times P(\text{Std. } Y_{min} \leq \text{b}) \tag{16}$$

4.2. Multivariate Copula Analysis Using Vine Copula

We employed vine copulas also known as pair-copula constructions (PCC) to model the dependence of strawberry yield and temperature anomaly. To model the dependence using vine copula, the data from the real domain were transformed to copula data, such that all data points lie inside [0, 1]. This was accomplished by computing the pseudo-observations of the data matrix by weighing the original variables based on the covariance structure of the data. The use of pseudo-observations helps to retain the essential structure of the data while making it easier to fit statistical models and interpret the results. Working with the pseudo-observations eliminates the marginal properties of the variables and retains the information about their interdependence [71].

The root node of a C-vine tree is identified by identifying the node with the strongest dependencies to all other nodes. The node with the maximum column sum in the empirical Kendall's tau matrix is selected as the root node. To do this, the Kendall correlation matrix for the variables was generated and the absolute value of the correlation coefficient of each variable was added together (Table 3). The variable with the highest sum is selected as the root node of the C-vine tree. In our case study, the root node of the C-vine tree is the minimum strawberry yield. The vine tree structure selected for this study is given in Figure 3.

Table 3. Kendall correlation matrix for the variables.

Variables	Std. Max Yield	Std. Min Yield	Temperature Anomaly	Sum
Standardised maximum yield	1	0.55	−0.29	1.84
Standardised minimum yield	0.55	1	−0.417	1.967
Temperature anomaly	−0.29	−0.417	1	1.707

We used bivariate copulas to formulate the PCC for the three links. Each link represents a bivariate copula that best describes the dependence between the corresponding variables. The bivariate copulas selected for each of the links are also shown in Figure 3. The selection and fitting of the copula were performed in R 4.2.2 using the "VineCopula" package. Copulas are fitted against the data using maximum likelihood estimation, and then the Akaike information criteria (AIC) criterion is computed for different copula families in R package "VineCopula" [68]. The Tawn II copula (asymmetric logistic copula) [72,73] which is an extension of the Gumbel copula describes the dependence of the marginal distributions of standardised minimum yield and standardised maximum yield. BB8 (Joe–Frank) copula, which is a mixture model of the Joe and Frank copulas, describes the relation between standardised minimum yield and temperature anomaly, and the Joe copula describes the conditional relation with standardised maximum yield and temperature anomaly given standardised minimum yield.

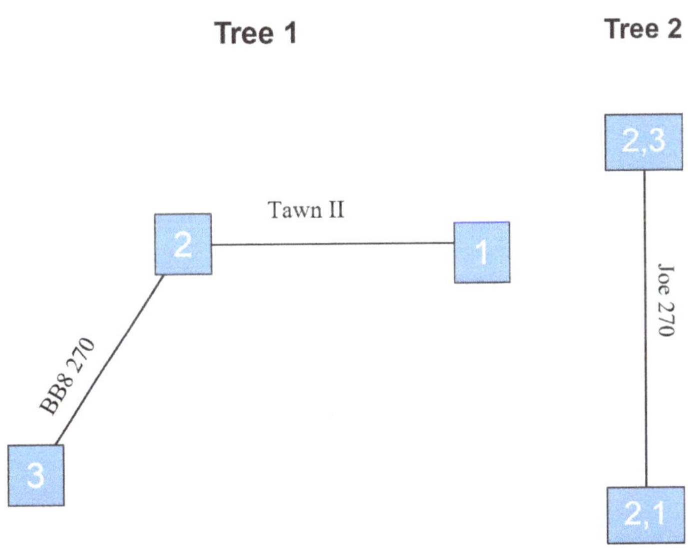

1. Standardised Max Yield, 2. Standardised Min Yield, 3. Temperature anomaly

Figure 3. C-Vine tree structure showing the dependencies and pair-copula structure.

Based on the fitted copula, the standardised maximum strawberry yield, standardised minimum strawberry yield, and temperature anomaly were simulated and compared with the observed data (Figure 4). It was observed that the copula could capture the pattern and behaviour of the yield and temperature data.

The joint probability analysis of standardised minimum and maximum strawberry yield and temperature anomaly was conducted using a vine copula. The focus of this study was to determine the impact of temperature anomaly on strawberry yield through conditional probability analysis. The conditional probability (CP) of standardised minimum and maximum strawberry yield (as given in Equation (3)) given the temperature anomaly was analyzed using the established vine copula. The results of the scenarios analyzed are presented in Table 4, which displays the conditional probability of various yield loss conditions for different temperature anomalies.

Table 4 demonstrates the impact of temperature anomalies on the standardised minimum and maximum strawberry yields. The table shows the probability of yield loss events (with its severity) for different temperature anomaly conditions. Our findings clearly demonstrate that temperature anomalies have a significant impact on strawberry yield in Santa Maria, California. By comparing the univariate and multivariate analysis results (as shown in Tables 2 and 4), we highlight the significance of multivariate probabilistic analysis in crop yield studies. The univariate analysis (Table 2), which focuses solely on marginal distributions, tends to underestimate the likelihood of yield loss events. This oversight occurs because it overlooks the intricate relationship between temperature anomalies and crop yields.

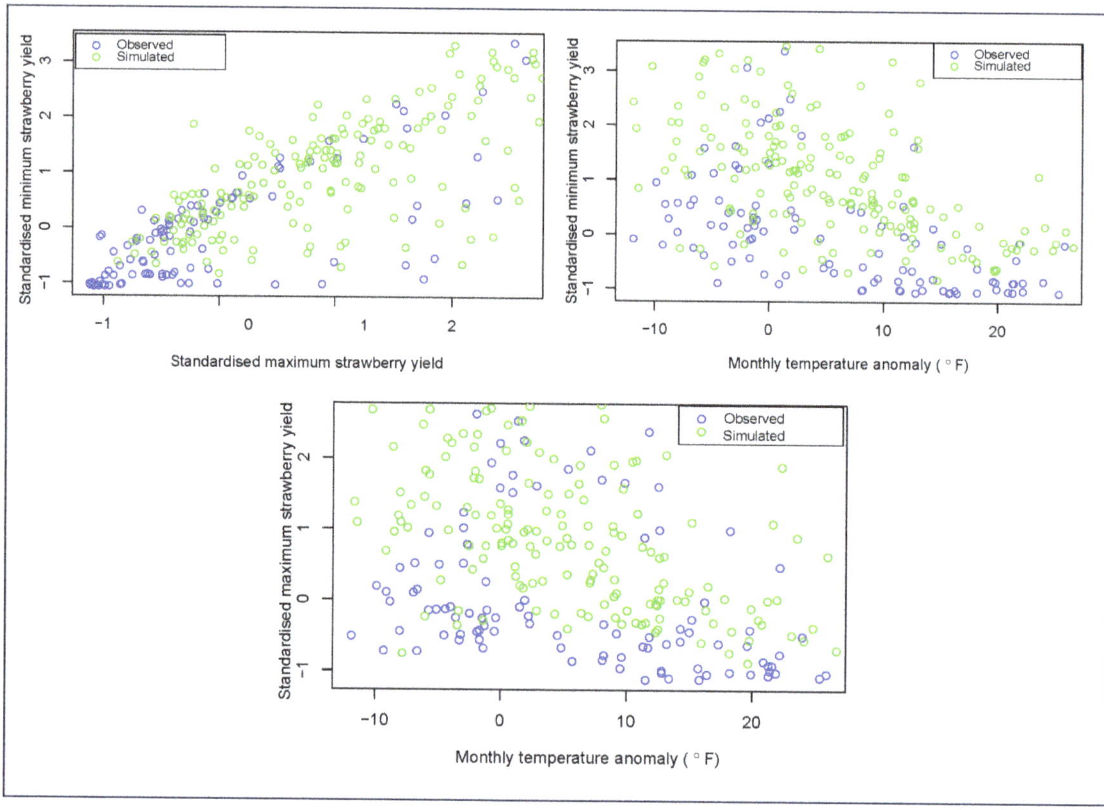

Figure 4. The standardised maximum and minimum strawberry yield and temperature anomaly along with the corresponding simulated data using the fitted copula.

Table 4. Multivariate probability of yield loss events for different temperature anomaly conditions.

No	Events	Conditional Probability of Occurrence of Events Given the Following T_A (°F)			Remarks
		$1 < T_A \le 2$	$2 < T_A \le 3$	$T_A > 3$	
1	Std. $Y_{max} \le -0.5$ & Std. $Y_{min} \le -0.5 \mid T_A$	0.584	0.56	0.8	Yield loss
2	Std. $Y_{max} \le -1$ & Std. $Y_{min} \le -1 \mid T_A$	0.039	0.037	0.66	Moderate or high yield loss
3	Std. $Y_{max} \le -1.5$ & Std. $Y_{min} \le -1.5 \mid T_A$	0.0006	0.0005	0.63	High yield loss

In contrast, the multivariate copula analysis, incorporating the joint dependence structure, allows for a more accurate assessment of the conditional probabilities of yield loss events for different temperature anomalies. This approach captures the interdependencies between temperature anomalies and crop yield, revealing the complex relationship between the variables. By considering the joint dependence structure, our analysis reveals that the conditional probability of various yield loss events is much higher for different possible temperature anomalies. For instance, we observe that when the temperature anomaly exceeds 3 °F, high yield loss events are highly probable. This highlights the critical importance of considering the intricate relationship between temperature anomalies and crop yields.

This comparison between the univariate and multivariate analyses underscores the importance of considering multivariate dependencies when studying the effects of temperature anomalies on crop yield, particularly in the context of climate change. Thus, copula-based multivariate analysis offers valuable insights by capturing the joint dependence structure between variables, enabling a more realistic assessment of the risks associated with temperature anomalies.

This study highlights the importance of understanding the influence of temperature anomalies on extreme crop yields and estimating the risk of yield reduction under extreme weather conditions to ensure food security under global warming. By applying copula analysis, we can accurately assess conditional probabilities and gain detailed insights into the influence of temperature anomalies on crop yields. It offers a more realistic assessment of the risks associated with temperature anomalies, enabling stakeholders to devise more effective risk management and climate change adaptation strategies. This becomes particularly relevant in the current climate change scenario, where global warming is projected and understanding the impact of temperature anomalies on agricultural production becomes crucial for effective adaptation and decision making. By understanding the complex relationship between temperature anomalies and crop yields, stakeholders can implement targeted measures to minimize yield losses and enhance the resilience of agricultural systems in the face of climate change.

The results of this study will be helpful in developing an integrated probabilistic crop climate model with improved prediction performance, especially in predicting the extreme crop yields and prices. The implications of our findings are significant for strawberry farmers in Santa Maria, California. Understanding the dependency patterns between temperature anomalies and crop yield can inform farmers about the risks associated with temperature extremes and help them make informed decisions regarding crop management practices, irrigation scheduling, and other adaptation measures. This knowledge can guide farmers and policymakers in implementing effective climate change adaptation strategies in strawberry production and potentially in other high-value agricultural crops.

5. Limitations of This Study

Despite the valuable insights gained from our study, it is important to acknowledge its limitations. These limitations include the following:

Generalizability: The findings of our study are specific to the context of Santa Maria, California, and the strawberry crop. Extrapolating these results to other regions or agricultural crops should be done with caution, as different crops and regions may exhibit unique climatic and agronomic characteristics. A significant area of future research could be exploring the potential compounding effects of extreme heat and drought on crops, especially in rain-fed regions, by considering other climate variables.

Biotic factors: In the current study, we considered commercial fully irrigated (pest controlled) crops, hence the influence of biotic factors, such as diseases, on crop yield was not explicitly considered in our analysis. Incorporating biotic variables into the copula modelling framework is a potential avenue for future research to improve the comprehensiveness of yield predictions especially for noncommercially grown crops.

Time scale: Another limitation of our study is the relatively short time scale of the research data, which ranges from 2011 to 2019. A longer time scale, ideally spanning at least 30 years, would be more suitable for studying the relationship between crop yield and climate change. This limitation should be acknowledged as it may affect the generalizability of our findings to longer-term climate variations. As future research, expanding the time frame can be considered, which will enhance our understanding of the changing climate–crop dynamics.

Despite these limitations, this study offers a valuable tool for analyzing the influence of temperature anomalies on strawberry yield. It provides a framework for studying the complex dependencies between climate variables and crop performance. Our results contribute to the growing understanding of climate–crop relationships and demonstrate the potential for utilizing multivariate copula analysis in agricultural decision making and climate change adaptation.

6. Conclusions

This study utilizes C-vine copula analysis to investigate the relationship between temperature anomalies and strawberry yield in Santa Maria, California. The findings demonstrate a significant dependence between temperature anomalies and crop yield, highlighting the importance of considering temperature variations in crop climate models.

While our study illustrates the influence of temperature on crop yield, integrating such analysis into broader crop models can yield more actionable insights for agricultural production practices. Such integrated models hold practical implications, offering guidance for decision-making processes aimed at mitigating the effects of temperature anomalies on crop production and contributing to the broader knowledge base on climate and agriculture. By incorporating temperature anomalies into crop models, resource allocation can be optimized, crop yield predictions can be improved, and farming system resilience can be enhanced.

This research contributes to the broader understanding of the intricate relationships between regional temperature anomalies and agricultural outcomes. By employing multivariate copula analysis, we showcase its effectiveness in quantifying the dependence between temperature anomalies and crop yield.

Future research can expand on our findings by including additional factors such as biotic factors and applying copula analysis to other high-value agricultural crops. This would further refine the predictive capabilities of crop climate models and provide a comprehensive understanding of temperature anomalies' impact on different agricultural systems.

Author Contributions: Conceptualization, P.U. and K.P.; methodology, P.U.; software, P.U.; validation, P.U., K.P. and F.K.; formal analysis, P.U.; investigation, P.U.; resources, P.U.; data curation, P.U.; writing—original draft preparation, P.U.; writing—review and editing, P.U., K.P. and F.K.; visualization, P.U.; supervision, K.P. and F.K.; project administration, P.U.; funding acquisition, K.P. and F.K. All authors have read and agreed to the published version of the manuscript.

Funding: This research was partially funded by NSERC and Loblaws NSERC-CRD.

Data Availability Statement: The raw data supporting the conclusions of this article will be made available by the authors on request.

Conflicts of Interest: The authors declare no conflict of interest.

References

1. Kumari, A.; Lakshmi, G.A.; Krishna, G.K.; Patni, B.; Prakash, S.; Bhattacharyya, M.; Singh, S.K.; Verma, K.K. Climate Change and Its Impact on Crops: A Comprehensive Investigation for Sustainable Agriculture. *Agronomy* **2022**, *12*, 3008. [CrossRef]
2. Elias, E.H.; Flynn, R.; Idowu, O.J.; Reyes, J.; Sanogo, S.; Schutte, B.J.; Smith, R.; Steele, C.; Sutherland, C. Crop Vulnerability to Weather and Climate Risk: Analysis of Interacting Systems and Adaptation Efficacy for Sustainable Crop Production. *Sustainability* **2019**, *11*, 6619. [CrossRef]
3. Malhi, G.S.; Kaur, M.; Kaushik, P. Impact of Climate Change on Agriculture and Its Mitigation Strategies: A Review. *Sustainability* **2021**, *13*, 1318. [CrossRef]
4. Talib, M.N.A.; Ahmed, M.; Naseer, M.M.; Slusarczyk, B.; Popp, J. The Long-Run Impacts of Temperature and Rainfall on Agricultural Growth in Sub-Saharan Africa. *Sustainability* **2021**, *13*, 595. [CrossRef]
5. Eck, M.A.; Murray, A.R.; Ward, A.R.; Konrad, C.E. Influence of Growing Season Temperature and Precipitation Anomalies on Crop Yield in the Southeastern United States. *Agric. For. Meteorol.* **2020**, *291*, 108053. [CrossRef]
6. John, R.P.; Megan, G. Temperatures and the Growth and Development of Wheat: A Review. *Eur. J. Agron.* **1999**, *10*, 23–36.
7. Ottman, M.J.; Kimball, B.A.; White, J.W.; Wall, G.W. Wheat Growth Response to Increased Temperature from Varied Planting Dates and Supplemental Infrared Heating. *Agron. J.* **2012**, *104*, 7–16. [CrossRef]

8. Zhao, C.; Liu, B.; Piao, S.; Wang, X.; Lobell, D.B.; Huang, Y.; Huang, M.; Yao, Y.; Bassu, S.; Ciais, P.; et al. Temperature Increase Reduces Global Yields of Major Crops in Four Independent Estimates. *Proc. Natl. Acad. Sci. USA* **2017**, *114*, 9326–9331. [CrossRef]
9. Asseng, S.; Ewert, F.; Martre, P.; Rötter, R.P.; Lobell, D.B.; Cammarano, D.; Kimball, B.A.; Ottman, M.J.; Wall, G.W.; White, J.W.; et al. Rising Temperatures Reduce Global Wheat Production. *Nat. Clim. Chang.* **2015**, *5*, 143–147. [CrossRef]
10. Challinor, A.J.; Wheeler, T.R.; Craufurd, P.Q.; Slingo, J.M. Simulation of the Impact of High Temperature Stress on Annual Crop Yields. *Agric. For. Meteorol.* **2005**, *135*, 180–189. [CrossRef]
11. Ferris, R.; Ellis, R.H.; Wheeler, T.R.; Hadley, P. Effect of High Temperature Stress at Anthesis on Grain Yield and Biomass of Field-Grown Crops of Wheat. *Ann. Bot.* **1998**, *82*, 631–639. [CrossRef]
12. Matsui, T.; Omasa, K.; Horie, T. The Difference in Sterility Due to High Temperatures during the Flowering Period among Japonica-Rice Varieties. *Plant Prod. Sci.* **2001**, *4*, 90–93. [CrossRef]
13. Vara Prasad, P.V.; Craufurd, P.Q.; Summerfield, R.J.; Wheeler, T.R. Effects of Short Episodes of Heat Stress on Flower Production and Fruit-set of Groundnut (*Arachis hypogaea* L.). *J. Exp. Bot.* **2000**, *51*, 777–784. [CrossRef] [PubMed]
14. Wang, X.; Peng, L.; Zhang, X.; Yin, G.; Zhao, C.; Piao, S. Divergence of Climate Impacts on Maize Yield in Northeast China. *Agric. Ecosyst. Environ.* **2014**, *196*, 51–58. [CrossRef]
15. Karimzadeh Soureshjani, H.; Ghorbani Dehkordi, A.; Bahador, M. Temperature Effect on Yield of Winter and Spring Irrigated Crops. *Agric. For. Meteorol.* **2019**, *279*, 107664. [CrossRef]
16. Grace, J. Temperature as a Determinant of Plant Productivity. *Symp. Soc. Exp. Biol.* **1988**, *42*, 91–107. [PubMed]
17. Goulart, H.M.D.; van der Wiel, K.; Folberth, C.; Balkovic, J.; van den Hurk, B. Storylines of Weather-Induced Crop Failure Events under Climate Change. *Earth Syst. Dyn.* **2021**, *12*, 1503–1527. [CrossRef]
18. Zhang, Y.; Hong, S.; Liu, D.; Piao, S. Susceptibility of Vegetation Low-Growth to Climate Extremes on Tibetan Plateau. *Agric For Meteorol.* **2023**, *331*, 109323. [CrossRef]
19. Feng, S.; Hao, Z.; Zhang, X.; Hao, F. Changes in Climate-Crop Yield Relationships Affect Risks of Crop Yield Reduction. *Agric. For. Meteorol.* **2021**, *304–305*, 108401. [CrossRef]
20. Powell, J.P.; Reinhard, S. Measuring the Effects of Extreme Weather Events on Yields. *Weather Clim. Extrem.* **2015**, *12*, 69–79. [CrossRef]
21. Shayanmehr, S.; Rastegari Henneberry, S.; Sabouhi Sabouni, M.; Shahnoushi Foroushani, N. Climate Change and Sustainability of Crop Yield in Dry Regions Food Insecurity. *Sustainability* **2020**, *12*, 9890. [CrossRef]
22. Schmitt, J.; Offermann, F.; Söder, M.; Frühauf, C.; Finger, R. Extreme Weather Events Cause Significant Crop Yield Losses at the Farm Level in German Agriculture. *Food Policy* **2022**, *112*, 102359. [CrossRef]
23. Zipper, S.C.; Qiu, J.; Kucharik, C.J. Drought Effects on US Maize and Soybean Production: Spatiotemporal Patterns and Historical Changes. *Environ. Res. Lett.* **2016**, *11*, 094021. [CrossRef]
24. Toreti, A.; Cronie, O.; Zampieri, M. Concurrent Climate Extremes in the Key Wheat Producing Regions of the World. *Sci Rep* **2019**, *9*, 5493. [CrossRef]
25. Lobell, D.B.; Cahill, K.N.; Field, C.B. Weather-Based Yield Forecasts Developed for 12 California Crops. *Calif. Agric.* **2006**, *60*, 211–215. [CrossRef]
26. Nassar, L.; Okwuchi, I.E.; Saad, M.; Karray, F.; Ponnambalam, K.; Agrawal, P. Prediction of Strawberry Yield and Farm Price Utilizing Deep Learning. In Proceedings of the 2020 International Joint Conference on Neural Networks (IJCNN), Glasgow, UK, 19–24 July 2020; IEEE: Piscataway, NJ, USA, 2020; pp. 1–7.
27. Newlands, N.K.; Zamar, D.S.; Kouadio, L.A.; Zhang, Y.; Chipanshi, A.; Potgieter, A.; Toure, S.; Hill, H.S.J. An Integrated, Probabilistic Model for Improved Seasonal Forecasting of Agricultural Crop Yield under Environmental Uncertainty. *Front. Environ. Sci.* **2014**, *2*, 17. [CrossRef]
28. California Department of Food and Agriculture (CDFA)—Statistics. Available online: https://www.cdfa.ca.gov/statistics/ (accessed on 9 October 2022).
29. Li, H.; Li, T.; Gordon, R.J.; Asiedu, S.K.; Hu, K. Strawberry Plant Fruiting Efficiency and Its Correlation with Solar Irradiance, Temperature and Reflectance Water Index Variation. *Environ. Exp. Bot.* **2010**, *68*, 165–174. [CrossRef]
30. Palencia, P.; Martínez, F.; Medina, J.J.; López-Medina, J. Strawberry Yield Efficiency and Its Correlation with Temperature and Solar Radiation. *Hortic. Bras.* **2013**, *31*, 93–99. [CrossRef]
31. Waister, P.D. Wind as a Limitation on the Growth and Yield of Strawberries. *J. Hortic. Sci.* **1972**, *47*, 411–418. [CrossRef]
32. Casierra-Posada, F.; Peña-Olmos, J.E.; Ulrichs, C. Basic Growth Analysis in Strawberry Plants (*Fragaria* sp.) Exposed to Different Radiation Environments. *Agron. Colomb.* **2012**, *30*, 25–33.
33. Ganmore-Neumann, R.; Kafkafi, U. The Effect of Root Temperature and Nitrate/Ammonium Ratio on Straw-berry Plants. II. Nitrogen Uptake, Mineral Ions, and Carboxylate Concentrations. *Agron J.* **1985**, *77*, 835–840. [CrossRef]
34. Heide, O.M. Photoperiod and Temperature Interactions in Growth and Flowering of Strawberry. *Physiol. Plant.* **1977**, *40*, 21–26. [CrossRef]
35. Webber, H.; Ewert, F.; Olesen, J.E.; Müller, C.; Fronzek, S.; Ruane, A.C.; Bourgault, M.; Martre, P.; Ababaei, B.; Bindi, M.; et al. Diverging Importance of Drought Stress for Maize and Winter Wheat in Europe. *Nat. Commun.* **2018**, *9*, 4249. [CrossRef] [PubMed]
36. Asseng, S.; Foster, I.; Turner, N.C. The Impact of Temperature Variability on Wheat Yields. *Glob. Chang. Biol.* **2011**, *17*, 997–1012. [CrossRef]

37. Zabel, F.; Müller, C.; Elliott, J.; Minoli, S.; Jägermeyr, J.; Schneider, J.M.; Franke, J.A.; Moyer, E.; Dury, M.; Francois, L.; et al. Large Potential for Crop Production Adaptation Depends on Available Future Varieties. *Glob. Chang. Biol.* **2021**, *27*, 3870–3882. [CrossRef] [PubMed]
38. Wang, Z.; Zhang, T.Q.; Tan, C.S.; Xue, L.; Bukovsky, M.; Qi, Z.M. Modeling Impacts of Climate Change on Crop Yield and Phosphorus Loss in a Subsurface Drained Field of Lake Erie Region, Canada. *Agric. Syst.* **2021**, *190*, 103110. [CrossRef]
39. Eyshi Rezaei, E.; Webber, H.; Gaiser, T.; Naab, J.; Ewert, F. Heat Stress in Cereals: Mechanisms and Modelling. *Eur. J. Agron.* **2015**, *64*, 98–113. [CrossRef]
40. United States Department of Agriculture. National Agricultural Statistics Service. Available online: https://www.nass.usda.gov/ (accessed on 9 April 2024).
41. Hanson, B.; Bendixen, W. Drip Irrigation Evaluated in Santa Maria Valley Strawberries. *Calif. Agric.* **2004**, *58*, 48–53. [CrossRef]
42. The California Strawberry Commission. Available online: https://www.calstrawberry.com/en-us/ (accessed on 9 January 2022).
43. The California Irrigation Management Information System. Available online: http://www.cimis.water.ca.gov/ (accessed on 10 January 2022).
44. Santa Maria Climate. Available online: https://en.climate-data.org/north-america/united-states-of-america/california/santa-maria-1488/ (accessed on 7 June 2022).
45. Sklar, A. *Fonctions de Repartition à n Dimensionls et Leurs Marges*; Publications de l'Institut Statistique de l'Université de Paris: Paris, France, 1959; pp. 229–231.
46. Ponnambalam, K.; Seifi, A.; Vlach, J. Yield Optimization with Correlated Design Parameters and Non-Symmetrical Marginal Distributions. In Proceedings of the 2003 International Symposium on Circuits and Systems, ISCAS '03, Bangkok, Thailand, 25–28 May 2003; IEEE: Piscataway, NJ, USA, 2020; Volume 4, pp. IV-736–IV-739.
47. Hofert, M.; Kojadinovic, I.; Machler, M.; Yan, J. *Elements of Copula Modeling with R*; Springer: Berlin/Heidelberg, Germany, 2018; ISBN 9783319896342.
48. Nelson, R.B. *An Introduction to Copulas*, 2nd ed.; Bickel, P., Diggle, P., Fienberg, S., Gather, U., Olkin, I., Zeger, S., Eds.; Springer: Berlin/Heidelberg, Germany, 2006; ISBN 13:978-0387-28659-4.
49. Zhang, L.; Singh, V.P. *Copulas and Their Applications in Water Resources Engineering*; Cambridge University Press: Cambridge, UK, 2019; ISBN 9781108565103.
50. Latif, S.; Mustafa, F. Bivariate Flood Distribution Analysis under Parametric Copula Framework: A Case Study for Kelantan River Basin in Malaysia. *Acta Geophys.* **2020**, *68*, 821–859. [CrossRef]
51. Roch, O.; Alegre, A. Testing the Bivariate Distribution of Daily Equity Returns Using Copulas. An Application to the Spanish Stock Market. *Comput. Stat. Data Anal.* **2006**, *51*, 1312–1329. [CrossRef]
52. Vaz de Melo Mendes, B.; Mendes Semeraro, M.; Câmara Leal, R.P. Pair-Copulas Modeling in Finance. *Financ. Mark. Portf. Manag.* **2010**, *24*, 193–213. [CrossRef]
53. Righi, M.B.; Ceretta, P.S. Analyzing the Dependence Structure of Various Sectors in the Brazilian Market: A Pair Copula Construction Approach. *Econ. Model* **2013**, *35*, 199–206. [CrossRef]
54. Hu, L. Dependence Patterns Across Financial Markets: A Mixed Copula Approach. *Appl. Financ. Econ.* **2006**, *16*, 717–729. [CrossRef]
55. Okhrin, O.; Ristig, A.; Xu, Y.-F. Copulae in High Dimensions: An Introduction. In *Applied Quantitative Finance*; Härdle, W., Chen, C.H., Overbeck, L., Eds.; Springer: Berlin/Heidelberg, Germany, 2017; ISBN 9783662544860.
56. Zhang, M.; Bedford, T. Vine Copula Approximation: A Generic Method for Coping with Conditional Dependence. *Stat. Comput.* **2018**, *28*, 219–237. [CrossRef]
57. Joe, H. *Multivariate Models and Multivariate Dependence Concepts*; Chapman and Hall/CRC: Boca Raton, FL, USA, 1997; ISBN 9780367803896.
58. Nazir, H.M.; Hussain, I.; Faisal, M.; Mohamd Shoukry, A.; Abdel Wahab Sharkawy, M.; Fawzi Al-Deek, F.; Ismail, M. Dependence Structure Analysis of Multisite River Inflow Data Using Vine Copula-CEEMDAN Based Hybrid Model. *PeerJ* **2020**, *8*, e10285. [CrossRef] [PubMed]
59. Spanhel, F.; Kurz, M.S. Simplified Vine Copula Models: Approximations Based on the Simplifying Assumption. *Electron. J. Stat.* **2019**, *13*, 1254–1291. [CrossRef]
60. Latif, S.; Mustafa, F. Parametric Vine Copula Construction for Flood Analysis for Kelantan River Basin in Malaysia. *Civ. Eng. J.* **2020**, *6*, 1470–1491. [CrossRef]
61. El Hannoun, W.; el Adlouni, S.-E.; Zoglat, A. Vine-Copula-Based Quantile Regression for Cascade Reservoirs Management. *Water* **2021**, *13*, 964. [CrossRef]
62. Xu, P.; Wang, D.; Wang, Y.; Singh, V.P. A Stepwise and Dynamic C-Vine Copula–Based Approach for Nonstationary Monthly Streamflow Forecasts. *J. Hydrol. Eng.* **2022**, *27*, 04021043. [CrossRef]
63. Chen, L.; Singh, V.P.; Shenglian, G.; Hao, Z.; Li, T. Flood Coincidence Risk Analysis Using Multivariate Copula Functions. *J. Hydrol. Eng.* **2012**, *17*, 742–755. [CrossRef]
64. Bedford, T.; Cooke, R.M. Vines: A New Graphical Model for Dependent Random Variables. *Ann. Stat.* **2002**, *30*, 1031–1068. [CrossRef]

65. Gräler, B.; van den Berg, M.J.; Vandenberghe, S.; Petroselli, A.; Grimaldi, S.; de Baets, B.; Verhoest, N.E.C. Multivariate Return Periods in Hydrology: A Critical and Practical Review Focusing on Synthetic Design Hydrograph Estimation. *Hydrol. Earth Syst. Sci.* **2013**, *17*, 1281–1296. [CrossRef]
66. Aas, K.; Czado, C.; Frigessi, A.; Bakken, H. Pair-Copula Constructions of Multiple Dependence. *Insur. Math. Econ.* **2009**, *44*, 182–198. [CrossRef]
67. Dißmann, J.; Brechmann, E.C.; Czado, C.; Kurowicka, D. Selecting and Estimating Regular Vine Copulae and Application to Financial Returns. *Comput. Stat. Data Anal.* **2013**, *59*, 52–69. [CrossRef]
68. Nagler, T.; Schepsmeier, U.; Stoeber, J.; Brechmann, E.C.; Graeler, B.; Erhardt, T. VineCopula: Statistical Inference of Vine Copulas. Available online: https://github.com/tnagler/VineCopula (accessed on 11 January 2022).
69. Rigby, R.; Stasinopoulos, M.; Heller, G.; De Bastiani, F. *Distributions for Modelling Location, Scale and Shape: Using GAMLSS in R*; CRC Press: Boca Raton, FL, USA, 2019.
70. Stasinopoulos, D.M.; Rigby, R.A. Generalized Additive Models for Location Scale and Shape (GAMLSS) in R. *J. Stat. Softw.* **2007**, *23*, 1–46. [CrossRef]
71. Jane, R.; Dalla Valle, L.; Simmonds, D.; Raby, A. A Copula-Based Approach for the Estimation of Wave Height Records Through Spatial Correlation. *Coast. Eng.* **2016**, *117*, 1–18. [CrossRef]
72. Eschenburg, P. *Properties of Extreme-Value Copulas*; Universitat Munchen: München, Germany, 2013.
73. Tawn, J.A. Bivariate Extreme Value Theory: Models and Estimation. *Biometrika* **1988**, *75*, 397–415. [CrossRef]

Disclaimer/Publisher's Note: The statements, opinions and data contained in all publications are solely those of the individual author(s) and contributor(s) and not of MDPI and/or the editor(s). MDPI and/or the editor(s) disclaim responsibility for any injury to people or property resulting from any ideas, methods, instructions or products referred to in the content.

sustainability

MDPI

Article

Optimizing Nitrogen Application for Enhanced Barley Resilience: A Comprehensive Study on Drought Stress and Nitrogen Supply for Sustainable Agriculture

Katarína Olšovská , Oksana Sytar * and Peter Kováčik

Institute of Agronomic Sciences, Slovak University of Agriculture in Nitra, 2, Tr. A. Hlinku, 94901 Nitra, Slovakia; katarina.olsovska@uniag.sk (K.O.); peter.kovacik@uniag.sk (P.K.)

* Correspondence: oksana.sytar@uniag.sk

Abstract: Soil water scarcity hinders crop productivity globally, emphasizing the imperative for sustainable agriculture. This study investigated the role of nitrogen in alleviating drought stress in barley. Parameters such as relative water content, photosynthetic rate, stomatal conductance, mesophyll concentration of CO_2, total leaf nitrogen, grain yield, total organic nitrogen content, starch content, and macronutrient concentrations (N, P, K, Ca, Mg) were examined. The optimal grain yield (3.73 $t \cdot ha^{-1}$) was achieved with 1 g of nitrogen per container (near 200 kg N $hectare^{-1}$) under ideal moisture conditions. However, under drought stress, nitrogen supply variants (1 g and 2 g per container) exhibited a significant decrease in photosynthetic rate (Pn), NRA activities, and a notable increase in Ci values. Stomatal conductance exhibited a substantial decrease by 84% in the early growth phase, especially with a 2 g dose of nitrogen supply. Nitrogen enhanced crude protein levels, yet both drought stress and nitrogen application reduced grain weight and starch content. Nitrogen effectively improved metabolic processes under drought, particularly in earlier growth stages (e.g., tillering). This research highlights the importance of sustainable agricultural practices related to the growth stage of barley, emphasizing nitrogen optimization to enhance crop resilience in water-scarce environments. The results underscore the intricate interplay between nitrogen fertilization, drought stress, and crop yield, indicating benefits during initial stress exposure but detrimental effects in subsequent growth stages.

Keywords: sustainable agriculture; relative water content; CO_2 assimilation; drought stress; nitrogen use efficiency

1. Introduction

The emerging impacts of climate change present a growing threat to agricultural crops, especially in tropical and sub-tropical regions. Drought stress stands out as a critical factor negatively influencing crop growth and productivity [1]. Simulation models play a crucial role in predicting the changing scenario of drought conditions due to climate change. The effort to enhance the resilience of strategic crops against the harmful effects of drought stress, inevitably resulting in reduced crop productivity, is essential for more sustainable agricultural practices [2].

The success of the Green Revolution can be primarily attributed to the development of superior plant varieties and the extensive application of fertilizers, particularly synthetic nitrogen fertilizers, to realize the increased yield potential of these new varieties. Approximately 50% of the nitrogen fertilizer applied to fields is assimilated and utilized by the designated crop, with the remainder being lost to the environment [3]. A small percentage is converted to the potent greenhouse gas N_2O, which contributes substantially to the total greenhouse gas emissions from agriculture. By the way, Green Economy thinking aims to promote sustainability through N_2O emission reduction in the context of the wider nitrogen cycle, with an emphasis on improving full-chain nitrogen use efficiency, optimal dose use for each crop plant and region, and exploiting a combination of technical measures in agriculture and other combustion sources [4,5]

Barley, a widely cultivated and economically valuable crop, stands at the forefront of the agricultural landscape [6]. The correlation between agronomic yield and nitrogen supply under well-watered conditions is well established [7]. Nitrogen, when efficiently managed, possesses the potential to alleviate water stress in crops by sustaining metabolic activities even under low tissue water potential. Establishing optimal fertilizer regimes is therefore essential to enhance metabolic and regulatory processes during kernel development in cereal crops [8]. Nitrogen application has been shown to influence starch synthesis and grain quality, particularly under drought stress conditions [9].

In the quest for sustainable solutions, the use of nitrogen supply not only improves water use efficiency but also supports the antioxidant system. This includes crucial enzymes like superoxide dismutase and catalase, which play a role in mitigating the stress associated with deficit irrigation [10]. Nitrogen serves as a regulator influencing the impact of short-term heat, drought, and combined stresses on diverse aspects of wheat physiology, encompassing photosynthesis, yield, nitrogen metabolism, and nitrogen use efficiency [11,12].

Despite extensive global research on plant responses to drought stress, studies on the role of nitrogen in mitigating drought stress in barley compared to other cereals crops are comparatively rare [13]. Barley, a resilient crop cultivated in both highly productive agricultural systems and marginal subsistence environments, holds the position of the fourth most important cereal crop globally, following wheat, maize, and rice [14,15]. Although its direct contribution to human food may be minor, the potential for new applications that leverage the health benefits of whole grains and beta-glucans is significant [16].

Drought stress significantly impacts barley physiology, affecting soluble and insoluble sugar levels, as well as the uptake of grain nitrogen (N), phosphorus (P), and potassium (K) [17]. The pursuit of climate-resilient and high-yielding barley varieties is ongoing, with researchers focusing on understanding the genetic controls influencing morphological and physiological responses to drought at different stages of plant growth [18].

In tandem with breeding strategies, exploring locally sourced barley genotypes becomes crucial to harness the potential of plant biodiversity in mitigating drought stress with the aid of nitrogen supply. Parameters such as relative water content, photosynthetic rate, stomatal conductance, mesophyll concentration of CO_2, total leaf nitrogen, grain yield, total organic nitrogen content, starch content, and macronutrient concentrations (N, P, K, Ca, Mg) are significant for optimizing agricultural practices. The current study aimed to investigate, through a long-term vegetation pot experiment, the above-mentioned parameters in spring barley measured under the influence of drought stress and different dosages of nitrogen supply. A focus on the potential of nitrogen supply in mitigating drought stress is important in developing more sustainable agricultural practices.

2. Materials and Methods

2.1. Plant Experiment

This three-year field experiment involved the cultivation of spring barley (variety Kompakt) plants in cylindrical plastic containers measuring 290 mm in diameter and 260 mm in height. Each container, as detailed in the preceding section, received 15 kg of soil, which was meticulously homogenized through thorough mixing. Soil was introduced into the containers ten days before sowing to facilitate proper settling. Before sowing, any germinating weeds on the soil surface were eliminated, and the top 30 mm of the soil was aerated by loosening.

The experiment aimed to investigate the impact of three different nutrition strategies under two levels of water regimes during three distinct growth phases of spring barley (shooting, stalk elongation, and budding flowering). An optimal water regime was set up and maintained at the level of 50 to 60% of field water capacity. Under drought stress, soil humidity was reduced to 15–20% of field water capacity.

The following fertilization methods were employed:

1st variant—without N fertilization (control);

2nd variant—N fertilization to the level of 1 g per pot + 0.33 g P + 1.1 g K (near 200 kg N hectare^{-1});

3rd variant—N fertilization to the level of 2 g per container + 0.33 g P + 1.1 g K (near 400 kg N hectare^{-1}).

In the context of the presented pot experiment, we consciously opted to maintain the original values of 1 g and 2 g of nitrogen per container, thereby preserving the experiment's initial framework. Simultaneously, we introduced the conversion to kg N hectare^{-1}, acknowledging its dependency on various factors in field experiments. This assumption was made with the intention of fertilizing one hectare of soil to a depth of 0.2 m. Considering an approximate weight of 3,000,000 kg per hectare of land and assuming uniform soil density (1.5 g/cm^3) in the Slovak Republic, this value was employed for calculations.

The nitrogen fertilizer utilized in this study was DAM 390 (containing 30% nitrogen). For phosphorus fertilization, we employed triple superphosphate (containing 20% phosphorus), and for potassium, we used potassium salt with a potassium content of 60% (equivalent to 49.8% K). These fertilizers were incorporated into the soil in predetermined amounts during the filling of the experimental containers. Each combination of experimental factors, defined by the interaction between moisture regime and fertilization, was replicated four times. The biological replication for each variant was n = 10 (10 containers for each variant, 29 plants in one container), and the analytical replication for the studied experimental parameters was n = 3.

The investigation focused on different fertilization variants applied under two distinct soil moisture regimes:

1. Optimal water regime (50–60% PVK);
2. Stress due to water deficit (15–20% PVK).

The growth and development of the spring barley plants was monitored during the growing season, and the onset of the main growth phases were recorded by date. The growth phases were rated using the DC scale:

1st branching (DC 21–DC 29);

2nd stabling end (DC 30–DC 49);

3rd blooming end (DC 50–DC 69).

DC is decimal code for the development stages of cereals (wheat, barley), corresponding to Zadoks et al. (1974) [19].

2.2. *Relative Water Content*

The relative water content (RWC), expressed as a percentage, reflects the relationship between the water content within a plant organ (e.g., leaf) and its water content under full turgor pressure conditions, as described by Turner (1981) and Olsovska et al. (2016) [20,21]. RWC values were calculated both before initiating the drought treatment (control, representing turgid plants) and after specific periods or stages of water deprivation (drought).

To determine RWC, leaf discs were extracted from the central region of an experimental leaf. Fresh weight (FW) was measured immediately after conducting gas exchange analyses. Turgid weight (TW) was obtained after hydrating the leaf disc in distilled water at 4 °C in the dark for 12 h. Dry weight (DW) was determined after drying the leaf disc at 80 °C for a 24 h period.

$$RWC = FW\text{-}DW/TW\text{-}DW \quad (1)$$

FW—fresh weight of the leaf segment;

TW—weight of the leaf segment in full turgor;

DW—dry weight of the analysed leaf segment.

2.3. Net Assimilation Rate of CO_2

The net assimilation rate of CO_2 (Pn—µmol·m^{-2}·s^{-1}) was determined by measuring the CO_2 consumption in intact photosynthesizing plant leaves enclosed within an assimilation chamber of the CIRAS-3 DC (PP Systems International, Inc., Amesbury, MA, USA) non-dispersive, open-type infra-red gas analyser, which contained, in the Ciras-3 console, four independent gas analysers simultaneously measuring absolute CO_2 and H_2O for both reference and analysis gas streams. All measurements were corrected for temperature and pressure. The external CO_2 concentration was maintained at 370 µL·L^{-1}, and the illumination level was set at 800 µmol·m^{-2}·s^{-1}.

2.4. Stomatal Conductance and Internal CO_2 Concentration

Stomatal conductance, denoted as g_s in mmol·m^{-2}·s^{-1}, and internal CO_2 concentration (Ci) in (µmol.m^{-2}.mol^{-1}) were derived from the same measurements of the CO_2 and H_2O flows in the CIRAS-3 photosynthetic system as Pn, as outlined by Olšovská and Brestic in 2001 [22].

2.5. Monitoring of Crop-Forming Elements

The following crop-forming elements were monitored in the container experiment:

(a) Grain yield per container (g);
(b) Thousand kernel (grain) weight—HTZ (g) (used DIPOS grain counter).

2.6. Assessment of Spring Barley Grain Quality

Following post-harvest maturation, the barley grain underwent evaluation based on the following parameters:

(a) Starch content (%): Determined using the Ewers polarimetric method.
(b) Measurement of total nitrogen: This was accomplished through Kjeldahl analysis, with nitrogen content being multiplied by 6.25 to calculate total nitrogen levels as per the method outlined by Kjeldahl in 1883 [23].

2.7. Agrochemical Analyses of Plants

Plant yield and the concentrations of nitrogen (N), phosphorus (P), potassium (K), calcium (Ca), and magnesium (Mg) were evaluated annually throughout the duration of the experimental study. The spring barley yield was determined by manually harvesting plants from two adjacent central rows at various growth stages: shoot emergence or when the second stem node became visible (DC 29), the second phase of stem elongation (DC 49), and the third phase of flowering (DC 69).

For the analysis of nutrient concentrations, six randomly selected spring barley plants (including both control and drought treatment groups) were divided into stem, leaf, husk, and grain components. These plant samples were dried at 65 °C to a constant weight and subsequently ground for further analyses. Nitrogen concentration in the plant material was determined using the Kjeldahl method [23]. Phosphorus and potassium concentrations were assessed in ground plant material after mineralization at 550 °C for 6 h. The resulting ash was mixed with 2 cm^3 of diluted HNO_3 (a 1:1 mixture of concentrated nitric acid and distilled water). Phosphorus was quantified calorimetrically using vanadium–ammonium molybdate, while potassium and calcium concentrations were determined via flame photometry (SpectrAA-250Plus, Varian, Markham, ON, Canada). Magnesium was quantified using atomic absorption spectrophotometry.

All macronutrient concentrations (N, P, K, Ca, Mg) were expressed as percentages based on their dry weight representation, and macronutrient uptakes by above-ground biomass were recalculated per kilogram per hectare (kg·ha^{-1}). This paper will present the results of the statistical analysis of the investigated agrochemical parameters.

2.8. Statistical Analyses

Statistical analyses were carried out using the Statistica v. 10 software (StatSoft Inc., Tulsa, OK, USA) and the graphics software SigmaPlot version 11.0 (Systat Software Inc., San Jose, CA, USA).

3. Results

Over the span of three experimental years, in conditions of optimal soil moisture, a positive impact of nitrogen fertilization on Nitrate Reductase Activity (NRA) was observed in all growth phases, regardless of the nitrogen dosage. NRA displayed a slight upward trend as the growth phases progressed, signifying an increase as the stand aged. The highest NRA value, reaching 45.8 nmol N-NO^{2-} g^{-1} fresh mass.min^{-1}, was observed during the scallion growth phase, when 2 g N per container was applied (Table 1).

Table 1. The physiological parameters of spring barley, as influenced by the growth phase, water regime, and fertilization treatments, averaged over a three-year period.

A	B	C	D	E	F	G	H	I
Stress in Growth. Phase	**Water Regime**	**g N. Container^{-1}**	**kg N hectare^{-1}**	**NRA (nmol N-NO_2^-.g^{-1} Fresh mass.min^{-1})**	**RWC (%)**	**P_n (µmol. m^{-2}.s^{-1})**	**C_i (µmol. m^{-2}.mol^{-1})**	**g_s (mmol. m^{-2}.s^{-1})**
tillering	control	0 g	0 kg	8.89	93.89	9.35	126.42	73.35
		1 g	200 kg	39.79	94.92	8.37	94.68	53.22
		2 g	400 kg	30.35	94.38	9.02	95.53	59.59
	stress	0 g	0 kg	5.01	83.57	5.82	120.21	34.09
		1 g	200 kg	16.83	65.31	3.52	145.94	18.43
		2 g	400 kg	15.61	69.28	0.98	183.33	9.09
shooting	control	0 g	0 kg	8.66	91.01	12.97	131.42	108.68
		1 g	200 kg	27.78	91.37	10.12	76.59	62.41
		2 g	400 kg	39.09	87.52	9.37	65.95	59.26
	stress	0 g	0 kg	7.50	79.40	10.84	98.54	85.43
		1 g	200 kg	4.68	63.90	2.83	224.48	19.09
		2 g	400 kg	12.91	62.05	3.62	202.00	22.11
blooming	control	0 g	0 kg	18.77	92.77	12.15	216.30	137.37
		1 g	200 kg	31.12	89.99	14.30	217.40	134.04
		2 g	400 kg	45.80	90.69	11.43	178.81	57.49
	stress	0 g	0 kg	9.22	77.60	7.00	213.59	76.57
		1 g	200 kg	7.01	64.95	2.77	438.49	15.17
		2 g	400 kg	19.77	63.14	4.23	382.00	15.36

Under drought conditions, significantly lower NRA values were consistently recorded at all growth stages and under various nutrition levels in comparison to plants grown under optimal soil moisture conditions. Among the stressed plants, NRA values decreased during the seedling phase in fertilized variants but increased during the tillering stage, particularly when a higher fertilization level (2 g N per container) was applied, compared to NRA values recorded during the tillering phase.

When averaged over three years, as well as within individual years, the relative water content (RWC) of the leaf tissues in optimally irrigated plants was found to be statistically significantly higher than that of leaves from plants subjected to soil drought.

In the nitrogen-fertilized treatments, a reduction in relative water content (RWC) was observed in all growth phases except for the tillering stage under optimal moisture conditions when compared to the unfertilized control. Over the average of the experimental years, net assimilation (Pn) in the main stem leaves of barley showed statistically significantly higher values in all monitored growth phases when grown under optimal irrigation conditions compared to the Pn values measured in the barley plants subjected to drought stress conditions. In all experimental variants with different nitrogen supplies (0 g, 1 g, 2 g of N), a significant decrease in Pn values was observed under the effects of drought stress, especially during the shooting phase of barley growth. Drought stress led to decreases in Pn values of 84% in the 0 g N supply variant, 72% in the 1 g nitrogen supply variant, and 61% in the 2 g nitrogen supply variant compared to the control (without N supply).

Under optimal soil moisture conditions, barley leaves displayed the highest intercellular CO_2 concentration (Ci) values on the non-fertilized treatments (ranging from 120 to 216 μmol $CO_2 \cdot m^{-2} \cdot s^{-1}$) in the individual growth phases, in contrast to the nitrogen-fertilized treatments, where a significant decrease in Ci values was evident (statistically supported). Under stress conditions, the barley plants responded differently to nitrogen fertilization concerning this parameter. The lowest Ci values were recorded in the unfertilized control treatment in all three growth phases. A particularly significant decrease of 25% in Ci value was observed during the shooting growth phase in the variant with 0 g N under drought stress. Nitrogen fertilization significantly increased the intercellular concentration of CO_2 in barley leaves, reaching peak values (438.49 or 382.00 μmol $CO_2 \cdot m^{-2} \cdot s^{-1}$) during the shooting growth phase, with a greater impact observed at the lower nitrogen dose (1 g N). At the lower N dose (1 g per container), the Ci values increased by 35% during the tillering growth phase, by 66% during the shooting growth phase, and by 50% during the blooming phase. Interestingly, higher increases in Ci values were observed in the variant with a higher nitrogen dose (2 g per container)—by 48% during the tillering growth phase, 67% during the shooting growth phase, and 52% during the blooming phase, respectively.

Overall, the stomatal conductance (g_s) values were higher in the unfertilized, optimally irrigated plants (ranging from 73.35 to 137.37 mmol$\cdot m^{-2} \cdot s^{-1}$) compared to the values obtained for the plants grown under stress conditions (ranging from 34.09 to 85.43 mmol$\cdot m^{-2} \cdot s^{-1}$), indicating that drought reduced stomatal conductance (through stomata closure). However, there was a substantial 84% reduction in stomatal conductance observed in the variant with a 2 g N supply during the early growth phase (tillering). In the variant with a 1 g N supply under drought stress, the barley plants exhibited a 65% decrease in stomatal conductance compared to the control variant without drought stress. As the growth stages progressed, both variants with nitrogen supply demonstrated noteworthy reductions in stomatal conductance values. Notably, the early stage of barley growth proved to be more sensitive to drought stress, displaying a significantly higher difference in the response between the variant with a 1 g N supply and the variant with a 2 g N supply—a 19% difference.

Nitrate reductase activity (NRA) was significantly decreased under drought stress in both experimental variants with N supply. At the tillering growth stage, NRA values were reduced by 58% in the variant with a nitrogen supply dose of 1 g and by 49% in the variant with a N supply dose of 1 g. At further growth stages, the variant with a nitrogen supply dose of 1 g under drought stress showed a significant decrease in NRA by 83% and 78% at the shooting and blooming growth phases, respectively. In the subsequent growth stages, the barley plants in the variant receiving 2 g of N supply under drought stress exhibited a noteworthy reduction in NRA, with decreases of 67% during the shooting growth phase and 57% during the blooming growth phase.

Across an average of three experimental years, all monitored factors, including year, growth phase, and fertilization, had a statistically significant effect on all of the observed physiological characteristics of barley, namely NRA, RWC, net CO_2 assimilation rate (Pn), intercellular CO_2 concentration (Ci), and stomatal conductance (g_s), under both optimal and stress moisture levels.

All experimental variables demonstrated a statistically significant influence on the nutrient concentration (N, P, K, Ca, Mg) within the dry matter of barley's above-ground biomass and the nutrient uptake by the above-ground plant biomass (Table 2). However, there were exceptions to this pattern. Notably, the impact of the year on the concentration and uptake of potassium (K) and magnesium (Mg) during barley cultivation under optimal moisture conditions did not reach statistical significance. In contrast, under drought stress conditions in barley plants, both the year's effect on Mg uptake and the impact of fertilization on the dry matter concentration of Mg were found to be statistically significant.

Table 2. Overview of the influence of experimental factors on the concentration of nutrients in the dry matter of the above-ground biomass and the uptake of nutrients by the above-ground mass of spring barley.

Monitored Parameter	Optimal Water Regime (n = 108)			Water Stress (n = 108)		
	Year	Growth Phase	Fertilization	Year	Growth Phase	Fertilization
Concentration N	++	++	++	++	++	++
Concentration P	++	++	++	++	++	++
Concentration K	-	++	++	++	++	++
Concentration Ca	++	++	++	++	++	++
Concentration Mg	++	++	++	++	++	+
Variant N	++	++	++	++	++	++
Variant P	++	++	++	++	++	++
Variant K	-	++	++	++	++	++
Variant Ca	++	++	++	++	++	++
Variant Mg	-	++	++	+	++	++

Note: ++ statistically highly demonstrable influence; + statistically demonstrable influence; - statistically unproven influence.

Over a three-year period, the nitrogen fertilization of spring barley resulted in a twofold increase in nitrogen (N) concentration within the dry matter of the above-ground biomass, both under optimal soil moisture conditions and in the plants subjected to stress compared to the unfertilized control. As the growth phases progressed, a dilution effect led to a reduction in N concentration in the dry matter for both the fertilized and unfertilized variants (Table 3).

In the final observed growth phase (heading), the N concentration in dry matter was higher for the fertilized variants of the stressed plants (2.19% and 2.71%, respectively) compared to the same fertilized plants grown under optimal conditions (1.85% and 2.41%, respectively). Nitrogen fertilization, whether under optimal or stress conditions, increased N uptake by the above-ground biomass; however, the rate of this increase was notably lower under stress conditions when compared to the plants grown under optimal moisture conditions. The application of 2 g N per container proved to be more effective in all cases than the half-dose of 1 g N per container.

Throughout all three years, the total harvested yield (HTZ) in the unfertilized treatments under stress conditions remained relatively consistent, occasionally even surpassing that of the unfertilized treatments under stress-free conditions (Table 3). Nitrogen fertilization further exacerbated this reduction. Conversely, when stress was applied during the different growth phases, nitrogen fertilization had the opposite effect, decreasing HTZ. At the tillering growth phase in the variants with N supply under drought stress, the HTZ values were mostly at the control level (without N supply). The HTZ values decreased under drought stress in barley plants of both nitrogen treatment variants (1 g and 2 g N) by 18% and 15%, respectively. This decreasing tendency regarding the HTZ level also continued at the shooting and blooming growth phase.

Nitrogen enhanced crude protein levels, yet both drought stress and nitrogen application reduced starch content. A significant increase in crude protein levels was observed in both the shooting and blooming growth phases. In the variant with a dose of 1 g of N supply, it increased by 43% in the shooting growth phase and by 34% in the blooming growth phase. Similar results with an increasing crude protein level were shown for the variant with a dose of N 2 g—by 43% in the shooting growth phase and by 33% in the blooming growth phase.

Table 3. Effect of stress on barley grain yield, HTZ, and quality parameters (average of three years).

A	B	C	D		E	F	G
stress in growth phase	dose N	dose N	Grain harvest		HTZ	Crude protein	Content of starch
	g.conteiner^{-1}	kg hectare^{-1}	(t·ha^{-1})	(g.container^{-1})	(g)	(%)	(%)
tillering	0 g	0 kg	1.01	6.69	32.84	12.17	63.14
	1 g	200 kg	3.14	20.72	30.86	13.99	60.65
	2 g	400 kg	3.07	20.27	34.05	19.50	59.11
shooting	0 g	0 kg	1.21	8.01	36.09	10.25	63.85
	1 g	200 kg	1.10	7.24	29.64	17.79	58.73
	2 g	400 kg	0.97	6.43	30.62	17.71	58.09
blooming	0 g	0 kg	0.95	6.29	29.95	11.72	62.17
	1 g	200 kg	2.32	15.31	21.16	17.63	61.34
	2 g	400 kg	0.98	6.45	21.90	17.58	56.21
optimum	0 g	0 kg	1.116	7.37	31.35	12.72	60.75
	1 g	200 kg	3.73	24.60	40.18	14.08	57.61
	2 g	400 kg	2.24	14.76	37.00	16.25	55.79

4. Discussion

4.1. The Physiological Parameters of Spring Barley, as Influenced by the Growth Phase, Drought Stress, and Nitrogen Fertilization Treatments

A significant increase in the intercellular CO_2 concentration in barley leaves, particularly during the steaming growth phase, attributed to nitrogen fertilization, with a more pronounced effect at a lower dose (1 g per container), was estimated. Concurrently, stomatal conductance (g_s) values displayed a noticeable trend, indicating higher values in the unfertilized, optimally irrigated plants compared to those under stress conditions, highlighting reduced g_s during drought. High levels of internal CO_2 and decreasing trends in stomatal conductance under drought stress are principal responses of C3 plants to help to reduce transpiration water loss by decreasing stomatal conductance and simultaneously increasing assimilation rates [24–26]. The changes in stomatal conductance indicate the adaptive changes of the experimental barley plants under different treatments, among which the dose of 2 g of N supply was evidently more stressful, presumably exposing the plant seedlings to soil osmotic stress.

Nitrogen fertilization had a negative impact on photosynthesis (Pn), resulting in reduced net assimilation values, though this impact was especially minor under irrigated conditions (except for the 1 g dose during the early growth phase). The high N application rate did not improve the net photosynthetic rate of the leaves but was able to inhibit it to some extent [27]. Specifically, this was visible in our experiment at the early stage of barley growth (tillering stage) for the variant with a 2 g dose of N. However, approved markable reductions in the photosynthesis of the plants under water deficit and combined N and water deficit were attributed to an imbalanced ATP/NADPH ratio, linked to the photosynthetic light reactions influencing carbon metabolism in the Calvin cycle [28]. Antagonistic, synergetic, and neutral effects between nitrogen addition and drought on resource use efficiency were found as well [29].

Conversely, under drought stress growing conditions, CO_2 assimilation was greater due to fertilization than in the non-fertilized control, resulting in Pn values dropping by 84% in the 0 g N supply variant. The decrease of 72% in the 1 g N supply variant and the decrease of 61% in the 2 g N supply variant compared to the control variant (without N) was estimated as well. In support of our findings, Guo et al. (2021) demonstrated that coupling a 20% reduced irrigation with traditional nitrogen application significantly enhanced grain yield in medium-density planted maize. This improvement was associated with heightened photosynthesis (Pn) and Y(NPQ) and a reduction in Y(NO) [30].

NRA was significantly decreased under drought stress in both experimental variants with N supply. However, during the early stage of barley growth (tillering), the variants with N supply showed a more moderate decrease in NRA activity compared to the variant without N under drought stress. Drought stress elevated nitrate nitrogen (NO_3-N) while concurrently diminishing the activities of N metabolism enzymes and the transcriptional levels of nitrate reductase, glutamine synthetase, and glutamate synthase [31].

Furthermore, during the early stage of barley growth (tillering), the variants with N supply demonstrated a more moderate decrease in NRA activity compared to the variant without N under drought stress. This may serve as evidence of a possible mitigating effect of nitrogen supply on barley plants under the influence of drought stress. Ru et al. (2022) demonstrated that a moderate application of nitrogen enhanced the activities of nitrate reductase and glutamine synthase in grains under post-anthesis heat and drought stress alone. This provided a basis for the accumulation of nitrogen and protein in the grains at the later stage of growth [11,32]. Chang et al. observed that nitrogen supply significantly enhanced the drought tolerance of grass plants. This improvement was attributed to the promotion of antioxidant metabolism and nitrogen metabolism, thereby safeguarding cell membranes against oxidative damage [33].

4.2. The Influence of Experimental Factors on the Concentration of Nutrients and Their Uptake in the Dry Matter of the Above-Ground Biomass of Barley Plants

The accumulation of minerals in plants during drought stress plays a crucial role in enhancing drought tolerance [34]. Plants employ various strategies to mitigate the deleterious effects of drought stress, including the promotion of phosphorus (P) and potassium (K) accumulation in various organs. They also regulate mineral concentrations in the phloem and xylem to prevent xylem embolism. Additionally, a decrease in magnesium (Mg) uptake was observed under drought stress [35]. All experimental variants with N supply under drought stress demonstrated a statistically significant influence on the nutrient concentration (N, P, K, Ca, Mg) within the dry matter of barley's above-ground biomass and the nutrient uptake by the above-ground plant biomass. Plants adapt to the high N and drought environment by altering their N uptake preference, and mineral uptake may explain changes in biomass with crude protein levels, nitrogen deposition, and drought [36].

The impact of drought stress on barley plants revealed a statistically significant effect on the annual uptake of magnesium (Mg) and the positive influence of nitrogen supply on the dry matter concentration of Mg. Additionally, another study confirmed that under rained conditions, nitrogen (N) fertilization significantly enhances barley productivity through its indirect influence on nitrogen accumulation in grain and straw. This process concurrently leads to an enhancement in grain quality by augmenting the accumulation of micronutrients such as magnesium (Mg) [37].

4.3. Effect of Stress on Barley Grain Yield, HTZ, and Quality Parameters: Average of Three Years

After our experimental assessment, it became clear that early exposure to stress during tillering significantly benefited the barley grain yield in terms of the stands' growth status. Křen et al. (2014) acknowledged that the timely and accurate prediction of grain yield and quality in spring barley is an essential requirement for efficient crop management [38]. To mitigate yield losses in plant objects due to high temperatures, it is crucial to apply the optimal amount of total nitrogen [10,39,40]. However, in the present study, in the later growth phases, starting from tillering, stress had a detrimental effect on grain yields.

Tillering is recognized as a crucial agronomic trait influencing the quality of plant crop populations and grain productivity. Wang et al., in 2017, illustrated that increased N levels enhanced the quantity and productivity of late-emerging tillers in rice, although these tillers demonstrated significantly fewer spikelets per panicle and less efficient grain filling in comparison to the primary stem or early-emerging tillers across varying nitrogen levels [41]. In our research, nitrogen application during the tillering phase in spring barley partially alleviated the detrimental effects of stress on the crop but proved ineffective when applied under stressful conditions or even resulted in a negative impact on grain yield. Abid et al. (2016) demonstrated that nitrogen nutrition enhances the capacity of wheat plant to mitigate the impacts of drought stress during vegetative growth stages [42].

Throughout all three years, the thousand kernel weight (HTZ) in the unfertilized treatments under stress conditions remained relatively consistent, occasionally even surpassing that of the unfertilized treatments under stress-free conditions. In the first year of the field experiment, drought stress consistently led to a reduction in HTZ, regardless of when the stress occurred. Nitrogen fertilization further exacerbated this reduction. In the second and third years of the experiment, HTZ increased in plants exposed to stress during tillering as a result of nitrogen fertilization. Conversely, when stress was applied during the seedling or tillering phases, nitrogen fertilization had the opposite effect, decreasing HTZ.

Utilizing linear fitting analysis, Wang et al. (2023) demonstrated a parabolic trend in grain yield, biomass yield, hundred kernel weight, and the ear grain number of maize plants as the nitrogen fertilization rate increased [43]. Across the three-year average considered, it became apparent that a fertilization rate of 1 g N per container significantly boosted yield. The application of moderate nitrogen levels was demonstrated to mitigate yield loss and ameliorate the decline in maize grain quality caused by post-silking heat stress [40]. In contrast, a double dose was found to be ineffective for optimal crop development, demonstrating a suppressive effect on crop formation, leading to reduced grain yield compared to the lower nitrogen dosage.

The barley plants demonstrated the lowest sensitivity to drought stress with regard to grain yield when the stress coincided with the tillering growth phase. Previous studies have highlighted that drought leads to a reduction in the number of tillers, plant height, and grains per ear in barley [44]. Overall, drought induced alterations in the morphology and architecture of the barley roots, with the roots transmitting stress signals to the caryopses, triggering the expression of various genes associated with protein biosynthesis, ultimately resulting in an increased accumulation of endosperm protein [45]. In this scenario, nitrogen fertilization demonstrated a significant stress-mitigating effect, maintaining grain yields at a high level. Conversely, the plants displayed the greatest sensitivity to stress in conjunction with fertilization when subjected to stress during transplanting. In this case, neither basal nor double nitrogen fertilization countered the impact of drought stress. When stress was induced during heading, a 1 g N per pot dose exhibited relatively effective stress mitigation, while the double dose failed to demonstrate any stress-mitigating effect, with yields remaining comparable to the unfertilized control.

Nitrogen enhanced crude protein levels, yet both drought stress and nitrogen application reduced starch content. A significant increase in crude protein levels was observed in the shooting and blooming growth phases. In the variant with a dose of 1 g of nitrogen supply, it increased by 43% in the shooting growth phase and by 34% in the blooming growth phases. Li and Wang (2023) reported a reduction in protein synthesis in barley leaves under drought stress conditions with a deficient nitrogen supply. The degradation of proteins in leaves was intensified in the absence of nitrogen [46]. The application of nitrogen amplifies the impact of pre-drought priming by modulating starch and protein biosynthesis in wheat [47]. Therefore, based on our results, we suppose that additional nitrogen supply may have a mitigating effect on barley plants under drought stress.

An investigation into the combined effect of drought stress and nitrogen fertilization on soybean plants revealed that a high nitrogen (N) rate is not advisable in the absence of drought conditions. In comparison to a low rate, it led to a reduction in the number of flowers and pods per plant, plant height, and seed yield. Conversely, under drought stress conditions, a high N rate positively influenced most traits, with thousand kernel weight showing the strongest correlation with yield [48].

Simultaneously, a modest reduction in nitrogen fertilizer application leads to enhanced plant quality through the modulation of starch properties without inducing any discernible yield loss. This was confirmed by our results derived from the use of 1 g of N supply as well. However, the augmentation of nitrogen levels induces alterations in the structure and characteristics of starch [49].

Nitrogen supply has been reported to be especially important under abiotic stresses [50,51]. When considering the three-year average data, it is evident that lower harvest total yield (HTZ) consistently occurred under stress conditions in the nitrogen-fertilized treatments compared to their counterparts, with analogous fertilization under optimal water conditions. Notably, nitrogen supply led to an increase in crude protein level and the maintenance of a moderate level of total yield compared to the unfertilized control.

5. Conclusions

In summary, our study reveals the complexity of nitrogen's role in sustainable barley cultivation. The additional nitrogen supply during drought stress heightened CO_2 concentration throughout the barley growth period, affecting stomatal conductance. Positive effects on magnesium uptake enhanced barley productivity, but the temporal dynamics varied across growth phases.

The thousand kernel weight (HTZ) responded inconsistently over three years, influenced by nitrogen and stress timing. While nitrogen boosted HTZ during tillering under stress, the complexities in the outcomes highlight the challenge of balancing nitrogen, water stress, and barley yield. Further research is essential for refined, sustainable crop management strategies.

Our findings underscore the intricate interplay between nitrogen fertilization, drought stress, and crop yield. The benefits observed during the early exposure of the plants to stress contrasted with the adverse effects in the later growth phases, emphasizing the need for a nuanced approach in sustainable crop management. These insights contribute to evolving strategies for effective and sustainable barley cultivation in changing environmental conditions.

Author Contributions: K.O.: conceptualization, funding acquisition, resources, writing—review and editing. O.S.: formal analysis, writing—original draft, writing—review and editing, supervision, P.K.: investigation, conceptualization, methodology, statistical analysis. All authors have read and agreed to the published version of the manuscript.

Funding: This research received no external funding.

Data Availability Statement: All data generated or analysed during this study are included in this published article.

Acknowledgments: The authors express their sincere gratitude to Martin Krček from Green Point, spol.s.r.o. (Farna, SVK), who assisted with the analysis in this study and provided practical suggestions on sustainable crop production in the agro-ecological conditions of southwest Slovakia.

Conflicts of Interest: The authors declare that they have no known competing financial interests or personal relationships that could have appeared to influence the work reported in this paper.

References

1. Wach, D.; Skowron, P. An overview of plant responses to the drought stress at morphological, physiological and biochemical levels. *Pol. J. Agron.* **2022**, *50*, 25–34.
2. Seleiman, M.F.; Al-Suhaibani, N.; Ali, N.; Akmal, M.; Alotaibi, M.; Refay, Y.; Dindaroglu, T.; Abdul-Wajid, H.H.; Battaglia, M.L. Drought stress impacts on plants and different approaches to alleviate its adverse effects. *Plants* **2021**, *10*, 259. [CrossRef] [PubMed]
3. Udvardi, M.; Brodie, E.L.; Riley, W.; Kaeppler, S.; Lynch, J. Impacts of agricultural nitrogen on the environment and strategies to reduce these impacts. *Procedia Environ. Sci.* **2015**, *29*, 303. [CrossRef]
4. Giordano, M.; Petropoulos, S.A.; Rouphael, Y. The Fate of Nitrogen from Soil to Plants: Influence of Agricultural Practices in Modern Agriculture. *Agriculture* **2021**, *11*, 944. [CrossRef]
5. van Grinsven, H.J.M.; Ebanyat, P.; Glendining, M.; Gu, B.; Hijbeek, R.; Lam, S.K.; Lassaletta, L.; Mueller, N.D.; Pacheco, F.S.; Quemada, M.; et al. Establishing long-term nitrogen response of global cereals to assess sustainable fertilizer rates. *Nat. Food* **2022**, *3*, 122–132. [CrossRef] [PubMed]
6. Sallam, A.; Alqudah, A.M.; Dawood, M.F.A.; Baenziger, P.S.; Börner, A. Drought Stress Tolerance in Wheat and Barley: Advances in Physiology, Breeding and Genetics Research. *Int. J. Mol. Sci.* **2019**, *20*, 3137. [CrossRef] [PubMed]
7. Govindasamy, P.; Muthusamy, S.K.; Bagavathiannan, M.; Mowrer, J.; Jagannadham, P.T.K.; Maity, A.; Halli, H.M.; Vadivel, R.; Raj, R.T.K.D.; Pooniya, V.; et al. Nitrogen use efficiency-a key to enhance crop productivity under a changing climate. *Front. Plant Sci.* **2023**, *14*, 1121073. [CrossRef] [PubMed]
8. Sandhu, N.; Sethi, M.; Kumar, A.; Dang, D.; Singh, J.; Chhuneja, P. Biochemical and Genetic Approaches Improving Nitrogen Use Efficiency in Cereal Crops: A Review. *Front. Plant Sci.* **2021**, *12*, 657629. [CrossRef]
9. Gous, P.W.; Warren, F.; Mo, O.W.; Gilbert, R.G.; Fox, G.P. The effects of variable nitrogen application on barley starch structure under drought stress. *J. Inst. Brew.* **2015**, *121*, 502–509. [CrossRef]
10. Agami, R.A.; Alamri, S.A.M.; Abd El-Mageed, T.A.; Abousekken, M.S.M.; Hashem, M. Role of exogenous nitrogen supply in alleviating the deficit irrigation stress in wheat plants. *Agric. Water Manag.* **2018**, *210*, 261–270. [CrossRef]
11. Ru, C.; Hu, X.; Chen, D.; Song, T.; Wang, W.; Lv, M.; Hansen, N.C. Nitrogen modulates the effects of short-term heat, drought and combined stresses after anthesis on photosynthesis, nitrogen metabolism, yield, and water and nitrogen use efficiency of wheat. *Water* **2022**, *14*, 1407. [CrossRef]
12. Khan, M.I.R.; Nazir, F.; Maheshwari, C.; Chopra, P.; Chhillar, H.; Sreenivasulu, N. Mineral nutrients in plants under changing environments: A road to future food and nutrition security. *Plant Genome* **2023**, *16*, e20362. [CrossRef] [PubMed]
13. El Sabagh, A.; Hossain, A.; Islam, S.; Barutcular, C.; Hussain, S.; Hasanuzzaman, M.; Akram, T.; Mubeen, M.; Nasim, W.; Fahad, S.; et al. Drought and salinity stresses in barley: Consequences and mitigation strategies. *Aust. J. Crop Sci.* **2019**, *13*, 810–820. [CrossRef]
14. Lister, D.L.; Jones, H.; Oliveira, H.R.; Petrie, C.A.; Liu, X.; Cockram, J.; Kneale, C.J.; Kovaleva, O.; Jones, M.K. Barley heads east: Genetic analyses reveal routes of spread through diverse Eurasian landscapes. *PLoS ONE* **2018**, *13*, e0196652. [CrossRef] [PubMed]
15. Newton, A.C.; Flavell, A.J.; George, T.S.; Leat, P.; Mullholland, B.; Ramsay, L.; Revoredo-Giha, C.; Russell, J.; Steffenson, B.J.; Swanston, J.S.; et al. Crops that feed the world 4. Barley: A resilient crop? Strengths and weaknesses in the context of food security. *Food Secur.* **2011**, *3*, 141–178. [CrossRef]
16. Askarnejad, R.M.; Soleymani, A.; Reza Javanmard, H. Barley (*Hordeum vulgare* L.) physiology including nutrient uptake affected by plant growth regulators under field drought. *J. Plant Nutr.* **2021**, *44*, 2201–2217. [CrossRef]
17. Cai, K.; Chen, X.; Han, Z.; Wu, X.; Zhang, S.; Li, Q.; Nazir, M.M.; Zhang, G.; Zeng, F. Screening of worldwide barley collection for drought tolerance: The assessment of various physiological measures as the selection criteria. *Front. Plant Sci.* **2020**, *11*, 1159. [CrossRef]
18. Elakhdar, A.; Solanki, S.; Kubo, T.; Abed, A.; Elakhdar, I.; Khedr, R.; Hamwieh, A.; Capo-chichi, L.J.A.; Abdelsattar, M.; Franckowiak, J.D.; et al. Barley with improved drought tolerance: Challenges and perspectives. *Environ. Exp. Bot.* **2022**, *201*, 104965. [CrossRef]
19. Zadoks, J.C.; Chang, T.T.; Konzak, C.F. A decimal code for the growth stages of cereals. *Weed Res.* **1974**, *14*, 415–421. [CrossRef]
20. Turner, N.C. Techniques and experimental approaches for the measurement of plant water status. *Plant Soil* **1981**, *58*, 339–366. [CrossRef]
21. Olsovska, K.; Kovar, M.; Brestic, M.; Zivcak, M.; Slamka, P.; Shao, H.B. Genotypically identifying wheat mesophyll conductance regulation under progressive drought stress. *Front. Plant Sci.* **2016**, *7*, 1111. [CrossRef] [PubMed]
22. Olšovská, K.; Brestič, M. Function of hydraulic and chemical water stress signalization in evaluation of drought resistance of juvenile plants. *J. Cent. Eur. Agric.* **2001**, *2*, 158–164.
23. Kjeldahl, J. Neue Methode zur Bestimmung des Stickstoffs in organischen Körpern" (New method for the determination of nitrogen in organic substances). *Z. Anal. Chem.* **1883**, *22*, 366–383. [CrossRef]
24. Wei, Z.; Abdelhakim, L.O.A.; Fang, L.; Peng, X.; Liu, J.; Liu, F. Elevated CO_2 effect on the response of stomatal control and water use efficiency in amaranth and maize plants to progressive drought stress. *Agric. Water Manag.* **2022**, *266*, 107609. [CrossRef]

25. Liu, X.; Zhang, H.; Wang, J.; Wu, X.; Ma, S.; Xu, Z.; An, B. Increased CO_2 concentrations increasing water use efficiency and improvement PSII function of mulberry seedling leaves under drought stress. *J. Plant Interact.* **2019**, *14*, 213–223. [CrossRef]
26. Guizani, A.; Askri, H.; Amenta, M.L.; Defez, R.; Babay, E.; Bianco, C.; Rapaná, N.; Finetti-Sialer, M.; Gharbi, F. Drought responsiveness in six wheat genotypes: Identification of stress resistance indicators. *Front. Plant Sci.* **2023**, *14*, 1232583. [CrossRef] [PubMed]
27. Noor, H.; Ding, P.; Ren, A.; Sun, M.; Gao, Z. Effects of Nitrogen Fertilizer on Photosynthetic Characteristics and Yield. *Agronomy* **2023**, *13*, 1550. [CrossRef]
28. Kang, J.; Chu, Y.; Ma, G.; Zhang, Y.; Zhang, X.; Wang, M.; Lu, H.; Wang, L.; Kang, G.; Ma, D.; et al. Physiological mechanisms underlying reduced photosynthesis in wheat leaves grown in the field under conditions of nitrogen and water deficiency. *Crop J.* **2023**, *11*, 638–650. [CrossRef]
29. Hu, J.; Ma, W.; Wang, Z. Effects of nitrogen addition and drought on the relationship between nitrogen- and water-use efficiency in a temperate grassland. *Ecol. Process.* **2023**, *12*, 36. [CrossRef]
30. Guo, Y.; Yin, W.; Fan, H.; Fan, Z.; Hu, F.; Yu, A.; Zhao, C.; Chai, Q.; Aziiba, E.A.; Zhang, X. Photosynthetic Physiological Characteristics of Water and Nitrogen Coupling for Enhanced High-Density Tolerance and Increased Yield of Maize in Arid Irrigation Regions. *Front. Plant Sci.* **2021**, *1*, 726568. [CrossRef]
31. Xia, H.; Xu, T.; Zhang, J.; Shen, K.; Li, Z.; Liu, J. Drought-Induced Responses of Nitrogen Metabolism in *Ipomoea batatas*. *Plants* **2020**, *9*, 1341. [CrossRef] [PubMed]
32. Statkevičiūtė, G.; Liatukas, Ž.; Cesevičienė, J.; Jaškūnė, K.; Armonienė, R.; Kuktaite, R.; Brazauskas, G. Impact of Combined Drought and Heat Stress and Nitrogen on Winter Wheat Productivity and End-Use Quality. *Agronomy* **2022**, *12*, 1452. [CrossRef]
33. Chang, Z.; Liu, Y.; Dong, H.; Teng, K.; Han, L.; Zhang, X. Effects of Cytokinin and Nitrogen on Drought Tolerance of Creeping Bentgrass. *PLoS ONE* **2016**, *11*, 0154005. [CrossRef] [PubMed]
34. Ahluwalia, O.; Singh, P.C.; Bhatia, R. A review on drought stress in plants: Implications, mitigation and the role of plant growth promoting rhizobacteria. *Resour. Environ. Sustain.* **2021**, *5*, 100032. [CrossRef]
35. Li, S.; Yang, L.; Huang, X.; Zou, Z.; Zhang, M.; Guo, W.; Addo-Danso, S.D.; Zhou, L. Mineral Nutrient Uptake, Accumulation, and Distribution in Cunninghamia lanceolata in Response to Drought Stress. *Plants* **2023**, *12*, 2140. [CrossRef] [PubMed]
36. Zhang, O.; Zhang, J.; Shi, Z.; Kang, B.; Tu, H.; Zhu, J.; Li, H. Nitrogen addition and drought affect nitrogen uptake patterns and biomass production of four urban greening tree species in North China. *Sci. Total Environ.* **2023**, *893*, 164893. [CrossRef] [PubMed]
37. Omirou, M.; Fasoula, D.; Stylianou, M.; Zorpas, A.A.; Ioannides, I.M. N-Source determines barley productivity, nutrient accumulation, and grain quality in Cyprus rainfed agricultural systems. *Int. J. Environ. Res. Public Health* **2023**, *20*, 3943. [CrossRef]
38. Křen, J.; Klem, K.; Svobodová, I.; Míša, P.; Neudert, L. Yield and grain quality of spring barley as affected by biomass formation at early growth stages. *Plant Soil Environ.* **2014**, *60*, 221–227. [CrossRef]
39. Li, Y.; Chen, J.; Tian, L.; Shen, Z.; Amby, D.B.; Liu, F.; Gao, Q.; Wang, Y. Seedling-Stage Deficit Irrigation with Nitrogen Application in Three-Year Field Study Provides Guidance for Improving Maize Yield, Water and Nitrogen Use Efficiencies. *Plants* **2022**, *11*, 3007. [CrossRef]
40. Yang, J.; Shi, W.; Xiao, G.; Zhang, X.; Wang, D.; Xu, H.; Wu, J.; Yang, Z.; Lai, Y.; Duan, M.; et al. Optimum total nitrogen application is required to reduce the yield loss of hybrid rice to high temperature. *Field Crops Res.* **2022**, *288*, 108696. [CrossRef]
41. Wang, J.; Fu, P.; Lu, W.; Lu, D. Application of moderate nitrogen levels alleviates yield loss and grain quality deterioration caused by post-silking heat stress in fresh waxy maize. *Crop J.* **2020**, *8*, 1081–1092. [CrossRef]
42. Abid, M.; Tian, Z.; Ata-Ul-Karim, S.T.; Cui, Y.; Liu, Y.; Zahoor, R.; Jiang, D.; Dai, T. Nitrogen Nutrition Improves the Potential of Wheat (*Triticum aestivum* L.) to Alleviate the Effects of Drought Stress during Vegetative Growth Periods. *Front. Plant Sci.* **2016**, *7*, 981. [CrossRef] [PubMed]
43. Wang, J.; Liu, G.; Cui, N.; Liu, E.; Zhang, Y.; Liu, D.; Ren, X.; Jia, Z.; Zhang, P. Suitable fertilization can improve maize growth and nutrient utilization in ridge-furrow rainfall harvesting cropland in semiarid area. *Front. Plant Sci.* **2023**, *14*, 1198366. [CrossRef] [PubMed]
44. Bartosiewicz, B.; Jadczyszyn, J. The impact of drought stress on the production of spring barley in Poland. *Pol. J. Agron.* **2021**, *45*, 3–11.
45. Li, F.; Chen, X.; Yu, X.; Chen, M.; Lu, W.; Wu, Y.; Xiong, F. Novel insights into the effect of drought stress on the development of root and caryopsis in barley. *PeerJ* **2023**, *8*, e8469. [CrossRef]
46. Li, L.; Wang, Y. Independent and combined influence of drought stress and nitrogen deficiency on physiological and proteomic changes of barley leaves. *Environ. Exp. Bot.* **2023**, *210*, 105346. [CrossRef]
47. Ullah, A.; Zhao, C.; Zhang, M.; Sun, C.; Liu, X.; Hu, J.; Zeeshan, M.; Zaid, A.; Dai, T.; Tian, Z. Nitrogen enhances the effect of pre-drought priming against post-anthesis drought stress by regulating starch and protein formation in wheat. *Physiol. Plant* **2023**, *175*, e13907. [CrossRef] [PubMed]
48. Basal, O.; Szabó, A. The Combined Effect of Drought Stress and Nitrogen Fertilization on Soybean. *Agronomy* **2020**, *10*, 384. [CrossRef]
49. Li, Y.; Liang, C.; Liu, J.; Zhou, C.; Wu, Z.; Guo, S.; Liu, J.; Na, A.; Wang, S.; Xin, G.; et al. Moderate Reduction in Nitrogen Fertilizer Results in Improved Rice Quality by Affecting Starch Properties without Causing Yield Loss. *Foods* **2023**, *12*, 2601. [CrossRef]

50. Ye, J.Y.; Tian, W.H.; Jin, C.W. Nitrogen in plants: From nutrition to the modulation of abiotic stress adaptation. *Stress Biol.* **2022**, *2*, 4. [CrossRef]
51. Kumari, V.V.; Banerjee, P.; Verma, V.C.; Sukumaran, S.; Chandran, M.A.S.; Gopinath, K.A.; Venkatesh, G.; Yadav, S.K.; Singh, V.K.; Awasthi, N.K. Plant nutrition: An effective way to alleviate abiotic stress in agricultural crops. *Int. J. Mol. Sci.* **2022**, *23*, 8519. [CrossRef]

Disclaimer/Publisher's Note: The statements, opinions and data contained in all publications are solely those of the individual author(s) and contributor(s) and not of MDPI and/or the editor(s). MDPI and/or the editor(s) disclaim responsibility for any injury to people or property resulting from any ideas, methods, instructions or products referred to in the content.

sustainability

MDPI

Article

Assessment of Suitability Area for Maize Production in Poland Related to the Climate Change and Water Stress

Aleksandra Król-Badziak [1,*], Jerzy Kozyra [1] and Stelios Rozakis [2]

[1] Institute of Soil Science and Plant Cultivation—State Research Institute in Puławy, 24-100 Puławy, Poland; kozyr@iung.pulawy.pl

[2] Bioeconomy and Biosystems Economics Lab, School of Chemical and Environmental Engineering, Technical University of Crete, 73100 Chania, Greece; srozakis@tuc.gr

* Correspondence: akrol@iung.pulawy.pl

Abstract: In this study, we identify the spatial distribution of water deficits in Poland. The analyses considered expert knowledge in soil categories importance in water stress evaluation influencing the climate suitability for maize production using the analytical hierarchy process (AHP). The Climatic Water Balance was calculated from April to September, for the baseline (BL) period (1981–2010) and two future periods of 2041–2070 (2050s) and 2071–2100 (2080s) using a six-member ensemble of GCM-RCM chain simulations under two representative concentration pathways (RCP) scenarios: low emissions (LE, RCP4.5) and high emissions (HE, RCP8.5). Taking into consideration water deficiency for the BL period, about 81% of Poland proved highly suitable, 18% moderately suitable and 1% marginally suitable for maize cultivation. According to LE and HE scenarios, the area of Poland that is highly suitable for maize production would decrease to 67 and 69% by the 2050s, and to 64 and 44% by the 2080s. By the 2080s, under the HE scenario, rain-fed maize production would become risky, as 21% of Poland would be marginally suitable, while 11% would not be suitable. According to our findings, supplemental irrigation is one of the effective adaptation strategies to maintain the production potential of maize in Poland.

Keywords: analytic hierarchy process; water deficit; climatic water balance; irrigation

1. Introduction

In Poland, the area under grain maize has increased significantly over the last 30 years, from 56 thousand ha in 1992 to 998 thousand ha in 2021, accounting for 11% of the total grain maize area in the EU [1]. The average grain maize yield over the last 30 years (1992–2021) was 5.9 t per 1 ha with a stable yield increase trend (0.7 t/10 years) but characterised by relatively high yield variability from year to year [2], from 3.7 t in 1992 to 7.3 t in 2012 [1].

Previous studies on land suitability for maize cultivation in Poland have mainly focused on thermal conditions [3]; however, in recent years, there has also been growing interest in other factors, mainly drought [2]. Climate change, with projected temperature increases, will increase the number of years with extremely unfavourable drought conditions, which might result in higher interannual yield variability and constitute a challenge for crop management [4]. In view of the increasing risk associated with drought and heat stress, the suitable production area for certain crops will change [5]. Previous studies have highlighted the need for adaptation measures to optimise water use on arable land [6]. The introduction of climate adaptation measures is necessary due to the predicted increase of droughts in Central Europe in the future [7]. Our study contributes to increasing the knowledge of adaptation practices in maize cultivation in the face of climate change. The methodology used in this work will be applied for further research aimed at assessing the suitability of land for maize cultivation in Poland, considering other climate change hazards such as spring frost stress, heat stress, etc.

In the context of climate change, it is essential to assess shifts in land suitability for crop production to minimise crop production risks [8]. Land suitability assessments may also help to indicate proper adaptation strategies to optimise agricultural production for improving the resilience of farming systems to climate change [9,10]. Land suitability assessment is a technique used to determine how suitable the land is for growing a given crop in a given region [11]. It involves the combination of several parameters, such as topographic, soil, climatic and management parameters, and is, therefore, a multi-dimensional problem [12]. Various multi-criteria evaluation techniques have been used to address crop suitability, such as Analytical Hierarchy Process (AHP) [12,13], Fuzzy Analytical Hierarchy Process (FAHP) [14–16], Analytic Network Process (ANP) [14], Technique for Order of Preference by Similarity to Ideal Solution (TOPSIS) [17,18] and ELimination Et Choice Translating REality (ELECTRE) [19] integrated with remote sensing and Geographical Information System [11]. Some studies [14,20–23] have carried out a comparative analysis in land suitability assessment. Seyedmohammadi et al. [14] compared two methods (square root and fuzzy set) for land suitability assessment for irrigated maize cultivation. Rodcha et al. [20] compared parametric, classical AHP, and fuzzy AHP (FAP) approaches in the suitability assessment of cash crops. Elaalem [21] compared Parametric and Fuzzy Multicriteria Methods to evaluate land suitability for olive production. Kalichkin et al. [22] obtained the criteria weights based on AHP and the direct ranking method in land suitability evaluation for spring wheat. Elaalem [23] compared two approaches, Boolean and Fuzzy MCDA for land suitability evaluation of sorghum production.

One of the challenges in water deficiency assessment for crops is considering the dependence of crop water requirements and different characteristics of soil retention related to granulometric composition according to soil categories [24,25]. Such analyses are conducted as part of the sustainability assessment of production by determining the criteria weights and can also be applied in the analysis of the suitability of areas. There are various methods that can be used to determine the criteria weights, one of which is the analytical hierarchy process (AHP) [13,26]. The AHP has been used to determine the weights of the parameters in the land suitability assessment for maize [13,27–29] and other crops such as tea [30,31], rice and soybean [29]. The AHP has the advantage of incorporating experts' opinion to prioritise the criteria [27]. Another advantage over other methods is the measurement and control of judgement inconsistency. One of the drawbacks of the AHP methodology is that it involves a subjective judgement of the relative importance of two criteria being compared, while there is a possibility of omitting interrelationships between other criteria [13].

As climate change may impact the land suitability for crop production, in some studies climate change projections with adaptation options were considered. According to an analysis in Europe, the most important factor for cultivated crops is the spatially distributed water availability from rainfall [32]; however, an increasing risk of late frost will also be very important [33]. Lopez-Blanco et al. [9] identified suitable areas for rainfed maize production in Mexico for current and future climate conditions projected under different climate change scenarios. Jia et al. [34] evaluated the suitable areas for maize cultivation in China and their change under future climate conditions. Tan et al. [10] evaluated suitable areas for maize and wheat in Ethiopia under current and future climate conditions using the GIS-MCE Planting Ecological Adaptability model.

In the last years, there has been a tendency for the growing intensity of the agricultural drought in Poland [35]. While assessing drought risk, it is important to distinguish between the different soil categories, crops and geographic regions [24]. Wójcik et al. [35] determined Central Poland (among others the Dolnośląskie Voivodeship) as an area with the highest frequencies of agricultural droughts in spring cereals, while in the Małopolskie Voivodeship drought occurred only twice. Łabędzki et al. [36] noted that the tendency of water shortage for crops will increase with the foreseen climate warming. Somorowska [37] indicated the increase of drought severity in relatively large area extending from the south-west to Central Poland over the 60-year period that was analysed. Among the adaptation strategies proposed to minimise possible climate related yield losses, there are various adaptation practices, including changing the timing of cultivation, introducing irrigation, selecting varieties with both a shorter and a longer maturity [38].

Therefore, this study attempts to analyse the distribution of water deficiency parameters that influence climate suitability for maize production in Poland using the analytical hierarchy process (AHP). The study considers irrigation as an agricultural practice for drought risk reduction. The next section presents the description of the study area and the climate data used in the study. The methodology for water condition evaluation and land suitability assessment is also presented. Section 3 shows temperature and water conditions in the current and future climate, as well as the results of the assessment of the climatic water balance in Poland. At the end of this section, the results of the sustainability assessment considering water deficiency on different soil categories are presented with testing the addition of different amounts of water supply through irrigation. Section 4 presents a discussion of the results obtained, which are summarised in Section 5.

2. Materials and Methods

2.1. Study Area

Poland is situated in a transitional warm temperate climate zone. A mild oceanic climate and a dry continental climate have an influence on the Polish climate. The mean annual air temperature varies from 6 to 9 °C and decreases from south-west to north-east [36,39]. The growing season in Poland varies from 180 to 225 days while in the mountains it is up to 150 days [40]. In Poland, there is spatial and temporal variability of precipitation [41]. The annual precipitation sum is about 550–650 mm in Central Poland, 650–850 mm in Northern Poland, 700–900 mm in Southern Poland and more than 1000 mm in the mountains [40]. Natural conditions in Poland are rather favourable for agriculture, but droughts may appear every 4–5 years and become an important risk factor [42].

2.2. Climate Data

The gridded climate data was derived from the regional downscaled dataset generated by the EURO-CORDEX project [43]. In the following study, we used EURO-CORDEX simulations for Europe with horizontal resolutions of 0.11° (EUR-11, ~12.5 km) [44]. The original gridded data were bilinearly interpolated to a regular grid by using CDO (climate data operators; v. 1.7.0) software [45]. In the study, a six-member ensemble of GCM-RCM chain simulations (Table 1) were used, and the ensemble mean was used as final indicator value. The bias adjustment model output performed under the CORDEX-Adjust project (SMHI-DBS45-MESAN, 1989–2010) was used [46,47]. The EURO-CORDEX simulations uses the RCP scenarios (Representative Concentration Pathways) used in the IPCC fifth assessment report [48]. For the present study, we applied two anthropogenic forcing scenarios, namely with low (LE, RCP4.5) and high emissions (HE, RCP8.5). The RCP4.5 is an intermediate stabilisation pathway that assumes radiative forcing stabilisation at 4.5 Wm^{-2} after year 2100 while RCP8.5 is a high pathway for which radiative forcing after year 2100 still continues to rise above 8.5 Wm^{-2} [49]. The analyses were performed for the three time periods: (1) baseline period (BL) 1981–2010, and two future periods, (2) 2050s (2041–2070) and (3) 2080s (2071–2100).

Table 1. GCM-RCM chain simulations used in the study.

N	GCMs	RCMs	Institution
1	CNRM-CERFACS-CNRM-CM5	SMHI-RCA4	SMHI
2	ICHEC-EC-EARTH	SMHI-RCA4	SMHI
3	ICHEC-EC-EARTH	KNMI-RACMO22E	KNMI
4	ICHEC-EC-EARTH	DMI-HIRHAM5	DMI
5	IPSL-IPSL-CM5A-MR	SMHI-RCA4	SMHI
6	MPI-M-MPI-ESM-LR	SMHI-RCA4	SMHI

While using climate change scenarios, it should be mentioned that the results of numerical simulations differ from observations, therefore it is necessary to compare them [50]. Ensemble mean of six bias-corrected simulations were evaluated for annual mean temperature and annual precipitation sum. Following other studies [51], the averaged error values and RMSEs were calculated for 42 Institute of Meteorology and Water Management (IMGW-PIB) meteorological stations [52] for 1981–2010 time frame (Figure 1). The RMSE in annual means was 0.3 °C for temperature and 36 mm/year (6%) for precipitation. For each meteorological station, the error for precipitation was less than 15% that is an acceptable model bias of up to 20% of true precipitation due to a systematic rain gauge undercatch [50,53].

Figure 1. Location of meteorological stations (green dot) of Institute of Meteorology and Water Management (IMGW-PIB) considered in comparison between observations and climate scenarios data.

2.3. Water Condition Evaluation for Maize Cultivation

Climatic water balance (CWB) is one of the indicators used in Poland for meteorological drought monitoring as well as for assessment of its intensity [24]. CWB is the difference between precipitation (P) and potential evapotranspiration (PET) [54]. Since there is a tendency for increasing the frequency of extreme weather phenomena such as drought [55], water deficiency indicated by climatic water balance sum for the period from April to September was indicated as suitability criteria for maize cultivation. In the literature, there are many methods for PET calculation, however, some of them require numerous input data. Depending on the application, a number of simplifications is possible to implement in PET formula [56]. Doroszewski and Górski [56] elaborated an index of PET as a function of sunshine duration, daylength and temperature. In the following study, the calculation of an index of potential evapotranspiration was carried out on the basis of the simplified Penman formula presented by Doroszewski and Górski [56] and Żyłowska et al. [57].We calculated CWB based on an indicator of moisture conditions as a function of mean monthly air temperature and monthly precipitation sum [57].

2.4. Suitability Assessment

AHP models were used in several studies for weighting factors in land suitability assessment [13,27,29,31]. Thus, the selected criteria were rated using the AHP methodology introduced by Saaty [58]. Steps of the AHP are given as follows:

Step 1: Construction of pairwise comparison matrix. In this step, two criteria are evaluated at the time in terms of their relative importance regarding the suitability of the land for maize cultivation, considering water deficiency occurrence [59]. In AHP 9 points importance scale is used to express individual preferences or judgments (Table 2) [31,58]. The pairwise comparison matrixes were created by consulting the opinions of the team experts from the Institute of Soil Science and Plant Cultivation (IUNG) and workers of Agricultural Experimental Stations (RZD IUNG) leading long-term experiments.

Table 2. Scale of relative importance of selected criteria based on Saaty [58].

Intensity of Importance	Definition
1	Equal importance
3	Moderate importance
5	Strong importance
7	Very strong importance
9	Extreme importance
2, 4, 6, 8	Compromise between above values
Reciprocals of above	Values for inverse comparison

Step 2: Calculation of the eigenvectors and eigenvalue of the judgment matrix. After construction of pairwise comparison matrix, the eigenvectors w_i (weights of criteria) and eigenvalue λ_{max} were calculated as follows [29]:

$$w_i = \frac{\sqrt[n]{\prod_{j=1}^{n} a_{ij}}}{\sum_{i=1}^{n} \sqrt[n]{\prod_{j=1}^{n} a_{ij}}}, \tag{1}$$

$$\lambda_{max} = \frac{1}{n} \cdot \sum_{i=1}^{n} \frac{(AW)_i}{w_i}, \tag{2}$$

where i and *j* is number of rank and column of A.

Step 3: Consistency test. By using AHP methodology, the inconsistency in pairwise comparison matrix is identified and taken into account [60,61]. The consistency ratio (CR) proposed by Saaty [58] is used to test the consistency of pairwise comparisons. If the CR is below 0.10, it is considered that the inconsistency of the judgments is acceptable [31,58,60]. The consistency index (CI) and the consistency ratio (CR) is given by [62]:

$$CI = \frac{\lambda_{max} - n}{n - 1}, \tag{3}$$

$$CR = \frac{CI}{RI}, \tag{4}$$

where RI is Random Consistency Index which depends on n (number of elements), listed in Table 3.

Table 3. Random Consistency Index (RI) values [58].

n	1	2	3	4	5	6	7	8	9	10
RI	0.00	0.00	0.52	0.89	1.11	1.25	1.35	1.40	1.45	1.49

Step 4: Group decision making. Since decision making frequently involves contribution of many respondents with different individual opinions, group decision making process has been applied to aggregate them [63,64]. There are alternative methods suggested by the literature to aggregate individual opinions applied in AHP methodology, depending on whether the decision group is assumed to be a synergistic unit or simply a collection of individuals. When the individuals act in concert, judgments should be pooled in such a way that the group 'merges' the different individuals into a new 'individual' representing it. Aggregation of Individual Judgments (AIJ) is suggested for this purpose, that requires satisfaction of reciprocity condition for the synthesis alike the individual judgments. This is provided by the geometric mean, thus in AIJ, the elements of aggregated comparison matrix $(a_{ij})_g$ would be calculated as follows [63–65]:

$$(a_{ij})_g = \sqrt[N]{\prod_{n=1}^{N} (a_{ij})_{Ind\_n}}, \tag{5}$$

where $(a_{ij})_{Ind\_n}$ is the judgment of individual n, N is number of individuals that belong to the group.

The selected suitability criteria (CWB values for four soil types) were reclassified into four suitability classes (S1, S2, S3, N) (Tables 4 and 5). The soil types reflect their agronomic category according to their granulometric composition [66,67]. The soils are divided into four categories reflecting their different susceptibility to drought in relation to the amount of available water (AW): very light soils—the most susceptible (AW less than 127.5 mm), light soils—susceptible soils (AW between 127.5 and 169.9 mm), medium soils—moderately susceptible soils (AW between 170 and 202.5 mm) and heavy soils—soils with low susceptibility to drought (AW more than 202.5 mm) [66]. The range of field water capacity in the soil profile from 0 to 100 cm of these categories of soils is as follows: 110–145 mm, 146–210 mm, 211–270 mm and 271–460 mm [67]. The final suitability map was generated by using weighted overlay method, by combining the standardized criterion maps with the criterion weights [59,61] using QGIS 3.28.4.

$$\text{Weighted Overlay} = \sum_{i=1}^{n} C_i \cdot W_i, \quad (6)$$

where C_i is each reclassified criterion (*i*), W_i is the weight of the selected criteria, n is the total number of criteria.

The final suitability map was divided into four suitability classes according to FAO [29,68]. The flowchart of the methodology is presented in Figure 2.

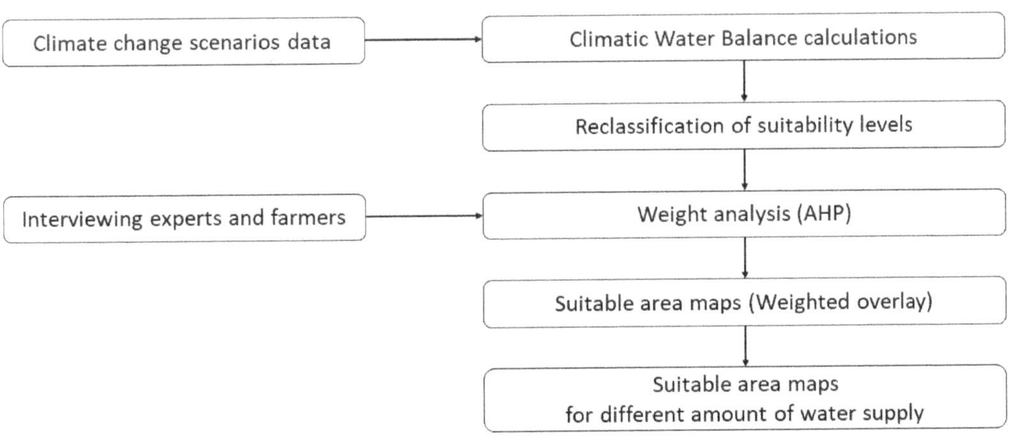

Figure 2. Flowchart of the methodology for assessing the suitability area for maize production in relation to the climate change and water deficiency.

Table 4. Suitability level per selected criteria [68].

Suitability Class	Description
S1—highly suitable	No significant limitations or only minor limitations
S2—moderately suitable	Moderately severe limitations that reduce productivity or benefits and rise inputs
S3—marginally suitable	Severe limitations that reduce productivity or benefits and rise inputs to the extent of being marginally justifiable
N—not suitable	Severe limitations that may preclude sustained use of the land in a specific manner

Table 5. Factor ratings of climate suitability for maize regarding water deficiency.

Characteristic		Suitability Class [1]				References
	Criteria	S1	S2	S3	N	
	Rating scale	85–100	60–85	40–60	0–40	[29,68]
Climatic water balance	on very light soils [mm]	>−150	−200–−150	−250–−200	<−250	[69]
	on light soils [mm]	>−210	−260–−210	−310–−260	<−310	[69]
	on medium soils [mm]	>−250	−300–−250	−350–−300	<−350	[69]
	on heavy soils [mm]	>−290	−340–−290	−390–−340	<−390	[69]

[1] S1—highly suitable, S2—moderately suitable, S3—marginally suitable, N—not suitable.

2.5. Irrigation as Potential Adaptation Strategy

Nowadays irrigation in Poland is treated as a supplemental practice, it is usually used in short time spans during the vegetation period. However, it can mitigate the effects of droughts on crop production locally. During the period from April to September, potential evapotranspiration usually exceeds precipitation causing a water deficit. It is seen especially on light soils with low water-holding capacity. The net irrigation water requirement in the growing period in Central Poland for maize is 50–70 mm [70]. Łabędzki [71] identified the need to assess the impact of climate change on irrigation requirements in Poland. Due to temperature increases with lower precipitation during summer, it is expected that irrigation needs will increase [71]. Therefore, in the current study, we tested irrigation by 30, 80 and 140 mm by adding three different amounts of irrigation water that impacted CWB for the growing season from April to September. The chosen amount of water supply (30, 80 and 140 mm) during irrigation is an example to test three types of irrigation requirements: (1) small (less than 50 mm—it is about 3 irrigation treatments per season), (2) medium (between 50 and 100 mm) and (3) high (more than 100 mm) [72].

3. Results

3.1. Mean Annual Temperature and Mean Annual Precipitation Sum for Baseline Period and by 2050s and 2080s

According to the ensemble mean of six analysed RCMs, the mean annual air temperature in Poland for 1981–2010 (BL) was 8.2 °C, while the mean annual precipitation sum was about 622 mm (Table 6). The spatial distribution of mean annual air temperature (Figure 3) shows its decrease from the southwest to the northeast of Poland under both current and future climate conditions. The projected increase in the annual temperature from the BL to the 2050s according to the LE scenario is about 1.5 °C and 2.1 °C to the 2080s (Table 6). For the HE scenario, projected mean annual temperature changes can reach about 2.1 °C and 3.8 °C by the 2050s and 2080s, respectively. The spatial distribution of mean annual precipitation is presented in Figure 4. Both in the baseline period and under future climate conditions, the highest annual precipitation occurs in mountainous areas (southern Poland) and the lowest in central Poland. According to the LE scenario, the annual precipitation will not change significantly in the 2050 perspective, whereas the increase in precipitation will be greater in the 2080 perspective. The HE scenario predicts a higher increase in precipitation than the LE scenario. Projected changes in mean annual precipitation sum according to the analysed ensembles of models are about +9% and +14% in the 2050s and +11% and +19% in the 2080s for the LE and HE scenarios, respectively.

Table 6. Mean values of annual temperature and annual precipitation sum and their changes (in brackets) in Poland.

Period [1]	**BL**	**2050s**		**2080s**	
Emission Scenario [2]		**LE**	**HE**	**LE**	**HE**
Mean annual temperature [°C]	8.2	9.7 (+1.5)	10.3 (+2.1)	10.3 (+2.1)	12.0 (+3.8)
Mean annual precipitation [mm]	622	676 (+9%)	710 (+14%)	692 (+11%)	741 (+19%)

[1] BL—baseline; [2] LE—low emissions scenario; HE—high emissions scenario.

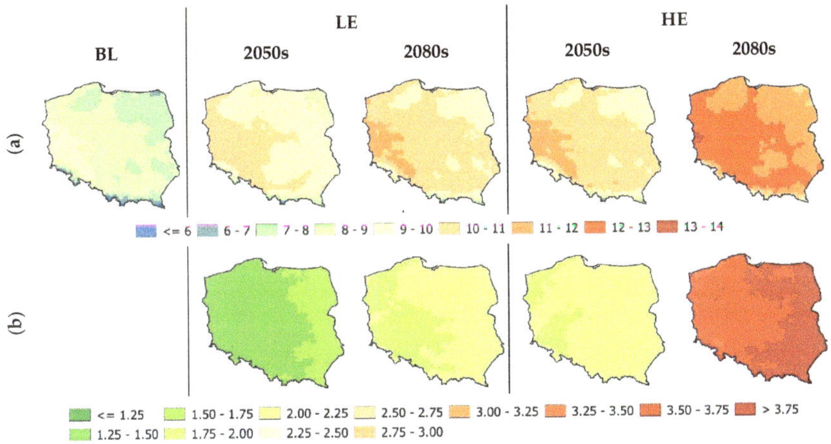

Figure 3. (**a**) Multimodel ensemble mean annual air temperature [°C]; (**b**) Changes in annual air temperature [°C] in Poland: baseline (BL, 1981–2010) and projected temperatures according to low emissions (LE, RCP4.5) and high emissions (HE, RCP8.5) scenarios for the 2050s (2041–2070) and the 2080s (2071–2100).

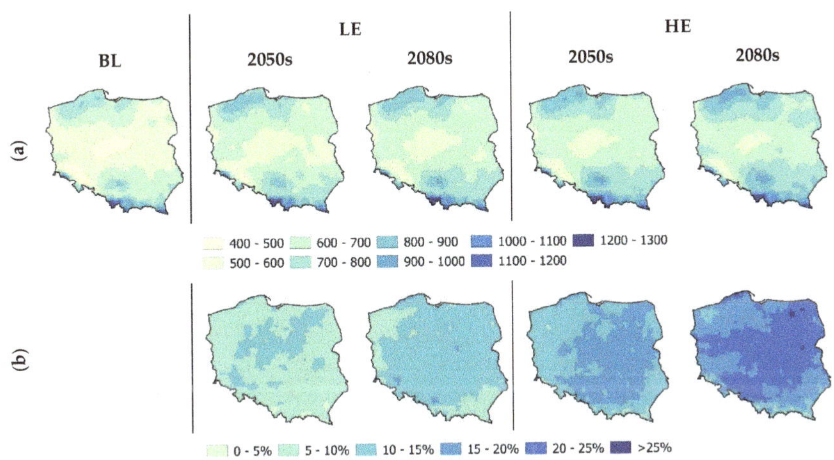

Figure 4. (**a**) Multimodel ensemble mean of annual precipitation sum [mm]; (**b**) Relative changes in annual precipitation sum [%] in Poland: baseline (BL, 1981–2010) and projected precipitation according to low emissions (LE, RCP4.5) and high emissions (HE, RCP8.5) scenarios for the 2050s (2041–2070) and the 2080s (2071–2100).

3.2. Mean Climatic Water Balance in the Period from April to September for Baseline Period and by 2050s and 2080s

The growing season sum of CWB (from April to September), averaged over Poland for the BL, was −213 mm (Table 7). According to the analysed LE scenario, the CWB will decrease by about 11% in the 2050s and will not change according to the HE scenario. In the 2080s, the seasonal CWB change will decrease by 13% according to the LE scenario and by 8% according to HE. The spatial distribution of the CWB for the growing season (from April to September) is shown in Figure 5. For most of the area of Poland, the CWB was negative as the PE exceeded the sum of precipitation. The highest positive values of CWB (more than 100 mm) were found in the south of Poland, where the sum of precipitation exceeded the PE. In north-western part of Poland (near the Baltic Sea) and in the southern part of Poland the values of CWB were more than −150 mm. The lowest values of the climatic water balance (less than −250 mm) were observed in the central Poland. The area with CWB less than −300 mm is projected to increase in the future, and according to the RCP8.5 scenario this area will dominate in 2071–2100.

Table 7. Mean values of the climatic water balance sum and its changes (in brackets) from April to September in Poland.

Period [1]	BL	2050s		2080s	
Emission Scenario [2]		LE	HE	LE	HE
CWB sum (Apr–Sept) [mm]	−213	−237 (−11%)	−214 (0%)	−240 (−13%)	−230 (−8%)

[1] BL—baseline; [2] LE—low emissions scenario; HE—high emissions scenario.

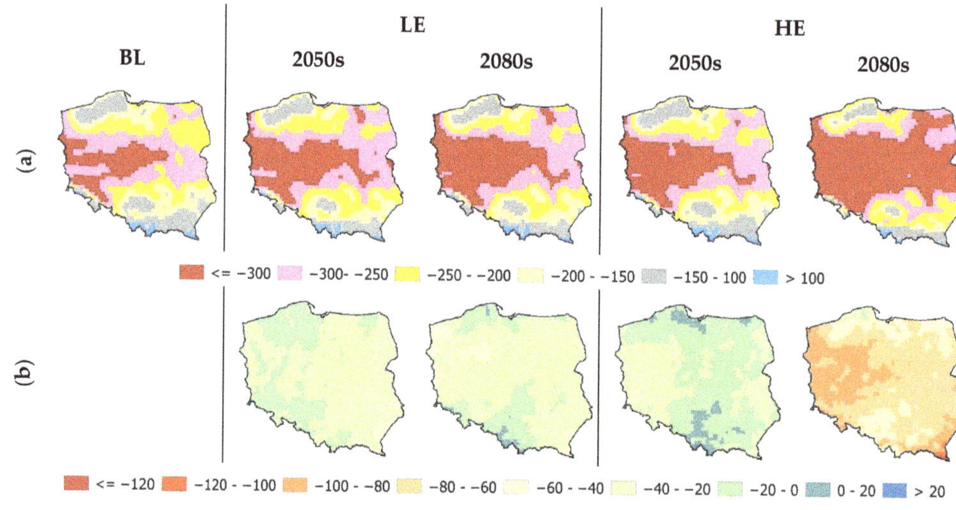

Figure 5. (**a**) Multimodel ensemble mean of the climatic water balance [mm]; (**b**) Changes in climatic water balance in the months of April to September: the baseline (BL, 1981–2010) and projected CWB according to the low emissions (LE, RCP4.5) and high emissions (HE, RCP8.5) scenario for the 2050s (2041–2070) and the 2080s (2071–2100).

3.3. Suitability Assessment

While assessing drought risk, it is important to distinguish water deficiency between the different soil categories. In the following study, we reclassified CWB values into four suitability classes varying according to soil types (Table 5, Figure 6). According to BL, the not suitable areas was observed for very light and light soils in central Poland. On very light soils, not suitable areas are going to increase in the future from 43% for BL to 57%

for LE and 72% of Poland for HE, while for light soil the area is projected to increase from 10% for BL to 26% for LE and 48% of Poland for the HE scenario. On heavy soils, there is no extreme deficiency for CWB observed in neither baseline nor futures scenarios apart from 2080s for the HE scenario. The highly suitable areas are observed on all kinds of soil in north-west and south of Poland, however the largest area is observed for medium and heavy soils.

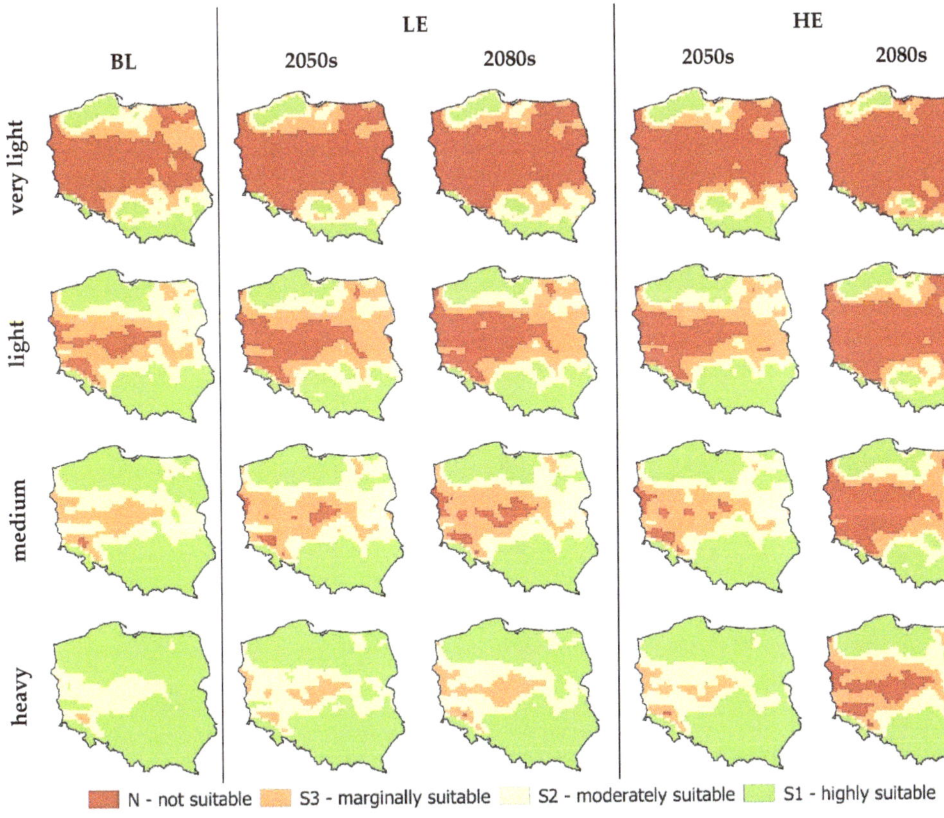

Figure 6. Water deficiency suitability classes on very light, light, medium and heavy soils for the baseline (1981–2010) and projected according to the low emissions (LE, RCP4.5) and high emissions (HE, RCP8.5) scenarios for the 2050s (2041–2070) and 2080s (2071–2100).

Four soil categories (very light, light, medium and heavy) were weighted using farmers' (5) and agricultural advisors/experts' (5) opinion using AHP pairwise comparison matrix (in total, 10 individuals). The purpose of the weighting is to determine the importance of each soil categories relative to others that influence land suitability for maize production, considering the occurrence of water deficiencys on these soil categories. The respondents indicated what type of soil they would choose for growing corn in the context of water stress (drought). The mean judgement of the individuals derived through pairwise comparisons is presented in Table 8. The weights of each soil categories calculated by the AHP method are listed in the last column of Table 8, with the higher value indicating that the criterion is more important. Based on the calculated values of the criteria weights (Table 8), medium soil categories are the most important factor, followed by heavy soils, while light soils are significantly less important. The least important factor in considering land suitability for maize cultivation due to water deficiency occurrence are very light soils.

In this suitability analysis, the consistency ratio (CR) value is 0.0535, which is below the threshold value of 0.1 indicating a reasonable level of consistency of the experts' judgment.

Table 8. AHP pairwise comparison matrix for the main criteria and resulted weight vector.

Criteria	Very Light	Light	Medium	Heavy	Weights
Very light	1.0000	0.2194	0.1305	0.1386	0.0433
Light	4.5581	1.0000	0.1909	0.2437	0.1169
Medium	7.6617	5.2382	1.0000	1.3286	0.4655
Heavy	7.2157	4.1038	0.7526	1.0000	0.3743

The final suitability map (Figure 7) was generated by using the weighted overlay method, by combining the reclassified criterion maps (Figure 6) with the criterion weights (Table 8). The suitability maps of the water deficiency indicators for maize cultivation were categorised into four suitability classes, namely: highly suitable (S1), moderately suitable (S2), marginally suitable (S3) and not suitable (N). The percentage of the suitability classes in Poland is presented in Table 9. Water deficiency is one of the key climate variables for maize, in consequence, climate change could significantly impact its production. Central Poland is the only area of Poland that has moderate suitability (18%) for maize cultivation due to the occurrence of water deficiency in the current climate. Subsequently, these conditions will extend to 22–26% of Poland depending on the period and scenario analysed. By the end of the century, water deficiency will increase and, according to HE, 11% of the area is projected to be not suitable while 21% will be only marginally suitable for rainfed maize cultivation (0 and 1% respectively in current conditions).

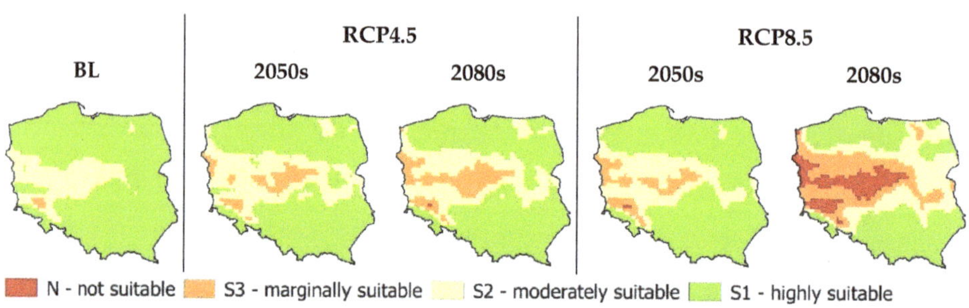

Figure 7. Final evaluation of climate suitability for maize regarding water deficiency for the baseline (1981–2010) and projected according to the low emissions (LE, RCP4.5) and high emissions (HE, RCP8.5) scenarios for the 2050s (2041–2070) and 2080s (2071–2100).

Table 9. Percentages of different water deficiency suitability levels for maize in Poland for the baseline (1981–2010) and projected according to low emissions (LE, RCP4.5) and high emissions (HE, RCP8.5) scenarios for the 2050s (2041–2070) and 2080s (2071–2100).

RCP					RCP4.5								RCP8.5							
Period	BL				2050s				2080s				2050s				2080s			
Rating scale [1]	S1	S2	S3	N	S1	S2	S3	N	S1	S2	S3	N	S1	S2	S3	N	S1	S2	S3	N
Percentage	81	18	1	0	67	26	7	0	64	24	12	0	69	22	8	0	44	24	21	11

[1] S1—highly suitable, S2—moderately suitable, S3—marginally suitable, N—not suitable.

3.4. *Adaptation Strategies Increasing Land Suitability for Maize—Irrigation*

Considering water deficiency, climate suitability for maize will decrease over time. The distribution with water limitations is shown in Figure 8. According to water stress, 19% of Poland, mainly in central Poland, was moderately or marginally suitable for maize cultivation under current climate conditions (Table 10). Furthermore, under future climate conditions, this area will increase to 33 and 30% in the 2050s and 36 and 45% in the 2080s for the LE and HE scenarios, respectively. However, in the 2080s for the RCP8.5 scenario according to water deficiency, 11% of the analysed area will not be suitable for maize production. Under current climate conditions, according to the results of reclassification of water deficiency, the highly suitable area would increase by 13%, 19% and 19% if 30, 80 and 140 mm of water were applied in irrigation, respectively. In the 2050s, the application of 30 mm of water will help to avoid marginally suitable areas for maize cultivation, while irrigation with 140 mm of water will create highly suitable conditions for the whole of Poland regarding water deficiency. In the 2080s, according to the HE scenario, the application of 30 mm of water will help to eliminate not suitable areas for maize cultivation, but to avoid marginally suitable areas, irrigation with 80 mm of water will be necessary. If 80 mm of water is applied by irrigation, about 18% of the analysed area will remain moderately suitable for maize production.

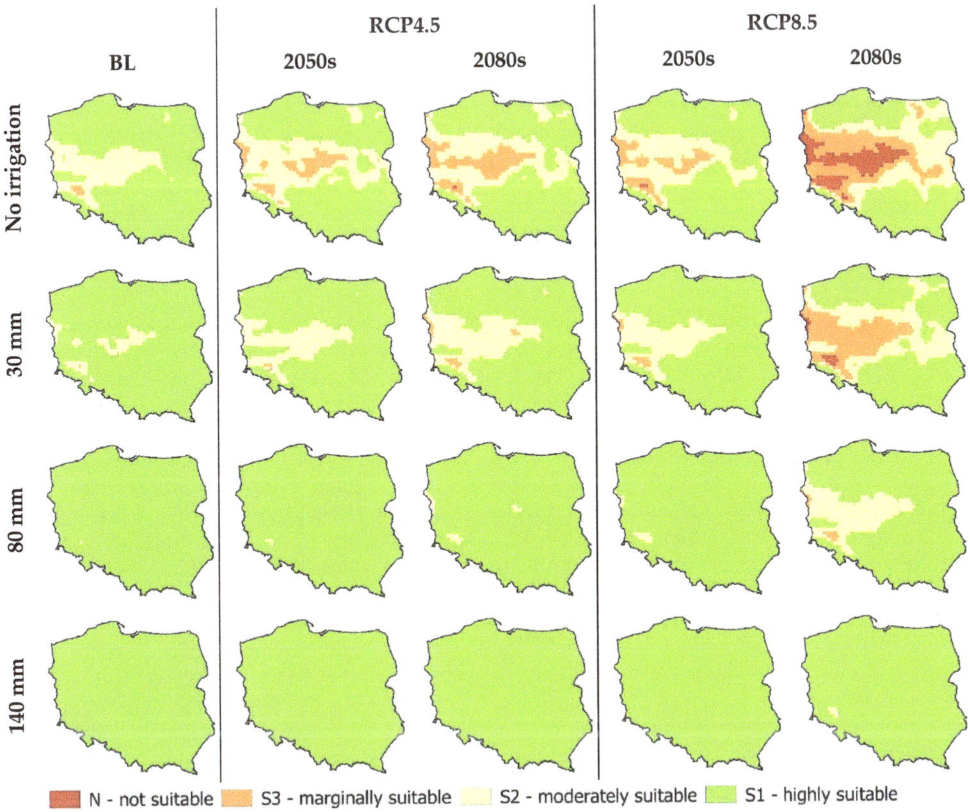

Figure 8. Climate suitability for different amount of water supply through irrigation (30, 80 and 140 mm per season) in maize cultivation for the baseline (1981–2010) and projected according to the low emissions (LE, RCP4.5) and high emissions (HE, RCP8.5) scenarios for the 2050s (2041–2070) and 2080s (2071–2100).

Table 10. Percentages of different water deficiency suitability levels for different amounts of water supply through irrigation (30, 80 and 140 mm per season) in maize cultivation in Poland for the baseline (1981–2010) and projected according to low emissions (LE, RCP4.5) and high emissions (HE, RCP8.5) scenarios for the 2050s (2041–2070) and 2080s (2071–2100).

RCP					**RCP4.5**								**RCP8.5**							
Period	**BL**				**2050s**				**2080s**				**2050s**				**2080s**			
Rating scale [1]	S1	S2	S3	N	S1	S2	S3	N	S1	S2	S3	N	S1	S2	S3	N	S1	S2	S3	N
0 mm	81	18	1	0	67	26	7	0	64	24	12	0	69	22	8	0	44	24	21	11
30 mm	94	6	0	0	84	16	0	0	79	20	2	0	82	16	1	0	57	23	18	1
80 mm	100	0	0	0	100	0	0	0	98	2	0	0	99	1	0	0	80	18	1	0
140 mm	100	0	0	0	100	0	0	0	100	0	0	0	100	0	0	0	100	0	0	0

[1] S1—highly suitable, S2—moderately suitable, S3—marginally suitable, N—not suitable.

4. Discussion

4.1. Climate Change Impact on Maize

The growth and yield of maize is significantly affected by climate change. There has been a tendency for grain maize area and yield to increase in Poland over the last 30 years [1]. This study presents the spatial distribution of selected climatic variables and how they change over time related to climate change. According to the climate change scenarios analysed, the increase in both annual temperature and precipitation sum is projected for Poland. The highest increase (by almost 4 °C) in annual mean temperature is projected in HE scenario for 2080s. Similarly, warmer and wetter conditions are projected for Poland on an annual scale by Mezghani et al. [51] and Strużewska et al. [73] together with an intensification of the changes expected by the end of the 21st century.

The obtained results show that despite the fact that annual precipitation sum is projected to increase in both LE and HE scenarios, the growing season sum of CWB is projected to decrease. The potential evapotranspiration exceeds precipitation over most of Poland during the period of April to September, resulting in a water deficit. The highest water deficit was observed in the Central Poland and this area is projected to increase under future climate conditions. Similar spatial distribution of climatic water balance is presented in [36]. Other studies highlight the uncertainty of summer precipitation projections depending on the models analysed, which leads to considerable uncertainty in the assessment of the impacts of climate change on the water balance [74].

Ziernicka-Wojtaszek [75] evaluated changes in thermal resources and CWB in Poland for two time periods: 1971–2000 and 1981–2010. It was shown that the very warm and dry areas increased from 10% to 48% and from 34 to 48%, respectively. The novelty of this paper is that we have also considered the expert opinion on soil categories importance in water deficiency evaluation, influencing climate suitability for maize production, that was not concerned in previous research on adaptation of maize production in Poland. Our study focused on water deficiency determined based on the climatic water balance and tested an adaptive irrigation practice that can mitigate the effects of increasingly frequent droughts.

In our study, soil categories were weighted using AHP methodology based on farmers' and experts' opinions. The aim of weighting the soil categories was to determine their relative importance in influencing land suitability for maize production due to the occurrence of water deficiency. According to the mean opinion of farmers and experts, medium and heavy soils were defined as the most important soil category in assessing land suitability evaluation for maize production, while light and very light soils were much less important. However, in Poland, the share of light and very light soils with low water capacity is predominant (60%) [76] and water deficit is most frequently observed on these soils [35].

Observed changes in climatic conditions indicate that increasingly frequent water shortages will be the main constraint for maize production in Poland, while thermal conditions will no longer limit production [77]. Therefore, we decided to evaluate the climate suitability for maize production in relation to occurrence of water stress under current and future climate conditions. Under current climate conditions, Poland has highly and moderately suitable water conditions for maize cultivation. In the future, water conditions for maize cultivation will worsen and about 10% of Poland will become marginally suitable. However, the biggest changes are observed in 2080s according to HE scenario, where 21% of Poland will have marginally suitable and 11% not suitable water conditions for maize production. The worsening of water conditions for maize production will be in central Poland, where the climatic water balance is the lowest. The previous studies [77,78] show an improvement of climatic conditions for maize cultivation in Poland. With climate warming the areas suitable for maize cultivation have increased significantly [77]. Ziernicka-Wojtaszek [79] introduced new agroclimatic regionalisation of the Polish area. It was simulated that increase in temperature by 1 °C and 2 °C with unchanged precipitation will result in a decrease in the area of optimal humidity from 70 to 48% and 26% respectively [79]. Ramirez-Cabral et al. [80] modelled the potential climate distribution of maize at global level in current and future climate conditions using the CLIMEX model taking into account stress factors that reduce climate suitability. The medium suitability for maize cultivation in Poland was mostly projected under current climate conditions, except in central Poland where marginal climate suitability occurs. For further climate conditions, different results were obtained for different general circulation models (GCM) CSIRO and MIROC. CSIRO projected a slight increase in the marginal suitability category, especially for the year 2100. However, for 2050 MIROC projected a change from suitable to optimal suitability in the northern and southern part of Poland while in 2100 almost the whole of Poland is projected to be optimal for maize cultivation except for the central part where suitable areas occur [80].

It should be noted as limitation of this research that the analyses carried out mainly focused on water deficiency factors and only include climatic factors and their changes in the future climate conditions, while the potential genetic progress of the species was not considered. Other research [80] suggests that it would be desirable to include non-climatic factors such as topography, land use, soil taxonomy and its physico-chemical properties in further studies. While including expert opinion in the identification of criteria relevance, especially when many criteria are evaluated, it is important to provide simple and intuitive techniques to obtain it, due to the possibility of errors occurring in the evaluation of criteria importance. There is interest in the literature for methods that better handle expert hesitation that can arise in subjective weighting methods. Więckowski et al. [81] proposed the RANCOM method, which is based on the ranking of criteria to obtain criteria weights on the basis of expert knowledge. RANCOM could be a valuable method for evaluating criteria weights or its comparison with the AHP methodology used in our study for future research since RANCOM is a simple method which can be adapted for both less and more experienced users. The RANCOM method can provide more repeatable results than the AHP method. This is especially true for a larger number of criteria [81].

4.2. Adaptation to Climate Change

Irrigation is a treatment that can effectively prevent the impact of drought on crops [82]. The use of irrigation in the cultivation of grain maize results in high yield increases that are difficult to achieve with other farming technique treatments [82]. The effect of agricultural drought on crops mainly depends on its size, duration and intensity [24]. Irrigation may help stabilize crop yields [82]. In assessing the impact of climate change on agriculture due to the lack of a clear picture regarding the amount of precipitation and the increase in evapotranspiration resulting in a deterioration of climatic water balances, the need for irrigation is increasingly being indicated [83]. Hence, we decided to test irrigation with 30, 80 and 140 mm of water and its impact on land suitability for maize production in Poland. According to water deficiency, Poland is highly or moderately suitable for maize production under current climatic conditions, but suitability is projected to decrease over time. In the future, marginally suitable water conditions and even not suitable water conditions are going to appear in central Poland. Under current climatic conditions, 80 mm of water should be applied to achieve highly suitable water conditions for maize production in almost all of Poland, while under future climatic conditions, 140 mm should be applied. Thanks to the application of 30 mm of water, the water conditions for maize production can be improved in future climate conditions, so that no marginally suitable areas will appear, except in the 2080s, according to the HE scenario where 80 mm of water should be applied.

Despite a considerable number of studies, it is very difficult to determine the amount of irrigation required for a given crop due to the variation in water demand caused by soil and meteorological factors [25]. The use of irrigation in maize production is essential in an area of distinct water deficits with low precipitation and soils with very low retention capacity [84]. Water requirements also vary between crops and their varieties, depending on the length of the growing season and the sensitivity to water shortages at different stages of development [25]. In Poland, agricultural drought is determined by the Agricultural Drought Monitoring System (ADMS) based on the Climatic Water Balance (CWB). In the determination of agricultural drought, the characteristics of the soil retention related to its granulometric composition constitute an important factor that is considered in ADMS. In the system, critical values of climatic water balance are differentiated for particular groups and species of crop plants [54].

The literature suggests that some adaptive agricultural practices should be introduced to counteract the negative effects of climate change, such as variety renovation, adjusting sowing dates, improving irrigation and fertiliser efficiency [85]. According to some research, irrigation in maize production can increase maize yield as well as stabilise yield under Polish conditions [82,86–89]. However, the use of irrigation in maize production has not always been cost effective [86]. The production effect is the main indicator of the desirability of irrigation. The production results are strongly influenced by weather conditions, in particular the amount and distribution of rainfall [82]. The effect of irrigation depends on precipitation sum in July and August, the lower precipitation, the higher the yield increase [82,87]. In the study developed by Żarski et al. [89] during the wet season, the maize grain yield increase was non-significant while in dry seasons the yield increased by more than half. Due to increased drought under climate change, soil water conservation practices along with more drought-resistant varieties are also proposed to adapt to these changes in maize production [90]. Furthermore, Zhao et al. [90] identified the use of cultivars that are more drought-resistant and cultivars with higher temperature requirements as effective adaptations to climate change in maize production in Poland (Central Europe).

5. Conclusions

This study assessed the potential effects of climate change on climate suitability related to water deficiency for maize cultivation in Poland using multi-criteria evaluation through the Analytical Hierarchy Process (AHP). The weights of soil category were calculated by AHP to include the expert opinion in the evaluation.

The obtained results suggest that the water stress limitations depend on the soil category and under the current climate conditions the extreme deficiency of the climatic water balance was observed on very light and light soils in central Poland, but not on heavy soils. Considering all soil categories with different relative importance according to experts' opinion, central Poland was moderately suitable under current climate conditions, while the rest of Poland was highly suitable. However, in the future, the suitability according to water deficiency is predicted to decrease, and some areas in central Poland will become marginally suitable for maize cultivation or even not suitable in the 2080s according to the RCP8.5 scenario.

This study focused on the impact of climate change on maize suitability in Poland, therefore water deficiency factors that may harm maize production in the future were considered. Studying the climate suitability of a particular crop can help to manage the impact of climate change on crop production. However, in future studies, the combined consideration of topographic, climatic and soil factors would allow a more complex assessment of maize land suitability.

The obtained results show that due to climate change and higher temperatures, there will be problems related to water deficiency. This study indicated that water deficiency may limit climate suitability for maize cultivation in far future, but irrigation could mitigate this effect. The results of the study could be used to develop adaptation strategies to enhance maize productivity under future climatic conditions. Identification of climate limitations, their spatial variability and changes in the future may play an important role in keeping maize production risk at a minimum level. The methods applied in the study can be used as a tool for finding pathways to enhance the resilience of farming systems dealing with challenges.

Author Contributions: Conceptualization, A.K.-B., J.K. and S.R.; methodology, A.K.-B., J.K. and S.R; software, A.K.-B.; validation, A.K.-B.; formal analysis, A.K.-B.; investigation, A.K.-B. and J.K.; writing—original draft preparation, A.K.-B.; writing—review and editing, A.K.-B., J.K. and S.R.; visualization, A.K.-B.; funding acquisition, J.K. All authors have read and agreed to the published version of the manuscript.

Funding: The research was funded by the Ministry of Agriculture and Rural Development, project "Drought monitoring system in Poland" (contract no. DBD.fin.070.10.2023).

Institutional Review Board Statement: Not applicable.

Informed Consent Statement: Not applicable.

Data Availability Statement: The raw data supporting the conclusions of this article will be made available by the authors on request.

Acknowledgments: The authors would like to thank Małgorzata Wydra for the linguistic revision of the article.

Conflicts of Interest: The authors declare no conflict of interest.

References

1. The Food and Agriculture Organization (FAO). Crops and Livestock Products. Available online: https://www.fao.org/faostat/en/#data/QCL (accessed on 30 October 2023).
2. Piniewski, M.; Marcinkowski, P.; O'Keeffe, J.; Szcześniak, M.; Nieróbca, A.; Kozyra, J.; Kundzewicz, Z.W.; Okruszko, T. Model-Based Reconstruction and Projections of Soil Moisture Anomalies and Crop Losses in Poland. *Theor. Appl. Climatol.* **2020**, *140*, 691–708. [CrossRef]
3. Górski, T.; Kozyra, J. Agroklimatyczna Norma Średniej Temperatury Powietrza w Polsce Na Lata 2011–2020. *Pol. J. Agron.* **2011**, *5*, 21–28.
4. Trnka, M.; Olesen, J.E.; Kersebaum, K.C.; Skjelvåg, A.O.; Eitzinger, J.; Seguin, B.; Peltonen-Sainio, P.; Rötter, R.; Iglesias, A.; Orlandini, S. Agroclimatic Conditions in Europe under Climate Change. *Glob. Chang. Biol.* **2011**, *17*, 2298–2318. [CrossRef]
5. Nendel, C.; Reckling, M.; Debaeke, P.; Schulz, S.; Berg-Mohnicke, M.; Constantin, J.; Fronzek, S.; Hoffmann, M.; Jakšić, S.; Kersebaum, K.-C. Future Area Expansion Outweighs Increasing Drought Risk for Soybean in Europe. *Glob. Chang. Biol.* **2023**, *29*, 1340–1358. [CrossRef] [PubMed]
6. Duffková, R.; Holub, J.; Fučík, P.; Rožnovský, J.; Novotný, I. Long-Term Water Balance of Selected Field Crops in Different Agricultural Regions of the Czech Republic Using Fao-56 and Soil Hydrological Approaches. *Sustainability* **2019**, *11*, 5243. [CrossRef]
7. Hänsel, S.; Ustrnul, Z.; Łupikasza, E.; Skalak, P. Assessing Seasonal Drought Variations and Trends over Central Europe. *Adv. Water Resour.* **2019**, *127*, 53–75. [CrossRef]
8. Gao, Y.; Zhang, A.; Yue, Y.; Wang, J.; Su, P. Predicting Shifts in Land Suitability for Maize Cultivation Worldwide Due to Climate Change: A Modeling Approach. *Land* **2021**, *10*, 295. [CrossRef]
9. Lopez-Blanco, J.; Pérez-Damián, J.L.; Conde-Álvarez, A.C.; Gómez-Díaz, J.D.; Monterroso-Rivas, A.I. Land Suitability Levels for Rainfed Maize under Current Conditions and Climate Change Projections in Mexico. *Outlook Agric.* **2018**, *47*, 181–191. [CrossRef]
10. Tan, Z.; Yang, Y.; Wang, Y.; Wang, L.; Sun, G. The Decrease of Potential Suitable Areas and the Distribution Tendency of Staple Crops in Ethiopia under Future Climate Conditions. *Afr. J. Agric. Res.* **2016**, *11*, 2092–2101.
11. Shaloo, S.; Bisht, H.; Jain, R.; Singh, R.P. Cropland Suitability Assessment Using Multi Criteria Evaluation Techniques and Geo-Spatial Technology: A Review. *Indian J. Agric. Sci.* **2022**, *92*, 554–562. [CrossRef]
12. Seif-Ennasr, M.; Bouchaou, L.; El Morjani, Z.E.A.; Hirich, A.; Beraaouz, E.H.; Choukr-Allah, R. Gis-Based Land Suitability and Crop Vulnerability Assessment under Climate Change in Chtouka Ait Baha, Morocco. *Atmosphere* **2020**, *11*, 1167. [CrossRef]

13. Ramamurthy, V.; Reddy, G.O.; Kumar, N. Assessment of Land Suitability for Maize (*Zea mays* L.) in Semi-Arid Ecosystem of Southern India Using Integrated AHP and GIS Approach. *Comput. Electron. Agric.* **2020**, *179*, 105806. [CrossRef]
14. Seyedmohammadi, J.; Sarmadian, F.; Jafarzadeh, A.A.; McDowell, R.W. Integration of ANP and Fuzzy Set Techniques for Land Suitability Assessment Based on Remote Sensing and GIS for Irrigated Maize Cultivation. *Arch. Agron. Soil Sci.* **2019**, *65*, 1063–1079. [CrossRef]
15. Maddahi, Z.; Jalalian, A.; Zarkesh, M.K.; Honarjo, N. Land Suitability Analysis for Rice Cultivation Using a GIS-Based Fuzzy Multi-Criteria Decision Making Approach: Central Part of Amol District, Iran. *Soil Water Res.* **2017**, *12*, 29–38. [CrossRef]
16. Xue, L.; Cao, P.; Xu, D.; Guo, Y.; Wang, Q.; Zheng, X.; Han, R.; You, A. Agricultural Land Suitability Analysis for an Integrated Rice–Crayfish Culture Using a Fuzzy AHP and GIS in Central China. *Ecol. Indic.* **2023**, *148*, 109837. [CrossRef]
17. Sari, F.; Sari, K.F. Multi Criteria Decision Analysis to Determine the Suitability of Agricultural Crops for Land Consolidation Areas. *Int. J. Eng. Geosci.* **2021**, *6*, 64–73. [CrossRef]
18. Bagherzadeh, A.; Gholizadeh, A. Modeling Land Suitability Evaluation for Wheat Production by Parametric and TOPSIS Approaches Using GIS, Northeast of Iran. *Model. Earth Syst. Environ.* **2016**, *2*, 1–11. [CrossRef]
19. Ali, I.; Gunawan, V.; Adi, K. Decision Support Systems for Land Suitability Evaluation on Rice Cultivation Using ELECTRE Method. *E3S Web Conf.* **2020**, *202*, 14004. [CrossRef]
20. Rodcha, R.; Tripathi, N.K.; Prasad Shrestha, R. Comparison of Cash Crop Suitability Assessment Using Parametric, AHP, and FAHP Methods. *Land* **2019**, *8*, 79. [CrossRef]
21. Elaalem, M. A Comparison of Parametric and Fuzzy Multi-Criteria Methods for Evaluating Land Suitability for Olive in Jeffara Plain of Libya. *APCBEE Procedia* **2013**, *5*, 405–409. [CrossRef]
22. Kalichkin, V.K.; Pavlova, A.I.; Logachova, O.M. GIS-Based Multi-Criteria Analysis of the Suitability of Western Siberian Forest-Steppe Lands. *Ann. GIS* **2021**, *27*, 225–237. [CrossRef]
23. Elaalem, M. Land Suitability Evaluation for Sorghum Based on Boolean and Fuzzy-Multi-Criteria Decision Analysis Methods. *Int. J. Environ. Sci. Dev.* **2012**, *3*, 357–361. [CrossRef]
24. Łabędzki, L.; Bąk, B. Meteorological and Agricultural Drought Indices Used in Drought Monitoring in Poland: A Review. *Meteorol. Hydrol. Water Manag.* **2014**, *2*, 3–13. [CrossRef]
25. Rzekanowski, C.; Żarski, J.; Rolbiecki, S. Potrzeby, Efekty i Perspektywy Nawadniania Roślin Na Obszarach Szczególnie Deficytowych w Wodę. *Postępy Nauk Rol.* **2011**, *1*, 51–63.
26. Duc, T.T. Using GIS and AHP Technique for Land-Use Suitability Analysis. In Proceedings of the International Symposium on Geoinformatics for Spatial Infrastructure Development in Earth and Allied Sciences, Ho Chi Minh, Vietnam, 9–11 November 2006; pp. 1–6.
27. Habibie, M.I.; Noguchi, R.; Shusuke, M.; Ahamed, T. Land Suitability Analysis for Maize Production in Indonesia Using Satellite Remote Sensing and GIS-Based Multicriteria Decision Support System. *GeoJournal* **2021**, *86*, 777–807. [CrossRef]
28. Ceballos-Silva, A.; Lopez-Blanco, J. Delineation of Suitable Areas for Crops Using a Multi-Criteria Evaluation Approach and Land Use/Cover Mapping: A Case Study in Central Mexico. *Agric. Syst.* **2003**, *77*, 117–136. [CrossRef]
29. Han, C.; Chen, S.; Yu, Y.; Xu, Z.; Zhu, B.; Xu, X.; Wang, Z. Evaluation of Agricultural Land Suitability Based on RS, AHP, and MEA: A Case Study in Jilin Province, China. *Agriculture* **2021**, *11*, 370. [CrossRef]
30. Rahaman, S.A.; Aruchamy, S. Land Suitability Evaluation of Tea (*Camellia sinensis* L.) Plantation in Kallar Watershed of Nilgiri Bioreserve, India. *Geographies* **2022**, *2*, 701–723. [CrossRef]
31. Layomi Jayasinghe, S.; Kumar, L.; Sandamali, J. Assessment of Potential Land Suitability for Tea (*Camellia sinensis* (L.) O. Kuntze) in Sri Lanka Using a GIS-Based Multi-Criteria Approach. *Agriculture* **2019**, *9*, 148. [CrossRef]
32. Donatelli, M.; Srivastava, A.K.; Duveiller, G.; Niemeyer, S.; Fumagalli, D. Climate Change Impact and Potential Adaptation Strategies under Alternate Realizations of Climate Scenarios for Three Major Crops in Europe. *Environ. Res. Lett.* **2015**, *10*, 075005. [CrossRef]
33. Zohner, C.M.; Mo, L.; Renner, S.S.; Svenning, J.-C.; Vitasse, Y.; Benito, B.M.; Ordonez, A.; Baumgarten, F.; Bastin, J.-F.; Sebald, V. Late-Spring Frost Risk between 1959 and 2017 Decreased in North America but Increased in Europe and Asia. *Proc. Natl. Acad. Sci. USA* **2020**, *117*, 12192–12200. [CrossRef] [PubMed]
34. Jia, C.; Luo, X.; Zhou, W.; Chen, Y.; Sun, G. Evaluation of Suitability Areas for Maize in China Based on GIS and Its Variation Trend on the Future Climate Condition. In *Ecosystem Assessment and Fuzzy Systems Management. Advances in Intelligent Systems and Computing*; Cao, B.-Y., Ma, S.-Q., Cao, H.-H., Eds.; Springer: Cham, Switzerland, 2014; Volume 254, pp. 285–299.
35. Wójcik, I.; Doroszewski, A.; Wróblewska, E.; Koza, P. Susza Rolnicza w Uprawie Zbóż Jarych w Polsce w Latach 2006–2017. *Woda-Śr.-Obsz. Wiej.* **2019**, *19*, 77–95.
36. Łabędzki, L.; Kanecka-Geszke, E.; Bak, B.; Slowinska, S. Estimation of Reference Evapotranspiration Using the FAO Penman-Monteith Method for Climatic Conditions of Poland. In *Evapotranspiration*; Łabędzki, L., Ed.; IntechOpen: Rijeka, Croatia, 2011; pp. 275–294.
37. Somorowska, U. Changes in Drought Conditions in Poland over the Past 60 Years Evaluated by the Standardized Precipitation-Evapotranspiration Index. *Acta Geophys.* **2016**, *64*, 2530–2549. [CrossRef]
38. Olesen, J.E.; Trnka, M.; Kersebaum, K.C.; Skjelvåg, A.O.; Seguin, B.; Peltonen-Sainio, P.; Rossi, F.; Kozyra, J.; Micale, F. Impacts and Adaptation of European Crop Production Systems to Climate Change. *Eur. J. Agron.* **2011**, *34*, 96–112. [CrossRef]

39. Blazejczyk, K. Climate and Bioclimate of Poland. In *Natural and Human Environment of Poland—A Geographical Overview*; Degórski, M., Ed.; Polish Academy of Sciences, Institute of Geography and Spatial Organization Polish Geographical Society: Warsaw, Poland, 2006; pp. 31–48.
40. Kędziora, A. *Podstawy Agrometeorologii*; PWRiL: Poznań, Poland, 2008.
41. Grzywna, A.; Bochniak, A.; Ziernicka-Wojtaszek, A.; Krużel, J.; Jóźwiakowski, K.; Wałęga, A.; Cupak, A.; Mazur, A.; Obroślak, R.; Serafin, A. The Analysis of Spatial Variability of Precipitation in Poland in the Multiyears 1981–2010. *J. Water Land Dev.* **2020**, *46*, 105–111.
42. Łabędzki, L. Estimation of Local Drought Frequency in Central Poland Using the Standardized Precipitation Index SPI. *Irrig. Drain. J. Int. Comm. Irrig. Drain.* **2007**, *56*, 67–77. [CrossRef]
43. Jacob, D.; Petersen, J.; Eggert, B.; Alias, A.; Christensen, O.B.; Bouwer, L.M.; Braun, A.; Colette, A.; Déqué, M.; Georgievski, G. EURO-CORDEX: New High-Resolution Climate Change Projections for European Impact Research. *Reg. Environ. Chang.* **2014**, *14*, 563–578. [CrossRef]
44. Benestad, R.; Haensler, A.; Hennemuth, B.; Illy, T.; Jacob, D.; Keup-Thiel, E.; Katragkou, E.; Kotlarski, S.; Nikulin, G.; Otto, J.; et al. Guidance for EURO-CORDEX Climate Projections Data Use. Version1.0—2017.08. 2017. Available online: https://www.euro-cordex.net/imperia/md/content/csc/cordex/euro-cordex-guidelines-version1.0-2017.08.pdf (accessed on 13 January 2020).
45. Schulzweida, U.; Kornblueh, L.; Quast, R. *CDO User's Guide*; Climate Data Operation Version; MPI for Meteorology: Hamburg, Germany, 2006.
46. Landelius, T.; Dahlgren, P.; Gollvik, S.; Jansson, A.; Olsson, E. A High-resolution Regional Reanalysis for Europe. Part 2: 2D Analysis of Surface Temperature, Precipitation and Wind. *Q. J. R. Meteorol. Soc.* **2016**, *142*, 2132–2142. [CrossRef]
47. Yang, W.; Andréasson, J.; Phil Graham, L.; Olsson, J.; Rosberg, J.; Wetterhall, F. Distribution-Based Scaling to Improve Usability of Regional Climate Model Projections for Hydrological Climate Change Impacts Studies. *Hydrol. Res.* **2010**, *41*, 211–229. [CrossRef]
48. IPCC. *Climate Change 2013: The Physical Science Basis: Working Group I Contribution to the Fifth Assessment Report of the Intergovernmental Panel on Climate Change*; Stocker, T.F., Qin, D., Plattner, G.-K., Tignor, M., Allen, S.K., Boschung, J., Nauels, A., Xia, Y., Bex, V., Midgley, P.M., et al., Eds.; Cambridge University Press: Cambridge, UK; New York, NY, USA, 2013; 1535p.
49. Moss, R.H.; Edmonds, J.A.; Hibbard, K.A.; Manning, M.R.; Rose, S.K.; Van Vuuren, D.P.; Carter, T.R.; Emori, S.; Kainuma, M.; Kram, T. The next Generation of Scenarios for Climate Change Research and Assessment. *Nature* **2010**, *463*, 747–756. [CrossRef]
50. Brzóska, B.; Jaczewski, A. Przyszłe Zmiany Wybranych Wskaźników Klimatycznych Dla Polski Na Podstawie Wyników Dynamicznego Downscalingu. *Pr. Geogr.* **2017**, *149*, 7–14.
51. Mezghani, A.; Dobler, A.; Haugen, J.E.; Benestad, R.E.; Parding, K.M.; Piniewski, M.; Kardel, I.; Kundzewicz, Z.W. CHASE-PL Climate Projection Dataset over Poland–Bias Adjustment of EURO-CORDEX Simulations. *Earth Syst. Sci. Data* **2017**, *9*, 905–925. [CrossRef]
52. IMGW Instytut Meteorologii i Gospodarki Wodnej PIB. Available online: https://danepubliczne.imgw.pl/data/dane_pomiarowo_obserwacyjne (accessed on 21 April 2021).
53. Kotlarski, S.; Keuler, K.; Christensen, O.B.; Colette, A.; Déqué, M.; Gobiet, A.; Goergen, K.; Jacob, D.; Lüthi, D.; Van Meijgaard, E. Regional Climate Modeling on European Scales: A Joint Standard Evaluation of the EURO-CORDEX RCM Ensemble. *Geosci. Model Dev.* **2014**, *7*, 1297–1333. [CrossRef]
54. Doroszewski, A.; Jadczyszyn, J.; Kozyra, J.; Pudełko, R.; Stuczyński, T.; Mizak, K.; Łopatka, A.; Koza, P.; Górski, T.; Wróblewska, E. Podstawy Systemu Monitoringu Suszy Rolniczej. *Woda-Śr.-Obsz. Wiej.* **2012**, *12*, 77–91.
55. Doroszewski, A.; Jóźwicki, T.; Wróblewska, E.; Kozyra, J. *Susza Rolnicza w Polsce w Latach 1961–2010*; IUNG-PIB: Puławy, Poland, 2014.
56. Doroszewski, A.; Górski, T. Prosty Wskaznik Ewapotranspiracji Potencjalnej. *Rocz. Akad. Rol. W Poznaniu Melior. Inż. Śr.* **1995**, *16*, 3–8.
57. Żyłowska, K. *Raport Końcowy z Tematu Badawczego Nr.4.03. Pt Warunki Agroklimatyczne w Polsce Według Scenariuszy Klimatycznych Na Lata 2030, 2050, 2080 Zrealizowanego w Ramach Działalności Statutowej w Podprogramie Badawczym Nr. 4.03*; IUNG-PIB: Puławy, Poland, 2015.
58. Saaty, T.L. How to Make a Decision: The Analytic Hierarchy Process. *Interfaces* **1994**, *24*, 19–43. [CrossRef]
59. Muluneh, A.; Tadesse, T.; Girma, R. Assessing Potential Land Suitable for Surface Irrigation Using GIS and AHP Techniques in the Rift Valley Lakes Basin, Ethiopia. *Sustain. Water Resour. Manag.* **2022**, *8*, 46. [CrossRef]
60. Akıncı, H.; Özalp, A.Y.; Turgut, B. Agricultural Land Use Suitability Analysis Using GIS and AHP Technique. *Comput. Electron. Agric.* **2013**, *97*, 71–82. [CrossRef]
61. Mishra, A.K.; Deep, S.; Choudhary, A. Identification of Suitable Sites for Organic Farming Using AHP & GIS. *Egypt. J. Remote Sens. Space Sci.* **2015**, *18*, 181–193.
62. Saaty, T.L.; Sodenkamp, M. The Analytic Hierarchy and Analytic Network Measurement Processes: The Measurement of Intangibles: Decision Making under Benefits, Opportunities, Costs and Risks. In *Handbook of Multicriteria Analysis*; Zopounidis, C., Pardalos, P.M., Eds.; Springer: Berlin, Germany, 2010; pp. 91–166.
63. Król-Badziak, A.; Pishgar-Komleh, S.H.; Rozakis, S.; Księżak, J. Environmental and Socio-Economic Performance of Different Tillage Systems in Maize Grain Production: Application of Life Cycle Assessment and Multi-Criteria Decision Making. *J. Clean. Prod.* **2021**, *278*, 123792. [CrossRef]

64. Parra-Lopez, C.; Calatrava-Requena, J.; de-Haro-Gimenez, T. A Multi-Criteria Evaluation of the Environmental Performances of Conventional, Organic and Integrated Olive-Growing Systems in the South of Spain Based on Experts' Knowledge. *Renew. Agric. Food Syst.* **2007**, *22*, 189–203. [CrossRef]
65. Forman, E.; Peniwati, K. Aggregating Individual Judgments and Priorities with the Analytic Hierarchy Process. *Eur. J. Oper. Res.* **1998**, *108*, 165–169. [CrossRef]
66. ADMS—Soil Categories. Available online: https://susza.iung.pulawy.pl/en/kategorie/ (accessed on 21 December 2023).
67. Kopacz, M.; Twardy, S. Potencjalne Zagrożenie Suszą Rolniczą w Północnej Części Województwa Małopolskiego. *Woda-Śr.-Obsz. Wiej.* **2016**, *16*, 5–19.
68. FAO (Food and Agriculture Organization). *A Framework for Land Evaluation*; Soils Bulletin; FAO: Rome, Italy, 1976; Volume 32.
69. Łabędzki, L. Potrzeby i Stan Nawodnień w Województwie Kujawsko-Pomorskim [Irrigation Needs and Condition in the Kuyavian-Pomeranian Voivodeship]. In Proceedings of the Warsztaty dla Interesariuszy Projektu OPERA [Workshops for the Stakeholders of the OPERA Project], Minikowo, Poland, 27 November 2017.
70. Łabędzki, L. Irrigation in Poland-Current Status after Reforms in Agriculture and Future Development. *J. Water Land Dev.* **2007**, *11*, 3–16. [CrossRef]
71. Łabędzki, L. Expected Development of Irrigation in Poland in the Context of Climate Change. *J. Water Land Dev.* **2009**, *13b*, 17–29. [CrossRef]
72. Łabędzki, L. Agroklimatyczne Uwarunkowania Potrzeb Melioracji Nawadniających. *Inż. Ekol.* **2016**, *47*, 199–204. [CrossRef]
73. Strużewska, J.; Jefimow, M.; Jagiełło, P.; Kłeczek, M.; Sattari, A.; Gienibor, A.; Norowski, A.; Durka, P.; Walczak, B.; Drzewiecki, P. *Raport Skrócony: Zmiany Temperatury i Opadu Na Obszarze Polski w Warunkach Przyszłego Klimatu Do Roku 2100*; Instytut Ochrony Środowiska—Państwowy Instytut Badawczy: Warszawa, Poland, 2020.
74. Szwed, M. The Elements of Water Balance in the Changing Climate in Poland. *Adv. Meteorol.* **2015**, *2015*, 1–13. [CrossRef]
75. Ziernicka-Wojtaszek, A. Klimatyczny Bilans Wodny Na Obszarze Polski w Świetle Współczesnych Zmian Klimatu. *Woda-Śr.-Obsz. Wiej.* **2015**, *15*, 93–100.
76. Igras, J.; Lipiński, W. Regionalne Zróżnicowanie Stanu Agrochemicznego Gleb w Polsce. In *Regionalne Zróżnicowanie Produkcji Rolniczej w Polsce*; Harasim, A., Ed.; Studia i Raporty IUNG-PIB: Puławy, Poland, 2006; Volume 15, pp. 71–79.
77. Nieróbca, A.; Kozyra, J.; Mizak, K. Zmiany Warunków Termicznych Dla Uprawy Kukurydzy w Polsce. In *Problemy Agrotechniki oraz Wykorzystania Kukurydzy i Sorgo*; Michalski, T., Ed.; UP Poznań: Poznan, Poland, 2008; pp. 26–30.
78. Nieróbca, A.; Kozyra, J.; Pudełko, R. Evaluation of Meteorological Conditions for the Cultivation of Maize Grain in Poland. *Acta Agrophys.* **2010**, *6*, 45–55.
79. Ziernicka-Wojtaszek, A. Weryfikacja Rolniczo-Klimatycznych Regionalizacji Polski w Świetle Współczesnych Zmian Klimatu. *Acta Agrophys.* **2009**, *13*, 803–812.
80. Ramirez-Cabral, N.Y.; Kumar, L.; Shabani, F. Global Alterations in Areas of Suitability for Maize Production from Climate Change and Using a Mechanistic Species Distribution Model (CLIMEX). *Sci. Rep.* **2017**, *7*, 5910. [CrossRef]
81. Więckowski, J.; Kizielewicz, B.; Shekhovtsov, A.; Sałabun, W. RANCOM: A Novel Approach to Identifying Criteria Relevance Based on Inaccuracy Expert Judgments. *Eng. Appl. Artif. Intell.* **2023**, *122*, 106114. [CrossRef]
82. Żarski, J.; Dudek, S.; Kuśmierek-Tomaszewska, R.; Rolbiecki, R.; Rolbiecki, S. Prognozowanie Efektów Nawadniania Roślin Na Podstawie Wybranych Wskaźników Suszy Meteorologicznej i Rolniczej. *Rocz. Ochr. Śr.* **2013**, *15*, 2185–2203.
83. Żarski, J. Tendencje Zmian Klimatycznych Wskaźników Potrzeb Nawadniania Roślin w Rejonie Bydgoszczy. *Infrastrukt. Ekol. Teren. Wiej.* **2011**, *5*, 29–37.
84. Żarski, J.; Dudek, S.; Grzelak, B.; Kuśmierek-Tomaszewska, R.; Rolbiecki, R.; Rolbiecki, S. Wpływ Nawadniania i Fertygacji Kroplowej Azotem Na Plonowanie Kukurydzy Na Obszarze Szczególnie Deficytowym w Wodę. *Infrastrukt. Ekol. Teren. Wiej.* **2015**, *11*, 279–289.
85. Zhao, Y.; Xiao, D.; Bai, H.; Tang, J.; Liu, D. Future Projection for Climate Suitability of Summer Maize in the North China Plain. *Agriculture* **2022**, *12*, 348. [CrossRef]
86. Dudek, S.; Żarski, J. Ocena Efektów Zastosowania Nawadniania w Uprawie Kukurydzy Na Ziarno. *Inż. Rol.* **2005**, *3*, 159–164.
87. Dudek, S.; Zarski, J.; Kusmierek-Tomaszewska, R. Reakcja Kukurydzy Na Nawadnianie w Świetle Wyników Wieloletniego Eksperymentu Polowego. *Infrastrukt. Ekol. Teren. Wiej.* **2009**, *3*, 189–195.
88. Żarski, J.; Dudek, S.; Grzelak, B. Rola Czynnika Wodnego i Termicznego w Ksztaltowaniu Plonow Ziarna Kukurydzy. *Acta Agrophys.* **2004**, *3*, 189–195.
89. Żarski, J.; Dudek, S.; Kuśmierek-Tomaszewska, R. Drip Irrigation as a Factor Mitigating Drought Impact in Corn Cultivation in Central Poland. In Proceedings of the 8th International Scientific Conference Rural Development 2017, Akademija, Lithuania, 23–24 November 2017; Raupelienė, A., Ed.; Aleksandras Stulginskis University: Kaunas, Lithuania, 2018; pp. 182–186.
90. Zhao, J.; Bindi, M.; Eitzinger, J.; Ferrise, R.; Gaile, Z.; Gobin, A.; Holzkämper, A.; Kersebaum, K.-C.; Kozyra, J.; Kriaučiūnienė, Z.; et al. Priority for Climate Adaptation Measures in European Crop Production Systems. *Eur. J. Agron.* **2022**, *138*, 126516. [CrossRef]

Disclaimer/Publisher's Note: The statements, opinions and data contained in all publications are solely those of the individual author(s) and contributor(s) and not of MDPI and/or the editor(s). MDPI and/or the editor(s) disclaim responsibility for any injury to people or property resulting from any ideas, methods, instructions or products referred to in the content.

sustainability

MDPI

Article

Eco-Scheme—Carbon Farming and Nutrient Management—A New Tool to Support Sustainable Agriculture in Poland

Anna Rosa *, Aleksandra Pawłowska and Michał Dudek

Institute of Rural and Agricultural Development, Polish Academy of Sciences, 00-330 Warsaw, Poland; apawlowska@irwirpan.waw.pl (A.P.); mdudek@irwirpan.waw.pl (M.D.)
* Correspondence: arosa@irwirpan.waw.pl

Abstract: This study investigates the spatial dynamics of participation in the carbon farming eco-scheme in Poland under the EU CAP for 2023–2027. Addressing the broader context of sustainable agriculture and climate change mitigation, this research explores how farm size and structural characteristics influence the adoption of regenerative practices incentivised through this eco-scheme. Using spatial statistical methods, including the global Moran's I test and the global spatial cross-correlation index, this study analyses county-level data from 2023 to 2024 on farm size, the number of beneficiaries, and payment levels. The findings reveal distinct spatial clusters in eco-scheme participation, with larger farms showing greater regional concentration and smaller farms displaying stronger local clustering in payment distribution. The findings highlight varied spatial mechanisms that influence adoption and financial support patterns, indicating that both farm size and regional context play a significant role in shaping the uptake of eco-schemes. This study emphasises the significance of comprehensive spatial and socio-economic data in the formulation of effective, evidence-based policies pertaining to sustainable agriculture. It establishes a basis for more precisely targeted interventions and optimal resource allocation, thereby supporting both national and EU climate objectives while simultaneously enhancing the resilience and sustainability of rural regions.

Keywords: carbon farming; sustainable agriculture; EU Common Agricultural Policy; direct payments; eco-scheme

1. Introduction

Nowadays, the greatest threat to the development of civilisation is climate change. In response to this challenge, it is necessary to adapt to changing climatic conditions and take decisive action to reduce greenhouse gas emissions. One solution is sustainable agriculture, an approach to agricultural production that harmoniously combines economic and social goals with environmental requirements. Its main objective is the effective use of natural resources such as soil, water, and energy in a way that minimises negative environmental impacts while ensuring production profitability and social acceptance [1]. The intensification of agriculture over recent decades—driven by the increasing demand for food production—has led to significant changes in the natural environment. Intensive land use, focused mainly on maximising the yield of plant and animal production through the use of chemicals, cultivation of new crop varieties and mechanisation, has brought economic benefits but has also negatively impacted the environment. The consequences of these practices include soil degradation, reduced biodiversity, increased greenhouse gas emissions, and intensified erosion and landscape deprivation due to the introduction of so-called monocultures. This model of industrial agriculture is now heavily criticised, primarily because of its negative externalities [2].

Regenerative agriculture offers solutions aimed at reversing the negative trends associated with production intensification. Its core practices include maintaining plant cover throughout the year (e.g., by incorporationg cover crops), using natural fertilisers, implementing crop rotation, and reducing tillage. Promoting such sustainable farming practices and developing policies that support farmers in adopting them is essential—not only for ensuring food security but also for protecting our planet for future generations [3]. In the report "Climate Change and Land", this model of agriculture is defined as a "sustainable land management practice" that "can be effective in building resilience of agro-ecosystems" [4]. According to the Food and Land Use Coalition (FOLU), regenerative agriculture is one of the ten critical transitions needed to transform food and land use systems. Their definition includes practices that regenerate soil, reduce (though not necessarily eliminate) synthetic pesticides and fertilisers, and go beyond minimising negative impacts by ensuring that agriculture has a positive effect on the environment—while also maintaining high productivity levels [5]. Rhodos [6] defines regenerative agriculture as an alternative method of food production that may have a lower—or even net positive—impact on the environment and/or society. It is described as a holistic design that aims to grow as much food as possible using the least amount of resources in a way that revitalises rather than depletes soil, while offering a solution to the problem of carbon sequestration. Rodale [7], on the other hand, defines it as agriculture that "increases productivity levels, enhances the biological base of our land and soil and has high built-in economic and biological stability". By applying regenerative agricultural practices, not only can positive environmental outcomes be achieved, but the resulting soil conditions also allow farmers to attain satisfying production and economic results [3,8]. This type of farming creates opportunities to reduce input costs, especially for energy, labour, fertilisers, and crop protection products [9].

An integral part of regenerative agriculture is carbon farming. Through restoring and maintaining the yield potential of soil—which is a natural CO_2 sink—farmers can counteract adverse climate changes by maximising organic matter content in the soil. Actions taken in this regard are referred to as carbon farming. Some use the terms regenerative agriculture and carbon farming interchangeably (although the authors argue that carbon farming would not exist without regenerative practices). Undoubtedly, carbon farming is based on regenerative agriculture principles, with its key goal being to increase CO_2 sequestration in soil. In practice, this means that the use of regenerative methods not only helps mitigate climate change but also improves soil structure and productivity (via increased organic matter), enhances agroecosystem resilience to extreme weather conditions, and improves the quality of produced food. The development of carbon farming is a result of the EU policy direction, which has adopted regulations aimed at achieving climate neutrality. According to the European Climate Law, greenhouse gas emissions and absorption in the European Union (EU) must be balanced no later than 2050 [10]. Farmers will play a significant role in achieving these goals, as they are now included in the system for "managing" carbon dioxide and other greenhouse gases. In light of the additional benefits that farmers may receive for applying carbon farming practices, the demand for knowledge in this area is expected to grow in the coming years.

In Poland, carbon farming activities align with the EU's Common Agricultural Policy (CAP) for 2023–2027. The key institutional tool of this policy is eco-schemes, voluntary financial mechanisms encouraging farmers to adopt environmentally and climate-friendly practices. The total budget for eco-schemes amounts to EUR 44.7 billion, compared to EUR 190 billion allocated to the entire CAP Pillar I for 2023–2027 [11]. According to the European Commission's recommendations, EU member states are required to allocate at least 25% of Pillar I funds for the years 2023–2027 to payments under eco-schemes. However, member states may reduce this share, provided that they increase financial allocations for environmental actions under Pillar II, which focuses on environmental protection, supporting structural changes in agriculture and rural development [12].

In Poland, as per the Strategic Plan for the CAP 2023–2027, financial resources amounting to EUR 25.2 billion are allocated for activities under both pillars. Of this, EUR 17.3 billion from the EU budget is designated for direct payments (Pillar I), while EUR 4.7 billion is allocated to rural development (Pillar II), supplemented with additional national contributions [13].

The so-called area-based eco-schemes are annual environmental interventions implemented under the direct payment system. In 2023, Poland introduced five area-based eco-schemes:

1. Carbon farming and nutrient management;
2. Areas with melliferous plants;
3. Water retention on permanent grasslands;
4. Plant production under the Integrated Plant Production scheme;
5. Organic crop protection.

Additionally, from 2024, farmers could apply for payments under the new eco-scheme "Decoupled land" and from 2025 under the eco-scheme "Seed material: elite and certified categories" [14]. Practices that directly support carbon farming have been included in the Strategic Plan of the CAP 2023–2027 under the eco-scheme "Carbon farming and nutrient management". It is one of the most comprehensive eco-schemes in Poland. This intervention supports nine practices that increase carbon storage in soil, capture atmospheric CO_2 in plants, and reduce its emissions.

Eco-schemes represent an innovative agricultural policy tool aimed at achieving the EU's climate and environmental goals, and their implementation started in 2023. Consequently, both in Poland and other European countries, a significant gap in research focused on identifying the determinants of the successful implementation of this initiative is observed. Understanding the various patterns and factors influencing the implementation of specific public interventions, such as carbon farming eco-schemes, across different levels of the CAP governance, is a topic that has not been extensively explored. This issue is particularly critical due to the highly spatially differentiated nature of agriculture, which varies significantly in terms of climate and environmental conditions, as well as the socio-economic structures in Poland and other EU member states. The empirical analyses presented here fit into a reflection on the dilemma between the rationale of achieving common environmental and climate objectives (fiscal vs. environmental federalism) and the need for solutions better targeted to local and regional needs (place-based policy). In addition to advancing theoretical knowledge, conducting such analyses serves to provide robust scientific arguments for more effectively aligning agricultural policy tools with the real needs of beneficiaries, increasing the efficiency of resource use, and supporting the sustainable development of rural regions. Moreover, insights presented here may contribute to the ongoing debate on the CAP post-2027 reform, particularly in efforts to enhance its decentralisation and regionalisation [15,16]. Providing robust spatial and socio-economic data supports the decision-making process, including the implementation of new programs aimed at achieving climate goals and advancing carbon farming in Poland and other EU countries.

The aim of this study is to identify selected determinants of the spatial distribution of participation in the carbon farming eco-scheme in Poland, taking into account differences resulting from the farm size structure. The analytical context also includes variations in agricultural structures, diverse farming conditions, and other socio-economic factors that may influence the spatial differentiation of farm participation in carbon farming practices. The conducted empirical research seeks to answer the following research questions:

1. Does farm size influence the occurrence of spatial clusters in terms of participation and payment levels within the carbon farming eco-scheme?
2. What are the spatial patterns of participation in the eco-scheme across different farm size groups, and how do these patterns change over time?
3. Is there a spatial correlation between farm size and payment levels? What does this imply about the direction and effectiveness of agricultural policy?
4. How do counties with the highest and lowest levels of agricultural support, number of beneficiaries, and farm size differ in terms of agricultural structure, socio-economic characteristics, and rural area typologies?

2. The Role of Carbon Farming in Sustainable Agriculture

Sustainable agriculture is a comprehensive concept based on three fundamental pillars: economic, social, and environmental [17–19]. This concept aims to strike a balance between agricultural production efficiency, concern for societal well-being, and the protection of the natural environment. Modern forms of sustainable agriculture began to take shape in Europe only in 1992 when the reform of the CAP placed emphasis on ecological aspects by integrating agri-environmental policy with farming practices. The CAP reform marked a breakthrough by linking rural development issues, ensuring appropriate economic conditions for farmers and integrating environmental protection principles into agricultural practices. These actions aimed to improve the quality of life in rural areas and ensure the long-term profitability of the agricultural economy while also working toward environmental protection. The CAP is one of main EU policy, managed and funded at the European level from the Union's budget. Its goals include supporting farmers and improving agricultural efficiency by ensuring a stable supply of food at affordable prices while also raising farmers' living standards—contributing to the development of rural areas. A key component of the CAP is the promotion of the sustainable management of natural resources, combating climate change, and protecting rural areas and landscapes across the Union, all of which support biodiversity conservation.

Sustainable agriculture is a term with various definitions in the literature, but a common denominator among most of them is the aim to organise agricultural production in a way that does not cause negative environmental changes—or seeks to minimise them—while eliminating resource degradation. The difficulty in defining this term stems from diverse methods of describing and measuring agricultural phenomena, as well as the considerable organisational and functional complexity of agriculture, production diversity, regional differences in management conditions, and the complexity of economic, demographic, and social processes—unparalleled in other types of economic activity [20].

In 1987, the Food and Agriculture Organization (FAO) of the United Nations adopted the following definition of sustainable agricultural development: "Sustainable development involves the utilisation and conservation of natural resources and the shaping of technologies and institutions in such a way as to meet the needs of present and future generations. Such development in agriculture, forestry and fisheries, through the conservation of soil, water resources, plants, and animal genetic resources, does not degrade the environment, uses appropriate technologies, is economically viable, and is socially acceptable" [21].

According to Stinner and Blair [22], sustainable agriculture differs from conventional high-input agriculture because it focuses on long-term yield stability with minimal environmental impact rather than short-term goals like yield maximisation. Kutkowska [23] emphasises that sustainable agriculture is a farming system that combines economic, social, and ethical priorities with ecological security. This is achieved through appropriate management, rational use of ecosystem self-regulation mechanisms, and scientific and technological progress. In practice, this primarily means reducing intensive production characterised by heavy chemical inputs and excessive mechanisation. At its core lies the principle of ensuring the longevity of agricultural systems by minimising negative environmental impacts, optimising resource utilisation, and enhancing resilience to external shocks [24].

Sustainable agricultural development has become a key component of the European Green Deal, a comprehensive policy agenda aimed at transitioning toward a climate-neutral, circular economy [25]. A crucial element of this strategy is the "Farm to Fork" initiative, which sets ambitious goals for reducing greenhouse gas emissions, promoting organic farming, and protecting biodiversity within the agricultural sector [26,27].

Carbon farming plays a key role in sustainable agricultural development, primarily through the sequestration of carbon dioxide (CO_2) in the soil, which contributes to mitigating climate change.

Contemporary global changes—such as advancing globalisation, technological development, and economic growth—are leading to significant transformations in the natural environment. One of the most significant consequences is the sharp increase in carbon dioxide emissions, which drives global warming. Data analysis shows that CO_2 emissions rose from about 5 billion tons per year in the mid-20th century to over 35 billion tons per year at the turn of the 21st century [28]. As a result, many institutions recognise the growing CO_2 emissions as the most urgent issue that must be addressed.

Carbon farming is an approach aimed at both reducing carbon dioxide emissions and storing this gas from the atmosphere in plants and soils using various agronomic techniques. Definitions of carbon farming vary slightly depending on the author, but they all focus on integrating agricultural methods that support carbon sequestration. Foley et al. [29] view carbon farming as part of a more complex agricultural strategy, which seeks to balance food production, environmental protection, and climate change by applying practices that reduce greenhouse gas emissions and actively contribute to the absorption of carbon dioxide from the atmosphere.

Lal [30], one of the pioneers of research on carbon farming, sees it as a set of practices that contribute to long-term carbon storage in the soil. He lists practices such as the use of compost, organic fertilisers, and minimal soil tillage, and the introduction of agroecosystems that enhance the efficiency of carbon sequestration. Similarly, according to Jarosz et al. [31], carbon farming consists of practices that minimise greenhouse gas emissions while maximising carbon uptake in the soil. These include techniques such as appropriate crop cultivation, implementation of strip farming, soil quality improvement, and crop rotation. The FAO defines carbon farming in a similar way. It understands carbon farming as a set of activities aimed at increasing carbon sequestration in the soil, improving its health, and enhancing the productivity of agricultural production [32].

Carbon-absorbing cultivation can also be defined as an ecological business model in which land managers are rewarded for adopting improved land management practices that lead to increased carbon sequestration in biomass, dead organic matter, and soil by enhancing CO_2 capture or limiting the release of carbon dioxide into the atmosphere. This is performed in accordance with ecological principles that support biodiversity and benefit overall natural capital [33].

In light of the expected outcomes of carbon farming—namely, carbon dioxide sequestration and the reduction of existing emissions, as well as the avoidance of future greenhouse gas emissions—specific agronomic and technological practices have been identified for support and promotion under the framework of carbon farming. Examples of supported practices include restoring wetlands and peatlands, establishing agroforestry systems, maintaining and increasing organic carbon content in soil and effective livestock and manure management, as well as nutrient management on arable land and grasslands. Through these practices, carbon farming becomes a foundation for achieving carbon neutrality goals, contributing both to climate change mitigation and the protection of natural resources [33].

3. Eco-Scheme—A Support Policy for Carbon Farming in Poland

The EU's climate goals are extremely ambitious and achieving climate neutrality by 2050 requires member states to adapt their strategies and implement effective methods for reducing greenhouse gas emissions across all sectors of the economy. The reduction in greenhouse gas emissions through the implementation of carbon farming practices in the EU has been estimated at 101–444 Mt CO_2-eq per year [34]. According to data from the Intergovernmental Panel on Climate Change (IPCC), one hectare of agricultural land where regenerative farming practices are applied can sequester between 0.4 and even 3 tons of CO_2 annually, whereas, in the case of peatlands and forests, sequestration may reach up to 8.6 Gt CO_2-eq per year [35].

Carbon farming plays a significant role in reducing greenhouse gas emissions in the agricultural sector. Therefore, governments and international organisations are introducing financial support programs for farmers who decide to adopt sustainable agricultural practices.

Eco-schemes are a new tool under Pillar I of the CAP, and there have been and continue to be great expectations that they will support the implementation of the goals of the European Green Deal as they pertain to agricultural management. These policy goals include, among others, supporting fair farmer incomes, increasing the competitiveness of farms, improving farmers' position in the value chain, combating climate change, and protecting the environment and biodiversity [36]. EU countries have developed national CAP Strategic Plans to help achieve these objectives, taking into account local needs and conditions [14].

Although eco-schemes are a new instrument in agricultural policy, their significance goes beyond the political sphere and has become a subject of interest to the scientific community. Researchers are analysing the effectiveness of these practices in terms of environmental protection and economic efficiency, and evaluating their impact on biodiversity, soil quality, and the adaptation of agriculture to climate change [37–45].

Eco-schemes are mandatory for all member states but voluntary for farmers [46,47]. They are annual, additionally paid practices tailored to national conditions and needs but evaluated by the European Commission in terms of achieving the environmental and climate objectives of the new CAP [48]. In Poland, eco-schemes have been designed to promote practices that translate into farm incomes by increasing soil fertility and rational fertilisation, and improving crop quality. The main mechanism for this is the carbon farming eco-scheme, under which farmers can choose from eight available practices that best suit their needs.

The practices covered by the eco-scheme include the following [49]:

1. Extensive use of permanent grasslands with livestock grazing—maintaining meadows and pastures through grazing, which supports the increase in organic matter content in the soil.
2. Winter catch crops or undersown crops—sowing cover crops during the autumn–winter period to protect the soil against erosion and increase its fertility.
3. Development and compliance with a fertilisation plan:
    a. Basic variant—preparation of a fertilisation plan tailored to crop needs.

    b. Variant with liming—additionally includes liming to improve soil pH.
4. Diverse crop structure—cultivating various crops to increase biodiversity and improve soil structure.
5. Incorporation of manure into arable land within 12 h of application—quick mixing of manure with soil limits ammonia emissions and improves fertilisation efficiency.
6. Application of natural liquid fertilisers using methods other than splashing—applying liquid fertilisers in ways that reduce nutrient loss.
7. Simplified tillage systems—using methods that minimise soil disturbance, such as no-till farming, which supports increased soil organic carbon content.
8. Incorporation of straw into the soil—leaving and mixing straw after harvest to enrich soil with organic matter.

Each of the above practices is assessed using a point system, where 1 point corresponds to a specified payment amount. To qualify for the eco-scheme, the farmer must obtain a minimum number of points, which equals the number of points awarded for implementing the highest-rated practice on at least 25% of the agricultural land on the farm. To benefit from the eco-schemes, a farmer must meet specific criteria (e.g., maintain appropriate documentation, such as an agrotechnical log, and submit statements or geotagged photos confirming the implementation of the practice) and submit an application for payment to the The Agency for Restructuring and Modernisation of Agriculture (ARMA), which is a public body implementing the instruments co-financed from the EU's budget as well as providing of aid from national funds.

Subsidies for implementing individual practices are calculated based on the number of points assigned to them. In 2023, the value of 1 point was PLN 104.89. It is noteworthy that the greatest interest was not directed towards the practices with the highest point values, but rather towards those that were the most straightforward for farmers to implement. At the start of the scheme "Carbon farming and nutrient management", 393,651 farmers in Poland submitted applications, covering nearly 8 million ha (7,648,652.39 ha) (ARMA, as of 11.2023) [50]. This represented as much as 79.6% of all eco-schemes submitted nationwide.

Based on the number of applications submitted, the three most popular practices nationwide were "incorporation of straw into the soil" (229,050), "winter catch crops/undersown crops" (153,814), and "incorporation of manure into arable land within 12 h of application" (106,869). It should be emphasised that the selected practices had already been widely used by farmers and the introduction of payments aimed to reward their earlier efforts in preserving soil structure, specifically through maintaining or increasing the levels of organic matter and nutrients in the soil. However, the results of the analysis of the area declared for individual practices present a different picture. Nearly 5 million ha were declared under two carbon farming practices: "simplified tillage systems" (2,683,495.40 ha) and "incorporation of straw into the soil" (2,180,127.04 ha). In third place was "diverse crop structure", with an area of 1,529,942.07 ha reported under this practice. The least interest among farmers was shown for the practice "development and compliance with a fertilisation plan", particularly its liming variant, for which only 5356 applications were submitted (covering 147,692.20 ha) (see Figure 1).

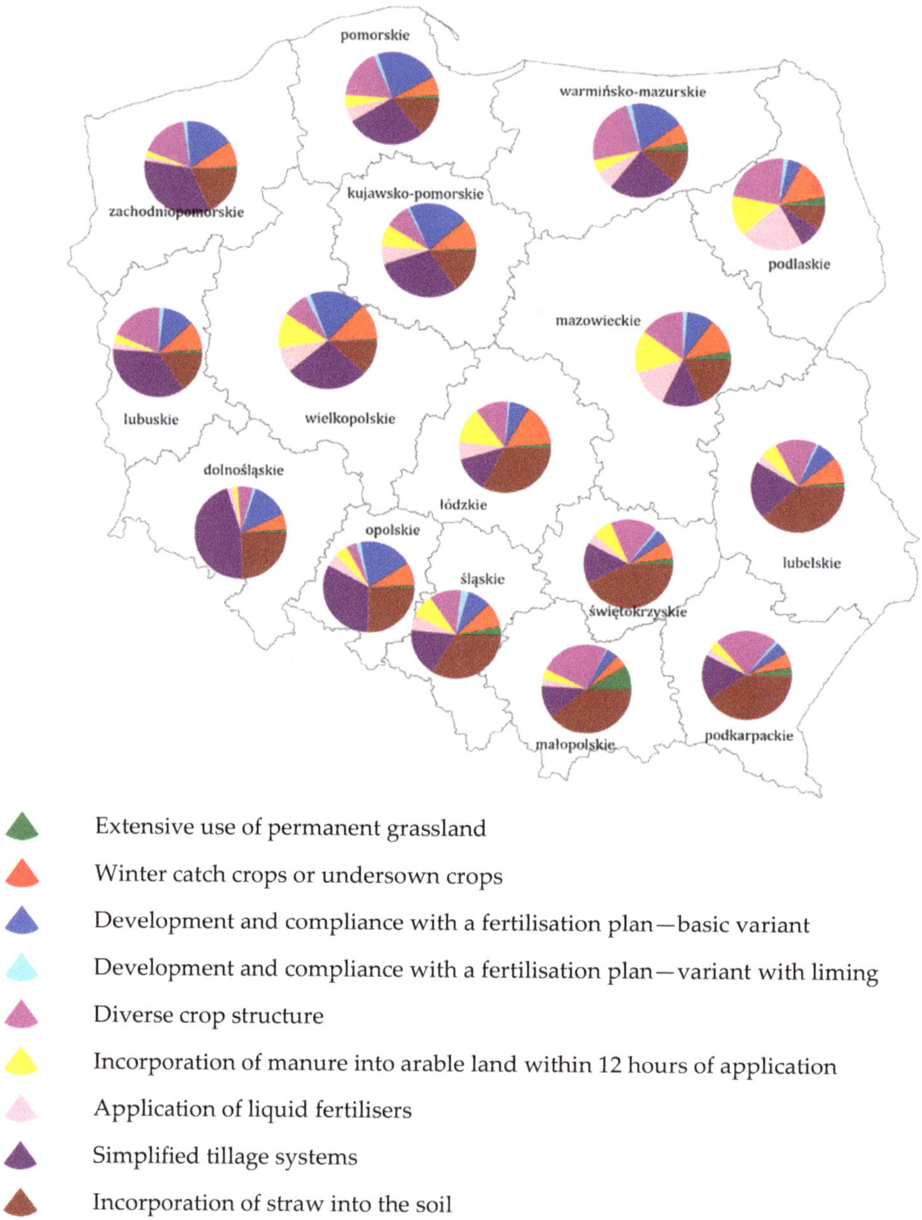

Figure 1. Structure of individual practices within the eco-scheme "Carbon farming and nutrient management" (based on the declared area). Source: own elaboration based on data from ARMA (as of 11.2023) [50].

An analysis of the data reveals clear regional variations, both in terms of the number of applications and the size of the declared area. The wielkopolskie and mazowieckie voivodeships stand out compared to other regions in terms of the area covered by eco-schemes (1,023,044.96 ha and 790,496.09 ha, respectively). In the case of the most popular practice, "simplified tillage systems", the leader was the wielkopolskie voivodeship

(407,379.33 ha), which is linked to its specific agrarian structure. Notably, the Polish CAP Strategic Plan allocated 800,000 ha for this practice using the simplified system, whereas farmers submitted applications for a total of 2,683,495.40 ha. This situation was related to a change in the payment rate. However, it should be noted that, according to the rules, farmers were allowed to combine several eco-schemes within a single application, which makes it difficult to assess the true scale of interest in this form of support.

4. Materials and Methods

4.1. Materials

An empirical analysis was conducted for Poland, one of the largest recipients of financial support from the European Union budget. This study focuses on regional disparities in agricultural support, using data aggregated at the county (powiat, NUTS 4) level. The county (powiat) is a second-level administrative unit in Poland, situated between the municipality (gmina, NUTS 5) and the voivodeship (województwo, NUTS 2). It plays a key role in the Poland's administrative structure, managing responsibilities that exceed the scope of individual municipalities but do not require intervention at the regional (voivodeship) level.

To achieve the stated research objective, a review of the relevant literature was conducted, supplemented by the use of data from several official national sources, covering both administrative and statistical domains. Information on the farm area, number of beneficiaries, and the amount of payments for the carbon farming eco-scheme for the years 2023 and 2024 was obtained from the ARMA, which manages the distribution of EU and national agricultural subsidies in Poland. Data on the total number of farms, dominant types of farming specialisation, and farm economic class sizes were sourced from the Central Statistical Office in Poland (Statistics Poland), based on the agricultural census conducted in 2020. These data provided a structural overview of Polish agriculture (see Appendix A). In addition, the typological classification of rural areas was derived from the Monitoring of Rural Areas Development based on data available for the year 2023 [51], which categorises municipalities according to functional, economic, and demographic characteristics.

4.2. Methods

In order to investigate the spatial dependence of farm area and payment amounts by farm size, we applied the global and local Moran's I statistics. These methods enable the detection of spatial autocorrelation, providing insight into whether the value of a variable in a given location is influenced by values in neighbouring areas [52,53]. Additionally, to explore the spatial relationship between these two variables, we used the global spatial cross-correlation index (GSCI), which allows for assessing the degree of co-dependence between spatial patterns of distinct variables.

4.2.1. Global and Local Autocorrelation

Global Moran's I is among the most commonly used measures of spatial autocorrelation. It is especially valuable for assessing whether the observed spatial distribution of a phenomenon is random, clustered, or dispersed. This measure was originally introduced by [54] and later formalised and expanded in the fields of spatial econometrics and geographic information science [55,56]. The statistic is defined as follows:

$$I = \frac{\sum_i \sum_j w_{ij}(x_i - \overline{x})(x_j - \overline{x})}{S^2 \sum_i \sum_j w_{ij}} \tag{1}$$

and

$$S^2 = \frac{1}{n} \sum_i (x_i - \overline{x})^2 \tag{2}$$

where x_i is the observed value in spatial unit I, $\overline{x}$ is the mean of all observed values, w_{ij} is the spatial weight between units I and j, and n is the number of spatial units. The spatial weights matrix $W = [w_{ij}]$ reflects the spatial structure and was row-standardised so that each row sums to 1.

Positive spatial autocorrelation ($I > 0$) means that similar values (either high or low) tend to cluster together. Negative spatial autocorrelation ($I < 0$) indicates that dissimilar values are neighbours, suggesting spatial dispersion (hot spots). As noted by [53], when the spatial weights matrix is appropriately defined, the Moran's I statistic can be interpreted as a form of a weighted correlation coefficient. It serves to identify spatial patterns in the distribution of variable x, specifically by testing whether adjacent regions exhibit greater similarity in their values than would be expected under a random, purely stochastic process. The significance of the Moran's I statistic can be assessed either through theoretical moments or a permutation-based approach. If the theoretical mean (expected value) of Moran's I approximates zero, it is interpreted as an indication of spatial randomness. To estimate the theoretical variance in the statistic, the randomisation method was employed [56,57].

To visualise spatial autocorrelation in each county, we employed a Moran scatterplot, displayed as a map based on the local Moran's I_i statistic. This map categorises data into four quarters (see Table 1), each indicating distinct spatial relationships:

Table 1. Relationships between regions and neighbours.

	Low Values in Region (L)	**High Values in Region (H)**
High values in neighbours (H)	LH: negative autocorrelation	HH: positive autocorrelation
Low values in neighbours (L)	LL: positive autocorrelation	HL: negative autocorrelation

Source: own elaboration based on Kopczewska [52].

This interpretation helps identify local clusters (HH—high value in the region with high values in neighbours, LL—low value in the region with low values in neighbours) and spatial outliers (HL—high value in the region with low values in neighbours, LH—low value in the region with high values in neighbours), providing a diagnostic tool for detecting spatial heterogeneity and potential spatial regimes.

The global Moran's I provides a synthetic measure indicating whether clustering exists across the entire study area, making it useful for general assessments. However, it does not identify where such clusters occur or their nature. In contrast, local Moran's I highlights specific areas of high or low spatial association, offering detailed insights into spatial patterns. It enables the detection of regions that strongly influence the formation of local clusters, the investigation of non-random local clusters, so-called hot spots, local instability, significant outlier observations, and spatial regimes [58]. Despite this advantage, local statistics are more sensitive to multiple testing issues and spatial outliers. Both methods depend heavily on the definition of spatial weights and assume stationarity, which can limit their interpretability in heterogeneous landscapes.

4.2.2. Global Spatial Cross-correlation Index

The GSCI is a measure designed to capture the spatial dependence between two variables across a given geographic area. Proposed by [59], the index extends the traditional concept of spatial autocorrelation by quantifying the cross-correlation between two different variables rather than a single variable's spatial distribution. The GSCI can be expressed as the sum of two components:

$$R_C = x^T W y \quad (3)$$

Chen [59] shows that the following relationship exists between Pearson's linear correlation coefficient and the GSCI:

$$R_0 = R_C + R_p \quad (4)$$

where R_0 is the overall spatial cross-correlation (Pearson linear correlation coefficient), R_C is the indirect component, capturing the influence of spatial effects, and R_p is the direct component, reflecting the correlation between variables that is not explained by the spatial structure. This index provides a global perspective on how strongly and in what manner two variables co-vary in space, offering insight into their spatial co-distribution patterns.

5. Results

5.1. Farm Size and Spatial Clustering in the Carbon Farming Eco-Scheme over Time

In order to examine the spatial patterns pertaining to the uptake and distribution of carbon farming payments (eco-scheme) in Poland, a spatial statistical analysis employing the Global Moran's I statistic was performed (see Table 2). This analysis was categorised by farm size to evaluate spatial dependencies across two dimensions: the total number of beneficiaries and the average payment amount (in EUR) received per beneficiary at the county (powiat) level for the years 2023 and 2024.

Table 2. Results of global Moran I test under randomisation.

Farm Size	Share of Beneficiaries in the Total Number of Farms (2023)	Share of Beneficiaries in the Total Number of Farms (2024)	Average Payment (2023)	Average Payment (2024)
<15 ha	0.277	0.306	0.339	0.395
15–30 ha	0.580	0.577	0.158	0.183
30–50 ha	0.633	0.612	0.151	0.200
50–100 ha	0.644	0.610	0.138	0.194
>100 ha	0.635	0.543	0.214	0.255

For all results of the global Moran I test, the p-values were below 0.01. Source: own calculation.

For farms with an area smaller than 15 ha (see Figure 2), the spatial autocorrelation of the number of beneficiaries demonstrated a positive and statistically significant correlation, registering at 0.28 in the year 2023, and experiencing a marginal increase to 0.31 in 2024. Notable low-value clusters were predominantly identified in the southern, north-western, and north-eastern regions of Poland in 2023; however, in 2024, these clusters extended into the central–eastern areas. The mean payment amount exhibited a heightened level of spatial autocorrelation, rising from 0.34 in 2023 to 0.39 in 2024, thereby distinctly highlighting high-value clusters in the northern regions and low-value clusters in the southern regions.

Farms with an area of 15 to 30 ha (see Figure 3) exhibited a considerably higher spatial autocorrelation in the number of beneficiaries, recorded at 0.58 in both 2023 and 2024. Notably, prominent low-value clusters were detected in the south-eastern and central–eastern regions, while high-value clusters emerged distinctly in western Poland. In contrast, the spatial autocorrelation pertaining to average payments was significantly weaker, measured at 0.16 in 2023 and 0.18 in 2024, lacking a clear cluster formation.

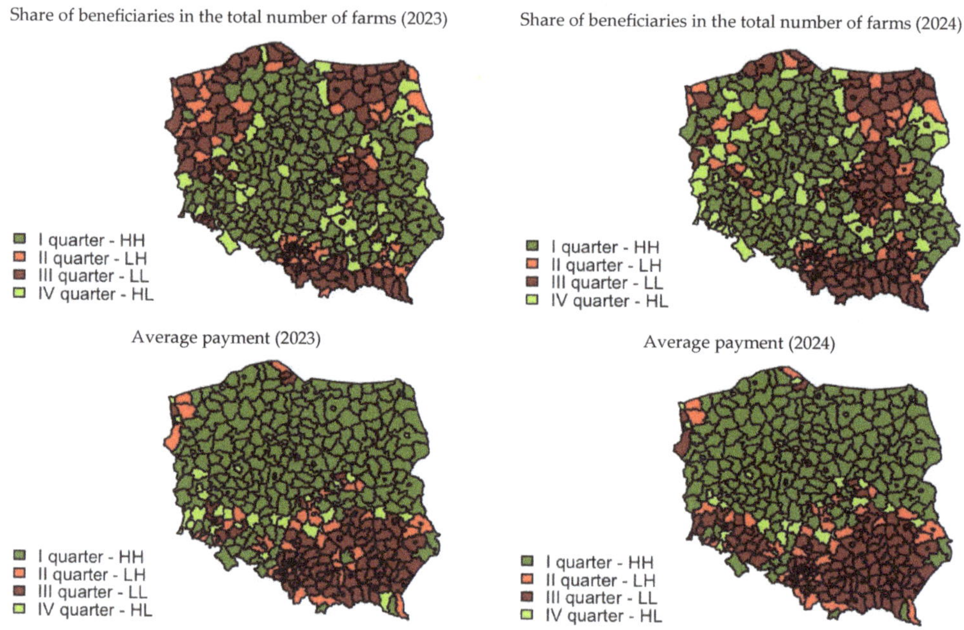

Figure 2. Spatial autocorrelation in the group of farms up to 15 ha. Source: the author's own elaboration.

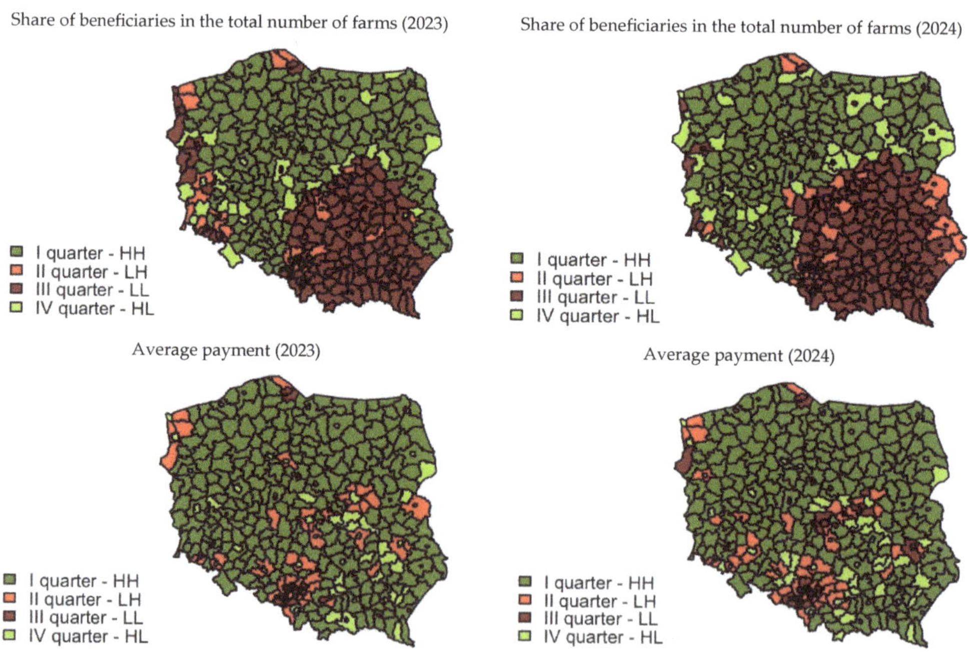

Figure 3. Spatial autocorrelation in the group of farms of 15–30 ha. Source: the author's own elaboration.

For farms ranging from 30 to 50 ha (see Figure 4), spatial autocorrelation among beneficiaries exhibited a pronounced strength, as evidenced by Moran's I values of

0.63 and 0.61 for the years 2023 and 2024, respectively. This phenomenon was typified by clear low-value clusters predominantly located in south-eastern Poland, while high-value clusters displayed a notable shift towards north-western regions in 2024. In contrast, the autocorrelation of average payments was markedly weaker, recording values of 0.15 in 2023 and 0.20 in 2024, and it exhibited no recognisable clustering patterns.

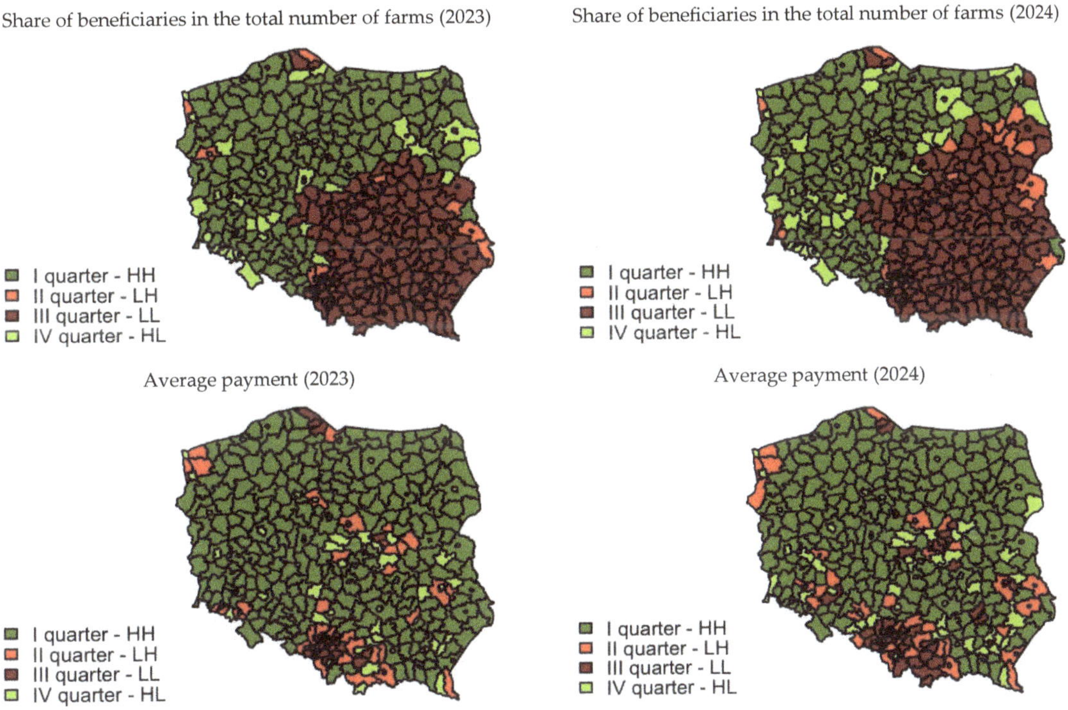

Figure 4. Spatial autocorrelation in the group of farms of 30–50 ha. Source: the author's own elaboration.

Within the category of farms ranging from 50 to 100 ha (see Figure 5), the Moran's I statistic for the number of beneficiaries peaked at 0.64 in 2023, subsequently experiencing a slight decline to 0.61 in 2024. These data reveal prominent low-value clusters situated in eastern Poland, alongside high-value clusters apparent in the western regions. Conversely, the spatial autocorrelation of average payments exhibited a modest value of 0.14 in 2023, with a slight increase to 0.19 in 2024, indicating the formation of low-value clusters in southern Poland.

In the case of the largest farms, exceeding 100 ha (see Figure 6), the spatial autocorrelation of beneficiary distribution demonstrated a notably strong correlation (0.63 in 2023, 0.54 in 2024). This observation highlights a distinct spatial regime characterised by high-value clusters situated in western Poland, in contrast to low-value clusters located in the eastern regions. Furthermore, the spatial autocorrelation of average payments, while relatively low, remained statistically significant (0.21 in 2023 and 0.26 in 2024), thereby indicating the existence of low-value clusters in the south-eastern region of Poland.

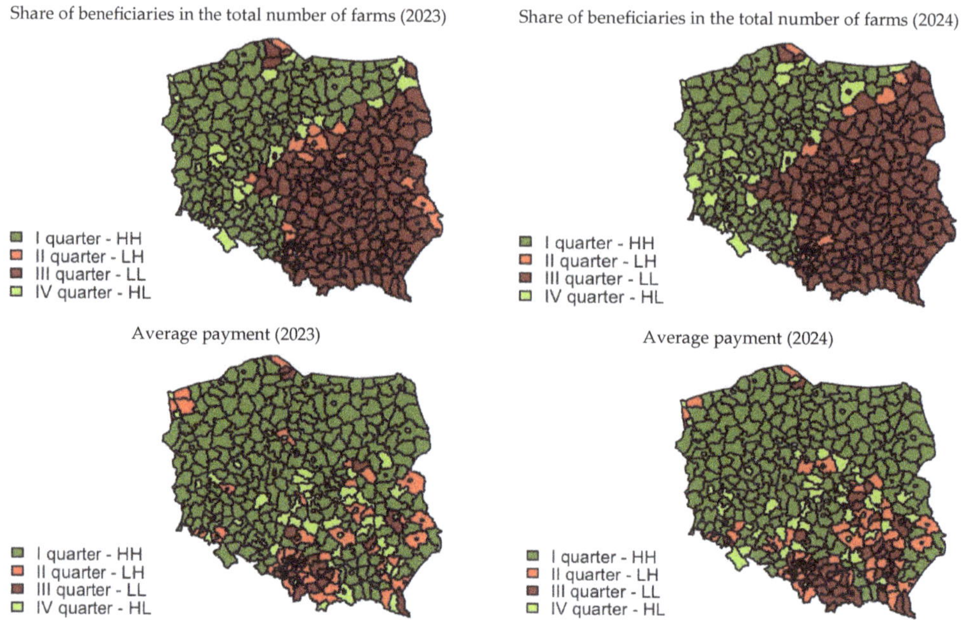

Figure 5. Spatial autocorrelation in the group of farms of 50–100 ha. Source: the author's own elaboration.

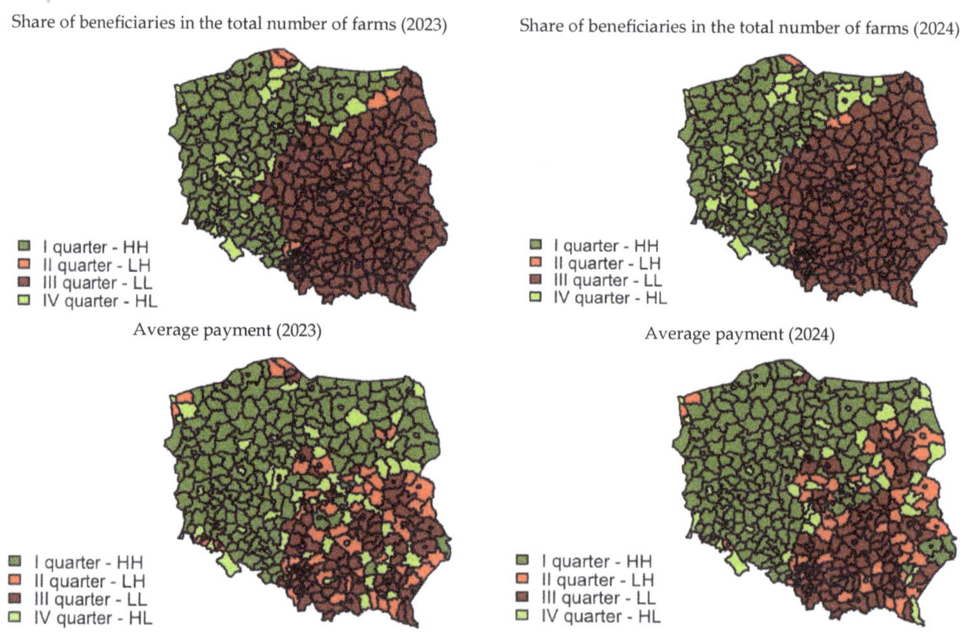

Figure 6. Spatial autocorrelation in the group of farms over 100 ha. Source: the author's own elaboration.

The analysis reveals that uptake clustering tends to increase with the size of the farm, indicating that larger farms demonstrate a greater regional concentration with respect to the adoption of eco-schemes in carbon farming, potentially reflecting the spatial distribution characteristic of farms in Poland. In contrast, payment clustering exhibits an inverse trend, diminishing as farm size increases. It is noteworthy that the strongest spatial dependency with regard to payment amounts is observed among the smallest farms, signifying a more localised spatial regime. Furthermore, an inverse relationship exists between uptake and payment spatial regimes: smaller farms exhibit more spatial structures in payments relative to uptake, whereas larger farms display the opposite pattern. These findings imply that diverse policy drivers and spatial mechanisms are influencing the adoption of eco-schemes in a manner distinct from the magnitude of financial benefits across different farm scales.

5.2. Spatial Correlation Between Farm Size and Payment Levels

Considering the emerging spatial patterns, we delved deeper into the connection between farm size and payment amounts by utilising the GSCI for 2023 and 2024 (see Table 3). This approach enabled us to assess the extent of spatial co-dependence between these two factors throughout all Polish counties, offering a further understanding of how the spatial arrangement of farm sizes may influence or relate to the spatial distribution of support under carbon farming.

Table 3. Results of spatial cross-correlation between total farm area and payments under carbon farming.

	2023	2024
GSCI (p-value)	0.096 (0.00)	0.102 (0.00)
Pearson correlation	0.983	0.884
Spatial effect	9.8%	11.5%
Cross-correlation effect	88.5%	87%

Source: own calculation.

The spatial cross-correlation results in Table 3 reveal that for both 2023 and 2024, the global spatial cross-correlation index is statistically significant (p-value = 0.00). This indicates a correlation between total farm area and payments under the eco-scheme on carbon farming, as well as spatial clustering. In 2023, the GSCI value stands at 0.096, with a Pearson correlation of 0.983, suggesting that approximately 9.8% of the total correlation between these variables is due to spatial relationships. Likewise, in 2024, this spatial contribution slightly rises to 11.5% (GSCI = 0.102, Pearson correlation = 0.884). This suggests that while the overall linear correlation declines slightly over time, spatial dependence increasingly influences this relationship. The cross-correlation effect, representing the portion of Pearson correlation not explained by spatial dependence, remains dominant in both years—at 88.5% in 2023 and 87% in 2024. This highlights that temporal or non-spatial factors are more significant in the observed correlation between farm area and payments. These findings support the idea that while spatial effects are meaningful, broader systemic or policy-driven influences are more critical. However, the rise in the spatial share from 2023 to 2024 may indicate emerging localised patterns or more spatially targeted policy adjustments.

5.3. Differences in Agricultural Structure and Rural Characteristics Across Counties with Varying Support Levels

In the final stage of the analysis, we examined the best- and worst-performing counties in terms of the number of beneficiaries, total farm area, and total payment amounts (see Figure 7 and Tables 4 and 5). This case-based approach enabled a deeper interpretation of spatial patterns and helped identify outliers in the dataset.

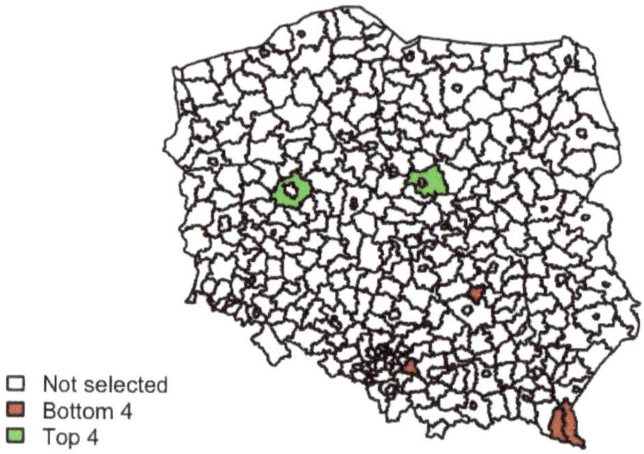

Figure 7. Top and bottom counties. Source: the author's own elaboration.

Table 4. Top and bottom counties by number of beneficiaries, area, and payments under the carbon farming eco-scheme.

County Name (Voivodeship Name)	Number of Producers in 2023	Total Area in 2023 in ha	Total Payments in 2023 in EUR	Number of Producers in 2024	Total Area in 2024 in ha	Total Payments in 2024 in EUR
			Bottom 4			
skarżyski (świętokrzyskie	53	489.54	48,971.6	53	500.55	17,000.9
chrzanowski (małopolskie)	53	540.66	45,474.8	57	674.24	9879.1
leski (podkarpackie)	57	390.48	33,654.5	53	374.21	14,974.1
bieszczadzki (podkarpackie)	70	649.42	71,362.8	51	519.29	18,083
			Top 4			
poznański (wielkopolskie)	2221	99,247.04	10,718,100	2223	101,108.14	4,400,240
płocki (mazowieckie)	4181	75,346.24	6,740,087	4086	77,087.66	1,268,832.9
bialski (lubelskie)	5070	69,450.05	6,584,446.2	5110	73,528.46	3,031,713.3
lubelski (lubelskie)	6269	68,620.47	5,201,413.3	6345	71,041.75	1,794,490

Source: Agriculture Restructurisation and Modernisation Agency in Poland.

A detailed comparison between the best- and worst-performing counties in terms of agricultural payments, farm area, and number of beneficiaries reveals a clear divide shaped by both structural and socio-economic factors. The worst-performing counties (skarżyski, chrzanowski, leski, and bieszczadzki) are all located in the less agriculturally developed voivodeships of south-eastern and central Poland. These areas exhibit low numbers of producers (ranging from 51 to 70 in both years), small total farm areas (between ~374 and 675 ha in 2024), and very modest total payments on carbon farming, with some counties showing a sharp decline in support—most notably skarżyski and chrzanowski, where payments in 2024 dropped by over 60% compared to 2023.

Table 5. Top and bottom counties by selected structural and socio-economic characteristics.

County Name (Voivodeship Name)	Total Number of Farms	Predominant Type of Farming	Predominant Economic Class Size in EUR Thousand of Standard Output (SO)	Monitoring of Rural Areas Development (Number of Municipalities)
Bottom 4				
skarżyski (świętokrzyskie)	812	field crops (60%), grazing livestock (13%), not classified (8%)	<2 (70%)	rural areas dominated by agriculture: intermediate (1), rural areas with dispersed agriculture and multiple sources of income (3)
chrzanowski (małopolskie)	1022	field crops (46%), grazing livestock (17%), not classified (14%)	<2 (64%)	urbanised rural areas with reduced agriculture (6)
leski (podkarpackie)	2274	field crops (51%), not classified (24%), grazing livestock (9%)	<2 (72%)	rural areas dominated by agriculture: intermediate (2), rural areas with dispersed agriculture and multiple sources of income (1), multifunctional rural areas: balance of sectors (1), urbanised rural areas with reduced agriculture (1)
bieszczadzki (podkarpackie)	1553	field crops (41%), not classified (26%), grazing livestock (18%)	<2 (56%)	rural areas dominated by large-scale agriculture (1), rural areas with dispersed agriculture and multiple sources of income (1), multifunctional rural areas: balance of sectors (1)
Top 4				
poznański (wielkopolskie)	4534	field crops (64%), mixed—crops and animals (14%)	<2 (15%), 8–15 (15%), 25–50 (14%)	multifunctional rural areas: balance of sectors (1), urbanised rural areas with reduced agriculture (5), highly urbanised rural areas (9)
płocki (mazowieckie)	9032	field crops (69%), mixed—crops and animals (12%), grazing livestock (11%)	4–8 (19%), <2 (18%)	rural areas dominated by traditional agriculture (6), rural areas dominated by large-scale agriculture (2), rural areas dominated by agriculture: intermediate (3), multifunctional rural areas: balance of sectors (3), urbanised rural areas with reduced agriculture (1)
bialski (lubelskie)	13,428	field crops (68%), mixed—crops and animals (13%), grazing livestock (8%)	<2 (22%), 4–8 (22%)	rural areas dominated by traditional agriculture (12), rural areas dominated by agriculture: intermediate (5)
lubelski (lubelskie)	15,248	field crops (71%), mixed—crops and animals (10%), fruit trees (8%)	<2 (22%), 2–4 (24%), 4–8 (22%)	rural areas dominated by traditional agriculture (2), rural areas dominated by agriculture: intermediate (7), rural areas with dispersed agriculture and multiple sources of income (2), urbanised rural areas with reduced agriculture (5)

Source: Central Statistical Office in Poland, Monitoring of Rural Areas Development.

These counties are characterised by dominant low-value farming systems, primarily field crops, grazing livestock, and a significant share of unclassified agricultural activity. The economic potential of farms is extremely limited—over 60% of producers in these counties belong to the lowest economic class (less than EUR 2000 SO). This reflects a highly fragmented agrarian structure and low market orientation. Additionally, these counties fall within rural typologies, such as rural areas with dispersed agriculture and multiple sources of income or urbanised rural areas with reduced agriculture, suggesting marginalisation in terms of agricultural function and a dependence on non-farm income sources.

In contrast, the best-performing counties (poznański, płocki, bialski and lubelski) are spread across agriculturally strong regions, primarily in central and eastern Poland. These counties report vastly higher numbers of producers (over 2000 and up to 6300), significantly larger agricultural areas (between 69,000 and 101,000 ha in total in 2024), and the highest payment on carbon farming, although some also experienced a drop in total payments (e.g., poznański). Farming systems in these counties are specialised and market-oriented, with a dominance of field crops and a more balanced inclusion of mixed and livestock farming. Economic class structures are significantly more varied, lacking a dominant class. From the perspective of a socio-economic development typology, the aforementioned leading counties are situated within more functionally dynamic rural regions, which include multifunctional rural areas characterised by a balance of sectors, highly urbanised rural areas, or regions predominantly dedicated to traditional agriculture.

6. Discussion

Eco-schemes designed to achieve climate and environmental objectives are a vital part of the EU's new and improved green agricultural policy. This policy aims to promote agriculture in the EU that is sustainable economically, socially, and environmentally. Eco-schemes account for one-quarter of the total allocation for direct payments, highlighting their importance in fostering sustainable agricultural practices in Europe [60]. Carbon farming and other proposed eco-schemes form a key element of the new reform of the EU CAP 2023–2027 and the implementation of EU strategies under the European Green Deal (the Farm to Fork and Biodiversity strategies) [48]. Our study showed that carbon farming turned out to be the relatively most popular instrument among all the eco-schemes implemented under the CAP in Poland during the analysed period. This eco-scheme involved the implementation of eight environmentally and climate-friendly farming practices, which provide compensation for incurred costs and rewards for generating public goods like carbon sequestration [29]. This mechanism was originally designed as a response to growing consumer expectations for agricultural policy to promote healthier and more environmentally, climate-friendly, and sustainable agriculture [1,6].

However, from the beginning of the implementation process of this instrument—intended to encourage voluntary but above-standard actions on farms—a number of controversies and practical difficulties have been faced. In the first two years of their perpetration (2023 and 2024), only a third of the total number of farms in Poland (around 390 thousand) used them. The new CAP proposal, which includes eco-schemes, has faced criticism from farmers not only in Poland but also across other EU countries [61]. Regardless of the various reasons for protest, the limited adoption of these instruments could be attributed to political factors, including farmers' resistance to new tighter solutions with greater environmental ambitions and errors in their creation and implementation at both the EU and member state levels [62]. The eco-schemes, along with other new components of the revised agricultural policy, were often perceived not as voluntary environmental solutions intended to improve farming conditions and the natural environment but as measures leading to income reductions and the next set of complex and burdensome administrative obligations, often described as an 'iron cage' of bureaucracy imposed on beneficiaries [63].

Comparative insights from other EU countries reinforce the need to revise the approach to eco-shemes. As highlighted in a recent cross-country analysis by BirdLife and NABU [64], the uptake and effectiveness of eco-schemes vary significantly across EU member states. In countries such as Spain or the Netherlands, nutrient management and carbon farming schemes recorded high levels of adoption—often surpassing 100% of planned areas. However, many of these schemes supported practices that were already widespread, resulting in limited additional environmental value. Similarly, in the Netherlands, the popularity of simple measures with a low environmental impact reflects a menu-based approach geared toward maximizing farmer participation. Conversely, biodiversity-oriented measures, like flower strips or landscape elements, although ecologically ambitious, faced low uptake due to their complexity and insufficient financial incentives. Poland demonstrates a similar pattern, where more demanding practices—despite their ecological potential—attracted fewer applicants. This evidence supports the argument that one-size-fits-all support mechanisms are suboptimal. Regionally differentiated eco-schemes, including tiered payments or region-specific practice menus, are increasingly recommended to better reflect agricultural realities.

Nevertheless, it is essential to interpret the effectiveness and reception of eco-schemes and the entire new CAP framework (including new conditionality standards, which tied the payment of direct subsidies to the fulfilment of specific requirements) not only in light of their design and implementation features, but also within the broader macroeconomic and geopolitical context. The scepticism toward eco-schemes and the broader CAP reform package, particularly the new conditionality rules linking direct payments to environmental compliance, was significantly reinforced by concurrent external shocks. These included the economic and political repercussions of Russia's aggression against Ukraine, as well as the deterioration of global agri-food markets, both of which substantially undermined farm income and production profitability. In such conditions, farmers' willingness to engage in new, often complex environmental schemes diminished further, regardless of their potential long-term benefits [65,66].

Although it is important to consider the general effects of political, administrative, and market conditions on the implementation of carbon farming eco-schemes, our analysis shows that in Poland, persistent spatial and structural factors at the local and regional level had an impact on this process as well. Among these structural determinants, the differentiated agrarian structure and the varying farm sizes, reflecting their economic potential, were of importance. Firstly, our analysis showed that uptake clustering tends to increase with the size of the farm, indicating that larger farms demonstrated a greater regional concentration with respect to the adoption of eco-schemes in carbon farming, potentially reflecting the spatial distribution characteristic of farms in Poland. The use of carbon farming eco-schemes was greater in the counties of western, northern, and south-western Poland—regions characterized by developed commodity agriculture and more concentrated agricultural structures [67]. In both years analysed (2023 and 2024), the observed spatial pattern showing greater participation in eco-schemes of group of larger farms did not change over time.

Secondly, in the analysis, an inverse relationship between uptake and payment spatial regimes was noted. In other words, the smaller farms exhibit more spatial structure in payments relative to uptake, whereas larger farms display the opposite pattern. Larger eco-scheme payments were generally received by small farms located in northern Poland compared to those in the south, which may indicate differences in the number of eco-schemes per farm. This trend is associated with the relatively lower significance of commodity agricultural production and farming activities for farm users in southern Poland [68].

Using the spatial cross-correlation, a strong, positive, and statistically significant relationship between farm size and the amount of support under carbon farming eco-schemes was confirmed. This observation was in line with other studies [69]. Nonetheless, the spatial effect for the correlation considered was relatively low, increasing from 9.8% to 11.5% in 2023–2024, indicating the importance of other determinants. In general, larger farms benefited more from carbon farming eco-schemes due to implementing more practices. However, the limited scale of use suggests that eco-schemes should be designed to be straightforward and tailored to the requirements of both larger and smaller farms. The case-based approach implemented in this paper enabled a deeper interpretation of spatial patterns and helped identify outliers in the dataset. A detailed comparison between the best- and worst-performing counties in terms of agricultural payments, farm area, and number of beneficiaries of carbon farming eco-schemes revealed a clear spatial divide shaped by both structural and socio-economic factors. In the case of small farms located in the eastern and particularly in south-eastern regions of Poland, the design, availability, and scale of support for carbon farming eco-schemes should consider the unique characteristics of agriculture in these areas. Often, this farming is not-marked oriented and constrained by natural conditions. While it is essential to meet EU climate and environmental objectives, the proposed support should also reflect a demand for the greater regionalisation of these payments (while not making it more complex) as well as offer more training and information activities for their users [37,70]. Better regionalised and tailored carbon farming eco-schemes to address different productions systems, agrarian structures, and natural constrains based on collaboration at different levels (farmers, advisors, central and regional authorities) could increase farmer and societal acceptance for this CAP support [71,72].

7. Conclusions

Eco-schemes, introduced under the CAP 2023–2027, constitute a modern tool supporting farmers in the adoption of environmentally and climate-friendly practices. It is estimated that only one-third of all farms in Poland benefited from eco-schemes, and the level of participation varied regionally. The relatively low and spatially varied interest in eco-schemes (including carbon farming practices) among Polish farmers during the initial phase of their implementation was due to complex factors. These were primarily related to the fragmented agrarian structure and diverse natural conditions for conducting agricultural activity.

It should be emphasised that the implementation of eco-schemes, including their components, such as carbon farming, posed a challenge for public institutions responsible for managing and executing the CAP in Poland (e.g., the Ministry of Agriculture and the paying agency), as these were new policy instruments. The process of designing and promoting eco-schemes (developing and disseminating the assumptions), followed by their implementation (creating an appropriate organisational system, including regulations and procedures) at the national level, progressed at a fast pace, dictated by legislative delays at the EU institutional level. Consequently, implementing eco-schemes required numerous actions and adjustments on the part of public administration bodies.

Our study of spatial patterns of uptake and payment distribution under the eco-scheme "Carbon farming and nutrient management" in Poland in 2023 and 2024 revealed significant regional differences in both the structure of beneficiaries and the distribution of payments, with the size of farms having a clear influence on these patterns. An analysis conducted at the county level using the Moran's I statistic showed that the larger the farm, the more pronounced the spatial concentration of both the number of beneficiaries and the amount of payments. The largest farms (over 100 ha) demonstrated strong concentration in the western regions of Poland, while the smallest farms (under 15 ha) showed more localised clustering, particularly in northern Poland.

An added value of the analysis was the consideration of differences in the spatial structure of payments. The results showed that small farms exhibited a higher spatial correlation regarding payment amounts, whereas larger farms had a less distinct concentration of payments. This phenomenon may result from different agricultural policy mechanisms that affect small and large farms in various ways.

This study also revealed differences between the best- and worst-performing counties. The best results were recorded in regions with high agricultural potential, such as poznański county or płocki county, characterised by large agricultural areas and diverse farming systems. In contrast, the worst results were observed in counties with less developed agriculture, such as skarżyski county or chrzanowski county, where smaller, less diversified farms with low agricultural income predominated.

This research highlighted the significant role of spatial and economic factors, particularly farm size, in the implementation of eco-schemes in Poland. Our study was preliminary, and further in-depth analyses are needed to better understand how these factors interact, using varied data and research methods.

The relevance of this research is further underscored by the specific nature of eco-schemes—while participation is voluntary for farmers and implementation is mandatory for member states. As such, the design and scope of eco-schemes reflect each country's adopted "green strategy". Member states are obliged to evaluate the effectiveness of their environmental and climate-related measures, making analytical studies like this particularly valuable. They offer critical insights into the actual uptake and perceived relevance of interventions such as carbon farming eco-schemes, thereby supporting evidence-based assessments of policy rationale.

Collaboration between public administration and the scientific community is crucial for the effective implementation of eco-schemes. Thanks to scientific research, it is possible to monitor the effects of these practices, identify the best solutions, and adapt agricultural policies to evolving environmental and social challenges.

Author Contributions: Conceptualization, A.R. and A.P.; methodology, A.R. and A.P.; software, A.P.; validation, A.R. and A.P.; formal analysis, A.R. and A.P.; investigation, A.R., A.P. and M.D.; resources, A.R., A.P. and M.D.; data curation, A.R. and A.P.; writing—original draft preparation, A.R., A.P. and M.D.; writing—review and editing, A.R., A.P. and M.D.; visualisation, A.R., A.P. and M.D.; supervision, A.R., A.P. and M.D.; project administration, A.R.; funding acquisition, A.R. All authors have read and agreed to the published version of the manuscript.

Funding: This research received no external funding.

Institutional Review Board Statement: Not applicable.

Informed Consent Statement: Not applicable.

Data Availability Statement: The data presented in this study are available on request from the corresponding author due to restrictions imposed by the Agency for Restructuring and Modernisation of Agriculture, as the data are not publicly available.

Conflicts of Interest: The authors declare no conflicts of interest.

Appendix A

To provide a more comprehensive context for the interpretation of spatial patterns observed in the results, the following figures present additional structural characteristics of agriculture at the county level. Specifically, they illustrate the total number of farms

(Figure A1), the number of farms by economic size class (Figure A2), the number of farms by type of farming (Figure A3), the number of beneficiaries of the carbon farming eco-scheme (Figure A4), the total area covered by the carbon farming eco-scheme (Figure A5), and the total payment for the carbon farming eco-scheme (Figure A6). These indicators help to better contextualise the variability in the uptake of the eco-scheme by reflecting the diversity of farming structures and production profiles across regions.

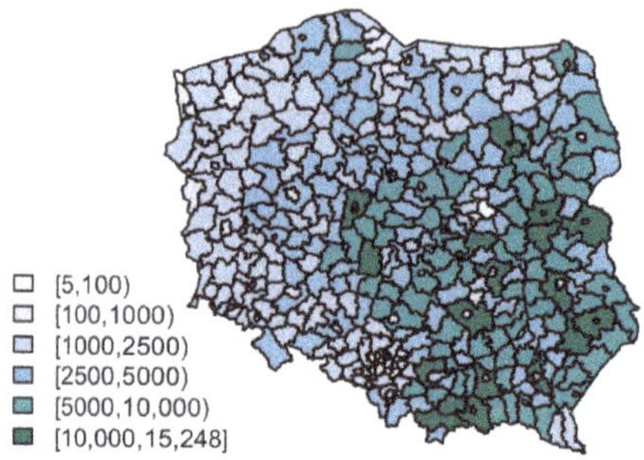

Figure A1. Number of farms in 2020.

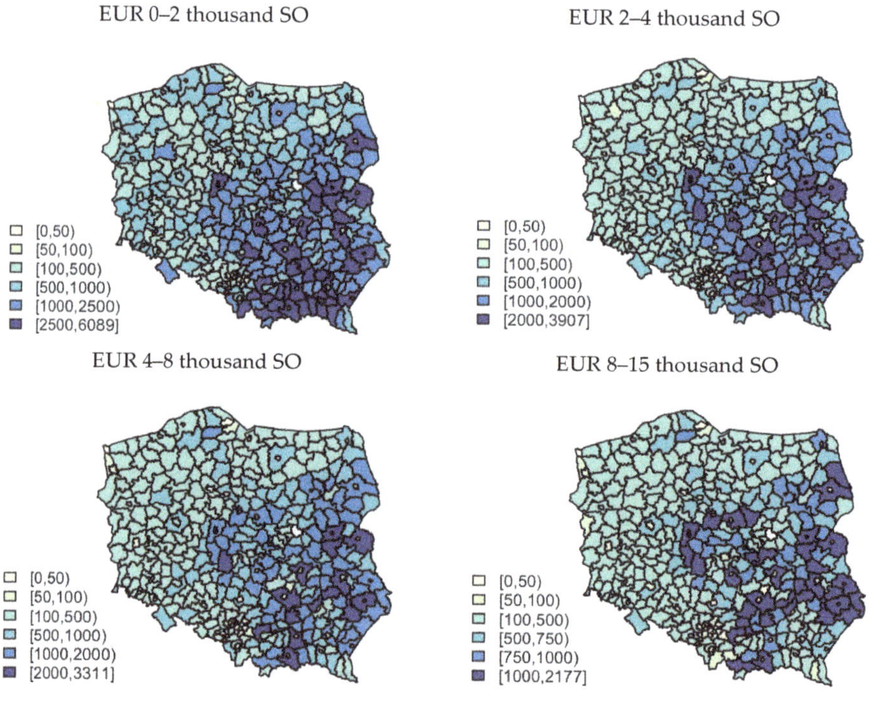

Figure A2. *Cont.*

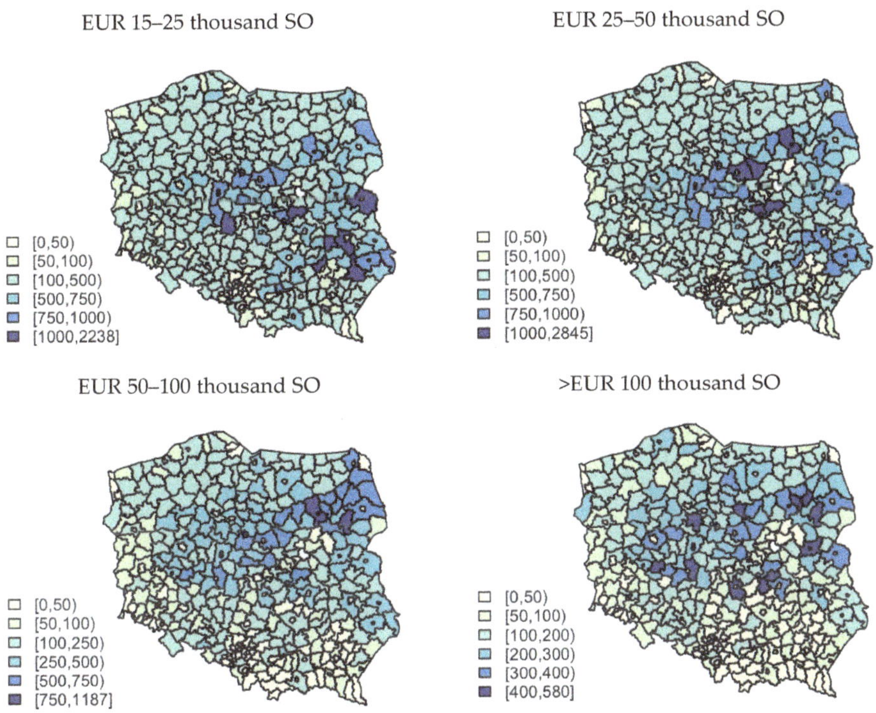

Figure A2. Number of farms by economic class size in 2020.

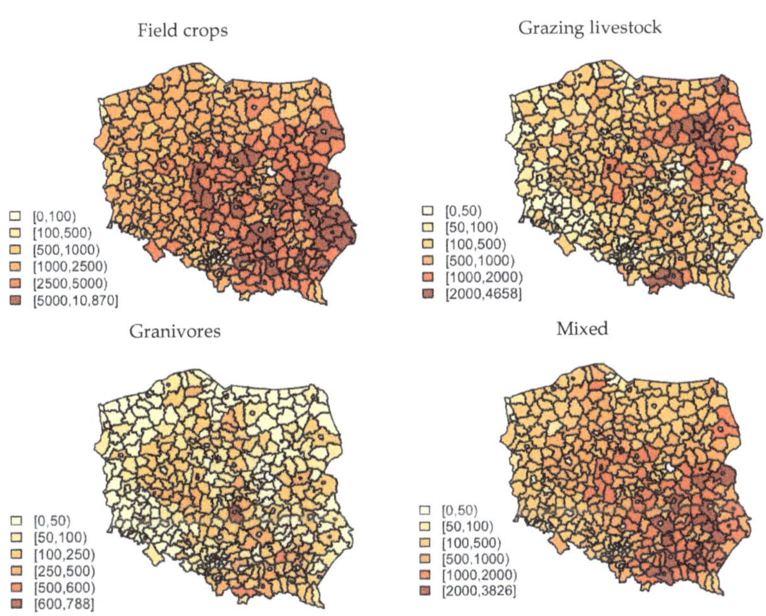

Figure A3. Number of farms by type of farming in 2020.

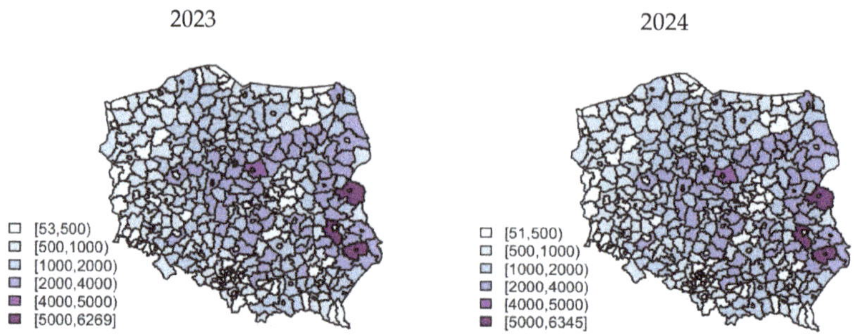

Figure A4. Number of beneficiaries of the carbon farming eco-scheme.

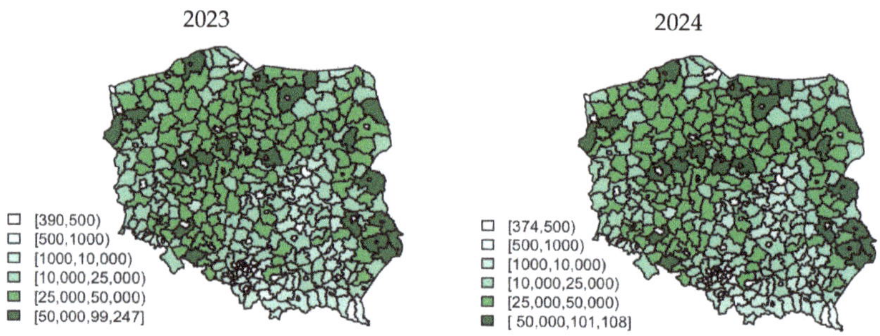

Figure A5. Total area covered by the carbon farming eco-scheme [in ha].

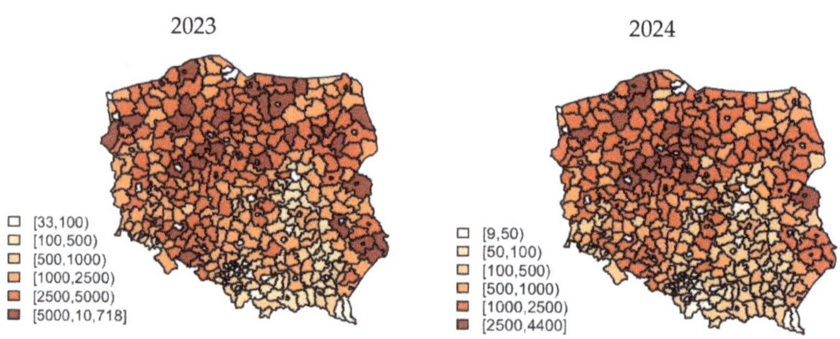

Figure A6. Total payment for the carbon farming eco-scheme [in EUR thousands].

References

1. Zegar, J.S.; Wrzaszcz, W. The holism principle in agriculture sustainable development. *Econ. Environ. Stud.* **2017**, *17*, 1051–1069. [CrossRef]
2. Zegar, J.S. *Premises of the New Agricultural Economics*; Zagadnienia Ekonomiki Rolnej: Warsaw, Poland, 2010.
3. Dudek, M.; Rosa, A. Regenerative Agriculture as a Sustainable System of Food Production: Concepts, Conditions, Perceptions and Initial Implementations in Poland, Czechia, and Slovakia. *Sustainability* **2023**, *15*, 15721. [CrossRef]
4. IPCC. Land degradation. In *Climate Change and Land: An IPCC Special Report on Climate Change, Desertification, Land Degradation, Sustainable Land Management, Food Security, and Greenhouse Gas Fluxes in Terrestrial Ecosystems*; Shukla, P.R., Skea, J., Calvo Buendia, E., Masson-Delmotte, V., Pörtner, H.-O., Roberts, D.C., Zhai, P., Slade, R., Connors, S., van Diemen, R., et al., Eds.; IPCC: Geneva, Switzerland, 2019; pp. 345–436. Available online: https://www.ipcc.ch/srccl/cite-report (accessed on 9 March 2025).
5. Food and Land Use Coalition (FOLU). Growing Better: Ten Critical Transitions to Transform Food and Land Use. The Global Consultation Report of the Food and Land Use Coalition, FOLU. 2019. Available online: https://www.foodandlandusecoalition.org/wp-content/uploads/2019/09/FOLU-GrowingBetterGlobalReport.pdf (accessed on 11 March 2025).
6. Rhodes, C.J. The imperative for regenerative agriculture. *Sci. Prog.* **2017**, *100*, 80–129. [CrossRef]

7. Giller, K.E.; Hijbeek, R.; Andersson, J.A.; Sumberg, J. Regenerative Agriculture: An Agronomic Perspective. *Outlook Agric.* **2021**, *50*, 13–25. [CrossRef] [PubMed]
8. Rosa, A.; Dudek, M.; Siemiński, P.; Sadowski, A.; Bartosik, S.; Kaczmarek, P.; Łata, K.; Markowicz, M.; Petrovic, J.; Dykes, I. *Biologization: The Key to Sustainable Agriculture. Catalogue of Good Practices*; Institute of Rural and Agricultural Development, Polish Academy of Sciences: Warsaw, Poland, 2022; p. 31.
9. Stanek, R. *Dobre Praktyki Rolnicze W Produkcji Integrowanej Z Zastosowaniem Pożytecznych Mikroorganizmów W Uprawie Buraka*; FDPA, MRiRW: Warsaw, Poland, 2012.
10. European Parliament and Council. Regulation (EU) 2021/2115 of the European Parliament and of the Council of 2 December 2021 Establishing Rules on Support for Strategic Plans to be Drawn up by Member States Under the Common Agricultural Policy (CAP Strategic Plans) and Financed by the European Agricultural Guarantee Fund (EAGF) and by the European Agricultural Fund for Rural Development (EAFRD) and Repealing Regulations (EU) No 1305/2013 and (EU) No 1307/2013. *Off. J. Eur. Union* **2021**. Available online: https://eur-lex.europa.eu/legal-content/EN/TXT/PDF/?uri=CELEX:32021R2115 (accessed on 11 March 2025).
11. Komisja Europejska. Zatwierdzono 28 planów strategicznych WPR (2023–2027). Available online: https://agriculture.ec.europa.eu/document/download/7b3a0485-c335-4e1b-a53a-9fe3733ca48f_en?filename=approved-28-cap-strategic-plans-2023-27.pdf (accessed on 11 January 2025).
12. European CAP Network. *Eco-Schemes: Background Paper*; European Union: Brussels, Belgium, 2023. Available online: https://eu-cap-network.ec.europa.eu/sites/default/files/2023-10/eco-schemes-background-paper.pdf (accessed on 3 April 2025).
13. Available online: https://www.gov.pl/web/rolnictwo/programowanie-ps-wpr (accessed on 9 March 2025).
14. MRiRW (Ministry of Agriculture and Rural Development). Plan Strategiczny dla Wspólnej Polityki Rolnej na lata 2023–2027 (PS WPR 2023–2027), Warsaw, Poland. 2023. Available online: https://www.gov.pl/web/rolnictwo/plan-strategiczny-dla-wspolnej-polityki-rolnej-na-lata--2023-2027 (accessed on 2 February 2025).
15. Wieliczko, B. Challenges of European integration–to what extent should the Common Agricultural Policy stay common? *Argum. Oeconom. Cracov.* **2019**, *2*, 97–109. [CrossRef]
16. Strategic Dialogue on the Future of EU Agriculture. *A Shared Prospect for Farming and Food in Europe*; European Commission: Brussels, Belgium, 2024. Available online: https://agriculture.ec.europa.eu/document/download/171329ff-0f50-4fa5-946f-aea11032172e_en?file-name=strategic-dialogue-report-2024_en.pdf (accessed on 11 March 2025).
17. Nistoroiu, B.F.; Privitera, D.; Prahoveanu, Ş.L.; Emek, M.L.; Condeianu, O. Enhancing Agricultural Progress via Sustainable Production and Consumption. *Ekon. Poljopriv.* **2024**, *71*, 973–985. [CrossRef]
18. Oleszko-Kurzyna, B. Rozwój zrównoważony rolnictwa wobec wymogów Unii Europejskiej w zakresie ochrony środowiska. *Zesz. Nauk. SGGW Probl. Rol. Świat.* **2008**, *4*, 326–336. [CrossRef]
19. Turner, S. *Koncepcja Zrównoważonego Rozwoju Rolnictwa*; ADAS Consulting: Warsaw, Poland, 2000.
20. Matuszczak, A. Wskaźniki wyników rozwoju: Przesłanki teoretyczne i wnioski o pomiar w regionach UE. *Wieś Rol. Rol. Wiej.* **2013**, *1*, 101–119. [CrossRef]
21. Wilkin, J. *Podstawy Strategii Zintegrowanego Rozwoju Rolnictwa I Obszarów Wiejskich W Polsce*; Warsaw University Press: Warsaw, Poland, 2003; p. 14.
22. Stinner, B.R.; Blair, J.M.; Edwards, C.; Lal, R.; Madden, P.; Miller, R.; House, G. Ecological and Agronomic Characteristics of Innovative Cropping Systems. In *Sustainable Agricultural Systems*, 1st ed.; Stinner, B.R., Blair, J.M., Eds.; CRC Press: Boca Raton, FL, USA, 1990. [CrossRef]
23. Kutkowska, B. *Wdrażanie Koncepcji Zrównoważonego Rozwoju Rolnictwa I Obszarów Wiejskich W Sudetach*; Self-Published: Warsaw, Poland, 2007; p. 90.
24. Leoveanu-Soare, B.E.; Petre, L.I.; Micu, M.M. Social and economic aspects regarding the development of agriculture in Romania. *Sci. Pap. Ser. Manag. Econ. Eng. Agric. Rural. Dev.* **2020**, *20*, 281–288.
25. European Commission. Komunikat Komisji do Parlamentu Europejskiego, Rady, Europejskiego Komitetu Ekonomiczno-Społecznego i Komitetu Regionów. Europejski Zielony Ład. Available online: https://eur-lex.europa.eu/legal-content/EN/TXT/?uri=CELEX:52019DC0640&qid=1726910790338 (accessed on 11 March 2025).
26. European Commission. Komunikat Komisji do Parlamentu Europejskiego, Rady, Europejskiego Komitetu Ekonomiczno-Społecznego i Komitetu Regionów. Strategia "Od Pola Do Stołu" na Rzecz Sprawiedliwego, Zdrowego i Przyjaznego dla Środowiska Systemu Żywnościowego. Available online: https://eur-lex.europa.eu/legal-content/EN/TXT/?uri=CELEX:52020DC0381 (accessed on 11 March 2025).
27. European Commission. Komunikat Komisji do Parlamentu Europejskiego, Rady, Europejskiego Komitetu Ekonomiczno-Społecznego i Komitetu Regionów. Unijna Strategia na Rzecz Bioróżnorodności 2030. Przywracamy Naturę do Naszego Życia. Available online: https://eur-lex.europa.eu/legal-content/EN/TXT/?uri=celex:52020DC0380 (accessed on 11 March 2025).
28. NOAA; ETH Zurich. Our World in Data. Available online: https://ourworldindata.org (accessed on 10 March 2025).

29. Foley, J.A.; Ramankutty, N.; Brauman, K.A.; Cassidy, E.S.; Gerber, J.S.; Johnston, M.; Mueller, N.D.; O'Connel, C.; Ray, D.K.; West, P.C.; et al. Solutions for a cultivated planet. *Nature* **2011**, *478*, 337–342. [CrossRef] [PubMed]
30. Lal, R. Soil carbon sequestration impacts on global climate change and food security. *Science* **2004**, *304*, 1623–1627. [CrossRef]
31. Jarosz, Z.; Faber, A. Praktyki rolnictwa węglowego. Praca przeglądowa. *Agron. Sci.* **2024**, *79*, 123–135. [CrossRef]
32. FAO. *The State of Food and Agriculture 2015: Social Protection and Agriculture: Breaking the Cycle of Rural Poverty*; Food and Agriculture Organization of the United Nations: Rome, Italy, 2015.
33. European Commission. *Communication from the Commission to the European Parliament and the Council Sustainable Carbon Cycles*; COM(2021) 800 final; European Commission: Brussels, Belgium, 2021.
34. McDonald, H.; Frelih-Larsen, A.; Lóránt, A.; Duin, L.; Pyndt Andersen, S.; Costa, G.; Bradley, H. Study for the Committee on Environment, Public Health and Food Safety (ENVI), Policy Department for Economic, Scientific and Quality of Life Policies. In *Carbon Farming–Making Agriculture Fit for 2030*; European Parliament: Luxembourg, 2021.
35. IPCC. Summary for Policymakers. In *Global Warming of 1.5 °C. An IPCC Special Report on the Impacts of Global Warming of 1.5 °C Above Pre-Industrial Levels and Related Global Greenhouse Gas Emission Pathways, in the Context of Strengthening the Global Response to the Threat of Climate Change, Sustainable Development, and Efforts to Eradicate Poverty*; Cambridge University Press: Cambridge, UK; New York, NY, USA, 2022; pp. 3–24. [CrossRef]
36. European Commission. Common Agricultural Policy Overview, CAP 2023–2027. Available online: https://agriculture.ec.europa.eu/common-agricultural-policy/cap-overview/cap-2023-27_pl?utm_source=chatgpt.com (accessed on 11 March 2025).
37. Zieliński, M.; Józwiak, W.; Żak, A.; Rokicki, T. Development of Eco-Schemes as an Important Environmental Measure in Areas Facing Natural or Other Specific Constraints Under the Common Agriculture Policy 2023–2027: Evidence from Poland. *Sustainability* **2025**, *17*, 2781. [CrossRef]
38. Cagliero, R.; Vassallo, M.; Pierangeli, F.; Pupo D'Andrea, M.R.; Monteleone, A.; Camaioni, B.; Tarangioli, S. The Common Agricultural Policy 2023–2027: How Member States Implement the New Delivery Model? *Ital. Rev. Agric. Econ.* **2023**, *78*, 49–66. [CrossRef]
39. Szymańska, M. Ekoschematy–nowa płatność bezpośrednia (Ecoschema–new direct payment). *Tyg. Porad. Rol.* **2020**, *23*, 14.
40. European Parliament. *European Parliamentary Research Service. Guide to EU Funding 2023 Edition*; European Parliament: Brussels, Belgium, 2023. Available online: https://www.europarl.europa.eu/thinktank/en/document/EPRS_STU(2023)747110 (accessed on 20 March 2025).
41. Latacz-Lohmann, U.; Termansen, M.; Nguyen, C. The New Eco-Schemes: Navigating a Narrow Fairway. *EuroChoices* **2022**, *21*, 4–10. [CrossRef]
42. Styburski, W.; Kozera-Kowalska, M.; Uglis, J. Ecoschemes As A New Tool To Support Agriculture–The State Of Implementation In Poland. *Ann. PAAAE* **2023**, *25*, 287–302. [CrossRef]
43. Włodarczyk, B. Prawne instrumenty ochrony środowiska i przeciwdziałania zmianom klimatu we Wspólnej Polityce Rolnej na lata 2023–2027. *Przegląd Prawa Rolnego* **2023**, *2*, 11–26. [CrossRef]
44. Niewiadomska, A. Krajowy Plan Strategiczny jako instrument prawny rozwoju rolnictwa. *Stud. Iurid.* **2022**, *91*, 242–255. [CrossRef]
45. Petsakos, A.; Ciaian, P.; Espinosa, M.; Perni, A.; Kremmydas, D. Farm-level impacts of the CAP post-2020 reform: A scenario-based analysis. *Appl. Econ. Perspect. Policy* **2023**, *45*, 1168–1188. [CrossRef]
46. Feindt, P.; Grohmann, P.; Hager, A. The CAP post-2020 reform and the EU budget process. In *EU Policymaking at a Crossroads. Negotiating the 2021–2027 Budget*; Munch, S., Heinelt, H., Eds.; Edward Elgar Publishing: Cheltenham, UK, 2022.
47. Castillo-Diaz, F.J.; Belmonte-Urena, L.J.; Alvarez-Rodriguez, J.F.; Cachao-Ferre, F. The Role of Sustainability and Circular Economy in Europe's Common Agricultural Policy. In *Environmentally Sustainable Production. Research for Sustainable Development*; Valls Martinez, M.A., Santos-Jaen, J.M., Eds.; Springer: Cham, Switzerland, 2024.
48. European Commission. *List of Potential Agricultural Practices That Eco-Schemes Could Support*; European Commission: Brussels, Belgium, 2021.
49. Ekoschematy. Available online: https://www.gov.pl/web/rolnictwo/ekoschematy3 (accessed on 11 March 2025).
50. Agencja Restrukturyzacji i Modernizacji Rolnictwa. *Ekoschemat Rolnictwo węglowe i zarządzanie składnikami odżywczymi*; Departament Płatności Bezpośrednich, November 2023. Available online: https://www.gov.pl/web/arimr (accessed on 11 March 2025).
51. Stanny, M.; Rosner, A.; Komorowski, Ł. *Monitoring Rozwoju Obszarów Wiejskich. Etap IV: Dekada Przemian Społeczno-Gospodarczych*; Fundacja Europejski Fundusz Rozwoju Wsi Polskiej; Instytut Rozwoju Wsi i Rolnictwa PAN: Warszawa, Poland, 2024. [CrossRef]
52. Kopczewska, K. *Przestrzenne Metody Ilościowe W R: Statystyka, Ekonometira, Uczenie Maszynowe, Analiza Danych*; CeDeWu: Warsaw, Poland, 2021.
53. Suchecki, B. *Ekonometria Przestrzenna. Metody I Modele Analizy Danych Przestrzennych*; C.H. Beck: Warsaw, Poland, 2010.
54. Moran, P.A.P. Notes on Continuous Stochastic Phenomena. *Biometrika* **1950**, *37*, 17–23. [CrossRef] [PubMed]
55. Cliff, A.C.; Ord, J.K. *Spatial Autocorrelation*; Pion Limited: London, UK, 1973.

56. Cliff, A.D.; Ord, J.K. *Spatial Processes: Models and Applications*; Pion Limited: London, UK, 1981.
57. Goodchild, M.F. *Spatial Autocorrelation*; Geo Books: Norwich, UK, 1986.
58. Ministeri, A. *Evolution of Spatial Distribution of Innovative Activity*; Working Paper; Università Ca' Foscari di Venezia: Venice, Italy, 2003.
59. Chen, Y. A New Methodology of Spatial Cross-Correlation Analysis. *PLoS ONE* **2015**, *10*, e0126158. [CrossRef]
60. European Commission; Joint Research Centre; Isbasoiu, A.; Fellmann, T. *CAP Strategic Plans of the EU Member States: CSPs Master File and Key Data*; Publications Office of the European Union: Luxembourg, 2024. Available online: https://data.europa.eu/doi/10.2760/3104919 (accessed on 2 April 2025).
61. Matthews, A. Farmer protests and the 2024 European Parliament elections. *Intereconomics* **2024**, *59*, 83–87. [CrossRef]
62. Finger, R.; Fabry, A.; Kammer, M.; Candel, J.; Dalhaus, T.; Meemken, E.M. Farmer Protests in Europe 2023–2024. *EuroChoices* **2024**, *23*, 59–63. [CrossRef]
63. Mc Loughlin, E. Protesting the future: The evolution of the European farmer. *Anthropol. Today* **2024**, *40*, 3–6. [CrossRef]
64. BirdLife Europe; NABU. *The Untapped Potential of Eco-Schemes: An Analysis of the Use of Eco-Schemes Across 12 Countries and Their Impact on Biodiversity*; BirdLife Europe and Central Asia & NABU: Bruxelles, Belgium, 2024.
65. Bułkowska, M.; Bazhenova, H. Direct and Indirect Consequences of the War in Ukraine for Polish Trade in Agri-Food Products. *Probl. Agric. Econ.* **2023**, *376*, 66–90. [CrossRef]
66. Czubak, W.; Kalinowski, S.; Pelpliński, B. *Ziarno Niezgody: Analiza Protestów Rolniczych*; Instytut Finansów Publicznych: Warsaw, Poland, 2024.
67. Wiśniewski, Ł.; Perdał, R.; Rudnicki, R. Proposed method for delimiting spatial structure on the example of agriculture types in Poland. *Bull. Geogr. Socio-Econ. Ser.* **2020**, *49*, 7–18. [CrossRef]
68. Czekaj, M.; Szafrańska, M.; Żmija, K.; Satoła, Ł.; Płonka, A.; Żmija, D.; Tyran, E.; Puchała, J. *Rola Małych Gospodarstw Rolnych. Diagnoza I Perspektywy Na Przyszłość Na Przykładzie Podregionu Rzeszowskiego*; Difin SA: Warsaw, Poland, 2020.
69. Zieliński, M.; Gołebiewska, B.; Adamski, M.; Sobierajewska, J.; Tyburski, J. Adaptation of eco-schemes to Polish agriculture in the first year of the EU CAP 2023–2027. *Econ. Environ.* **2024**, *89*, 817. [CrossRef]
70. Zieliński, M.; Józwiak, W.; Adamski, M.; Wrzaszcz, W.; Sobierajewska, J.; Mirkowska, Z. *Dochody Gospodarstw Rolniczych Versus Ekoschematy*; Fundacja Europejski Fundusz Rozwoju Wsi Polskiej: Warsaw, Poland, 2024.
71. EU CAP Network. Factsheet: Improving the Design and Implementation of Eco-Schemes in the New CAP Strategic Plans (CSPs)–Experiences from the First Year of Implementation. Available online: https://ruralnirazvoj.hr/files/eu-cap-network-factsheet-eco-schemes.pdf (accessed on 9 March 2025).
72. Drygas, M.; Nurzyńska, I. *Pożądana Struktura Wsparcia W Ramach Wspólnej Polityki Rolnej Unii Europejskiej Po 2020 Roku Oraz Cele Polityki Krajowej W Świetle Globalnych Wyzwań Modernizacyjnych Wobec Polskiego Rolnictwa*; Fundacja Europejski Fundusz Rozwoju Wsi Polskiej, IRWiR PAN: Warsaw, Poland, 2021.

Disclaimer/Publisher's Note: The statements, opinions and data contained in all publications are solely those of the individual author(s) and contributor(s) and not of MDPI and/or the editor(s). MDPI and/or the editor(s) disclaim responsibility for any injury to people or property resulting from any ideas, methods, instructions or products referred to in the content.

sustainability

MDPI

Article

An Evaluation of Dryland Ulluco Cultivation Yields in the Face of Climate Change Scenarios in the Central Andes of Peru by Using the AquaCrop Model

Ricardo Flores-Marquez [1,*], Jesús Vera-Vílchez [2], Patricia Verástegui-Martínez [2], Sphyros Lastra [1] and Richard Solórzano-Acosta [1,3]

1 Centro Experimental La Molina, Dirección de Supervisión y Monitoreo en las Estaciones Experimentales Agrarias, Instituto Nacional de Innovación Agraria (INIA), La Molina 1981, Lima 15024, Peru; slastrapaucar@gmail.com (S.L.); investigacion_labsaf@inia.gob.pe (R.S.-A.)

2 Estación Experimental Agraria Santa Ana, Dirección de Supervisión y Monitoreo en las Estaciones Experimentales Agrarias, Instituto Nacional de Innovación Agraria (INIA), Carretera Saños Grande—Hualahoyo Km 8, Huancayo 12007, Peru; jvera@lamolina.edu.pe (J.V.-V.); patymarve@gmail.com (P.V.-M.)

3 Facultad de Ciencias Ambientales, Universidad Científica del Sur (UCSUR), Lima 15067, Peru

* Correspondence: ricardo.floresm29@gmail.com

Abstract: *Ullucus tuberosus* is an Andean region crop adapted to high-altitude environments and dryland cultivation. It is an essential resource that guarantees food security due to its carbohydrate, protein, and low-fat content. However, current change patterns in precipitation and temperatures warn of complex scenarios where climate change will affect this crop. Therefore, predicting these effects through simulation is a valuable tool for evaluating this crop's sustainability. This study aims to evaluate ulluco's crop yield under dryland conditions at 3914 m.a.s.l. considering climate change scenarios from 2024 to 2100 by using the AquaCrop model. Simulations were carried out using current meteorological data, crop agronomic information, and simulations for SSP1-2.6, SSP3-7.0, and SSP5-8.5 of CMIP 6. The results indicate that minimum temperature increases and seasonal precipitation exacerbation will significantly influence yields. Increases in rainfall and environmental CO_2 concentrations show an opportunity window for yield increment in the early stages. However, a negative trend is observed for 2050–2100, mainly due to crop temperature stress. These findings highlight the importance of developing more resistant ulluco varieties to heat stress conditions, adapting water management practices, continuing modeling climate change effects on crops, and investing in research on smallholder agriculture to reach Sustainable Development Goals 1, 2, and 13.

Keywords: Andean tuber; *Ullucus*; climate change; AquaCrop; yield response

1. Introduction

Tubers are important crops within Andean societies and their agro-food systems. They are characterized by their resilience to the region's adverse conditions. Among these crops, ulluco is prominent in food security due to its fundamental role in the Andean population's diet [1,2]. It is cultivated primarily in Peru, Bolivia, and Ecuador and is grown between 2000 and 4000 m.a.s.l. at temperatures between 11 and 13 °C [1–3]. Its nutritional contribution is based on its high carbohydrate content (64.96–84.2%), between 8.5 and 15.7% protein content, low-fat content (0.1–1.4%), fiber ranging 0.5–5%, and vitamin C [1,4,5]. Generally, tubers are boiled before eating. Nonetheless, they can also be dried to turn into *chuño* or milled into flour [6]. It is grown by more than 60 thousand producers in Peru, and during 2023, a cropping area of 42,912 ha with a 7.62 $t{\cdot}ha^{-1}$ average yield was reported, generating a total production of 173,116 t [7]. On average, dryland ulluco cultivation represents 82% of Peru's cropping area of ulluco, widely located in the Quechua and Sunni regions (94%). Monoculture patterns (86%) prevail over associations (14%) with crops such as corn, mashua, potatoes, quinoa, etc. [8]. Even though its productive potential is less than other Andean tubers or roots [9], it is widely consumed in this region [1,10,11].

In the current context, climate change driven by carbon emissions from agriculture, livestock, industry, and households affects global agriculture through extreme weather events [12,13]. To research future impacts, the Working Group I (WGI) of the Intergovernmental Panel on Climate Change (IPCC) analyzed historical information and Global Climate Models (GCMs), generated as part of the Coupled Model Intercomparison Project (CMIP) to broaden our comprehension about the process and effects linked to climate change [14]. The fifth phase of the CMIP (CMIP5) analyzed future climatic projections based on Representative Concentration Pathways (RCPs), which express the increase in the most probable radiative forcing for 2100 [14]. Subsequently, the sixth phase's report provided different socioeconomic development pathways to achieve each RCP, thus proposing new levels of mitigation and adaptation actions to the effects of climate change [15]. These new scenarios are named Shared Socioeconomic Pathways (SSPs). SSP 1 and SSP 5 both propose increasing human development, differing in that the second proposes intensive fossil energy consumption as a means to achieve this; SSP 3 and SSP 4 propose pessimistic and inequitable development trends, while SSP 2 outlines continuity regarding historical patterns [14,15]. O'Neill et al. [15] identified priority analysis scenarios to update and cover information gaps of CMIP5 projections for RCPs—SSP1-2.6, SSP2-4.5, SSP3-7.0, and SSP5-8.5.

With all this information, it is possible to use existing models to explore potential yield changes in different crops. There are some experiences regarding modeling tuber and root growth such as potatoes, cassava, and sweet potatoes under diverse climates and water-available conditions [16–19]. Nevertheless, there is little research on crops in Andean regions [20–23] or on native crops such as ulluco [24–28], despite its importance for Andean food security and value in gourmet cuisine and industry.

AquaCrop is a model that simulates crop yields under different conditions based on plant water consumption; for this, it requires information about climate, crop, soil, and management conditions without considering pests' or diseases' potential effects [29]. Constant efforts are made to improve AquaCrop's simulation results while maintaining its simple modeling scope [30]. Due to its versatility, a wide range of research is linked to crop modeling through AquaCrop use for different climatic and management conditions, including climate change projections at different latitudes, longitudes, and altitudes [20,31,32]. At this point, the model's main limitations include its single-field-scale scope, assumptions of spatial uniformity inside the field, and the unidimensional vertical water fluxes simulated [33]. In addition, Vanuytrecht et al. [30] also outline the constant improvement of AquaCrop results to make them more accessible and understandable for agricultural extensionists, decision-makers, and stakeholders.

Several studies suggest that smallholding farming productivity in the Peruvian Andes is at imminent risk of decreasing due to water availability changes and increased temperatures [34–36]. The susceptibility of dryland crops to rainfall variability is exacerbated by changes in rainfall patterns due to climate change [34,37] which ultimately determines poverty conditions in rural areas [38]. Additionally, rising temperatures reduce the productive potential of crops adapted to cold regions [39]. This study focused on precipitation and temperature changes and their effect on ulluco yield.

Agronomic factors that influence tuber development have been investigated [40–43], as well as their potential food industry use due to their nutritional benefits [10,44–46] and their importance in local culture [47,48]. Ulluco's phenological development and productivity depend on climatic, edaphic, and genotypic factors, among others. Sowing dates are usually linked to rain presence in a crop rotation system with potatoes, oca, or barley [3,49].

Therefore, analyzing crop behavior facing extreme climate events is essential to develop adaptation strategies for different regional conditions [36]. This study evaluates the yield changes in *Ullucus tuberosus*, a key native Andean tuber, under dryland cultivation in the Peruvian central highlands, according to climate change scenarios from the Coupled Model Intercomparison Project (CMIP) Phase Six for the 2024–2100 period.

2. Materials and Methods

2.1. Studied Area

This study was conducted in the town of San José de Apata, district of Apata, Jauja province, department of Junín, Peru. The experimental field was installed at 75°19′30.41″ W; 11°47′53.73″ S at 3914 m.a.s.l. (Figure 1). The area's climatic conditions are average annual precipitation, 689 mm; minimum temperatures between −4 °C (June–July) and 4 °C (March); maximum temperatures between 20 °C (March) and 22 °C (November–December); and average relative humidity, 72%. Averages were calculated based on historical data from Peru's National Service of Meteorology and Hydrology (SENAMHI) [50].

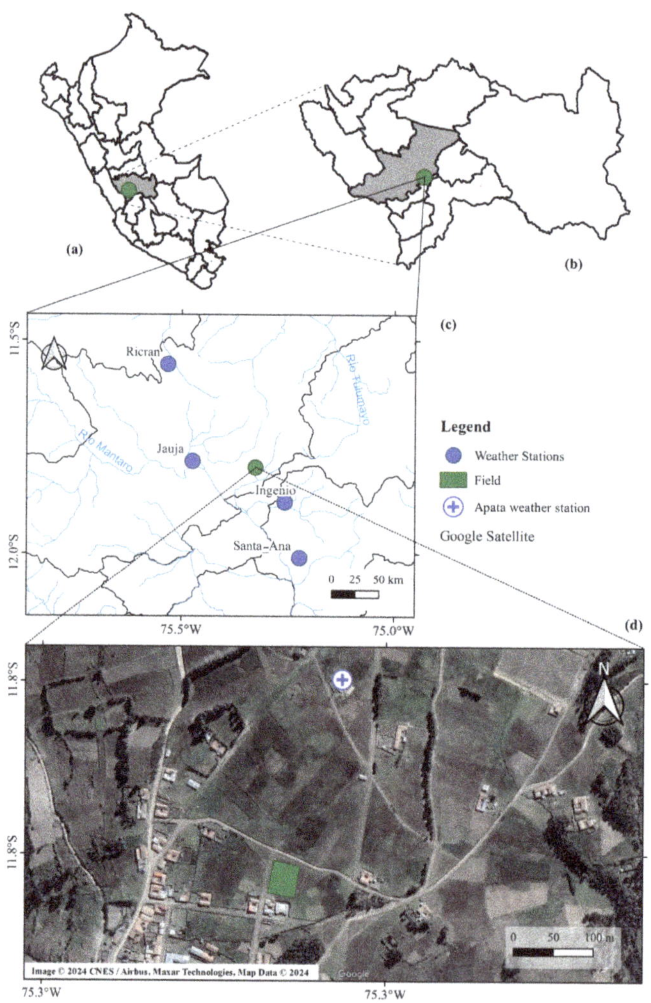

Figure 1. Studied area; (**a**) Junín department location within Peru; (**b**) Jauja province location; (**c**) location of the evaluation plot and meteorological stations used; (**d**) aerial satellite view of the evaluation plot.

2.2. Experimental Design

This study had a split-plot design in which the following two factors were tested: (i) sowing date (main plot with three levels) and (ii) the type of fertilization (sub-plot with two levels). Thus, six treatments with four replicates per treatment were used for 24 experimental units, as described in Table 1.

Table 1. Factors and levels tested.

Factor	Levels
Sowing date (F)	F1: 13 October 2022 F2: 28 October 2022 F3: 12 November 2022
Fertilization management (M)	M1: Traditional (sub-dose) * M2: Recommended (optimal dose) *

* Details of each fertilization management are described in Table of Section 2.5.

2.3. Experimental Plots Set-Up

Experimental plots were installed on a three-year fallow field where potatoes had previously been grown. *Ullucus tuberosus* cv. Canario was chosen due to its widespread cultivation in the central Peruvian Andes [51], sown at a density of 41,666 plants ha-1 (0.8 m × 0.3 m). A net plot consisting of 9 furrows, 6 m long, was established within each experimental unit. Data were collected from them. The spatial arrangement of the treatments can be found in Figure 2.

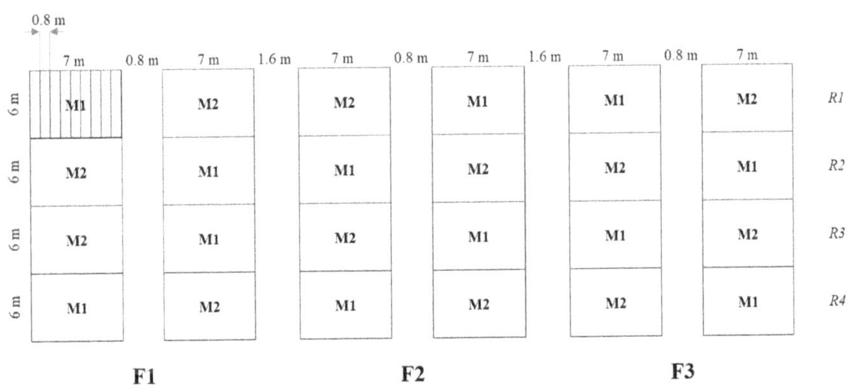

Figure 2. Field distribution of the experimental units.

2.4. Soil Characteristics

Before installation, a soil study was conducted in the experimental area. First, a soil pit was dug to a depth of 1.3 m, and, after observation, three horizons were determined, comprising topsoil (AP) and two strata of partially decomposed parent material (C1 and C2). For each of them, samples were taken and sent to the Soil, Water and Leaves Laboratory (LABSAF)—INIA Santa Ana to analyze bulk density (BD) [52], texture (Txt) [53], soil pH [54], electrical conductivity (EC) [55], organic matter (OM) content, nitrogen (N), phosphorus (P), potassium (K), and exchangeable aluminum (Al) [53] (Table 2). In addition, other soil samples were taken at the central zone of each evaluation unit, at 30 cm depth, to measure soil moisture content using the gravimetric method [56].

Table 2. Soil profile characterization.

Cod	Depth	pH	EC	OM	P	K	Al	N	Txt	BD
	m		d·Sm^{-1}	%	mg·kg^{-1}	mg·kg^{-1}	meq·(100 gr)$^{-1}$	%		g·(cm^3)$^{-1}$
AP	0.4	5.00	0.11	9.30	24.37	358.05	18.21	0.47	Loam	1.10
C1	0.65	4.92	0.07	1.15	2.24	54.61	19.50	0.06	Loam	1.38
C2	1.3	4.78	0.07	0.76	4.19	36.31	7.29	0.04	Sandy loam	1.80

Cod: horizon code, EC: electrical conductivity, OM: organic matter, P: phosphorus, K: potassium, Al: exchangeable aluminum, N: total nitrogen, Txt: texture, BD: bulk density.

2.5. Crop Fertilization

Fertilization was planned according to the second factor, proposed management (Table 1), considering the physical–chemical soil characterization results from the topsoil (Table 2). Details of the fertilization plan are shown in Table 3.

Table 3. Fertilization plan.

Management	Sheep Manure	Urea	DAF	ClK	Sulpomag	Agr Lime	Ammonium Nitrate
	t·ha^{-1}	kg·ha^{-1}	kg·ha^{-1}	kg·ha^{-1}	kg·ha^{-1}	kg·ha^{-1}	kg·ha^{-1}
M1: Traditional	5.00	0	0	0	0	0	0
M2: Recommended	5.00	91.36	222.67	138.10	55.55	807.14	91.36

DAF: Diammonium phosphate; ClK: potassium chloride; Sulpomag: potassium and magnesium sulfate; Agr: Agricultural.

2.6. Weather Data

Weather data were obtained from a Davis Vantage Pro2 station managed by AGRORURAL located at 75°19′28.32″ W; 11°47′45.59″ S at 3950 m.a.s.l. Records included the whole growing season (2022–2023) (Figure 3).

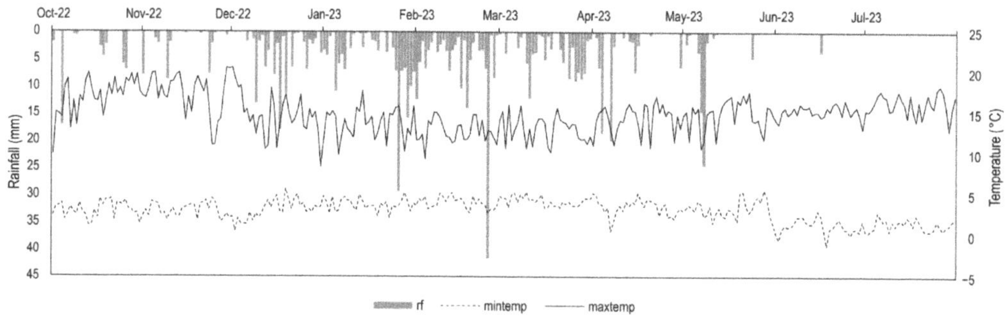

Figure 3. Apata station data for 2022–2023 on a daily scale. Rainfall (rf), maximum temperature (maxtemp), and minimum temperature (mintemp).

2.7. Data Collection of Morpho-Physiological Variables for Model Calibration

To calibrate the AquaCrop model, 9 evaluations were carried out during the whole growing season, starting from the phenological stage of emergence. In each one, five variables were assessed. In the phenological stage, direct observations of plants within furrow number 5 (the central 5 m of it) were performed. Plant canopy cover percentage was evaluated using RGB photographs taken with a NikonD3500 camera and processed with Image Canopy 3.6 and AutoCAD 2022 software. A total of 12 plants were assessed from each net plot (three plants from furrow numbers 2, 3, 7, and 8 each). The photographs were taken approximately 1 m above ground level. A square metal piece measuring 10 cm on a side was used as a measuring reference. Then, these plants were extracted from the ground to measure their height and root depth. Finally, plant leaves were separated and prepared for dry matter analysis [16].

At the end of the season, samples were taken from the central 9 m^2 (3 m $\times$ 3 m) of the net plot to estimate the following crop-yield-associated variables: crop fresh weight, tuber dry matter, aerial and total dry matter, tuber dry matter (10 representative tubers per unit of assessment), and aerial and total dry matter (the sampled plants for yield) [16].

2.8. Statistical Analysis

Morpho-physiological variables were analyzed using a two-way ANOVA (alpha = 0.05) to examine treatment differences. Model assumptions were verified graphically with the autoplot function from the ggfortify library for R [57]. Variables that showed significant differences were analyzed with the Least Significant Difference Test (alpha = 0.05), LSD.test function from agricolae library for R [58], corrected with the Benjamini–Hochberg procedure.

2.9. AquaCrop's Crop Development Modeling

AquaCrop's climate input came from the weather station. It included rainfall data, maximum and minimum temperatures, wind speed, and relative humidity on a daily scale, which enabled the calculation of reference evapotranspiration (ETo) through the Penman–Monteith equation using the EToCalculator 3.2. The cropping module was completed with the image processing results from the canopy coverage analysis. Soil input values were determined with the information from the physical-chemical analysis of the soil pit. A management module was configured for dry conditions, weed management, fertilization, and furrowing characteristics. The iterative calibration process was based on the recommendations of [29], using calibrated parameters for potato cultivation from AquaCrop, the literature, and experts' opinions. The first simulation was performed with field-collected data (soil and weather). Missing parameters (conservative and non-conservative) were completed with available values in the literature, including those from potato crop research, which is extensive. Canopy cover data were adjusted for each phenological stage, considering the potential effects of heat and water stress. The impact of traditional fertilization (Table 1) was also considered. Calibration was performed considering the data from sowing dates 1 and 2 (Table 1). Model efficiency was evaluated through Pearson's correlation index (r) and the Nash–Sutcliffe efficiency index (NSE) for plant canopy coverage and crop yield data. The model was then validated with the data from sowing date 3 (Table 1), without changing the conservative parameters but adapting the sowing date. The model's efficiency was assessed using both previously mentioned indexes.

2.10. Climate Change Explorations

Since the weather station used for AquaCrop climate input does not have data before 2019, we used data from the SENAMHI network of weather stations [50]. The criteria for choosing proper data involved proximity to the experimental location and accessibility to historical data for rainfall (rf), maximum temperature (maxtemp), and minimum temperature (mintemp). Records from 1986 to 2017 were found. Downloaded data from four weather stations (Jauja: 75°29′12.73″ W–11°47′11.87″ S; Ingenio: 75°17′47.9″ W–11°52′30.8″ S; Ricran: 75°31′38.29″ W–11°32′24.05″ S; Santa Ana: 75°13′17.7″ W–12°0′34.4″ S) were homogenized, and missing registers were filled by the Paulhus and Kohler [59] method included in the Climatol package in R [60]. Subsequently, homogenized weather data for the experiment's geographical location were calculated using the weighted inverse distance interpolation method (WID) in Excel. Regarding climate change models, a total of 7 General Circulation Models (GCMs) belonging to the Coupled Model Intercomparison Project Phase 6 (CMIP6) were chosen (Table 4). The selection criteria involved prioritizing those evaluated for Peru and South America's conditions, in terms of their statistical downscaling adjustment [61,62]. GCM data corresponded to the historical period (1980–2014). To avoid the innate bias problems of the GCMs (over-estimation of minimum temperatures and rainfall, as well as under-estimation of maximum temperatures) [62], after climate data extraction for plot location (using the terra package in R) [63], bias correction by quantile mapping was performed using the Qmap package in R [64]. Given our historical data, the model efficiency of all seven GCMs was evaluated to select the best-fitting model, based on Fernandez-Palomino et al. [62]. As a result of this analysis, the EC-Earth3 model was chosen to perform future projections, based on the Shared Socio-Economic Pathways' (SSPs) scenarios SSP1-2.6, SSP3-7.0 and SSP5-8.5, using the chosen model. The SSP1-2.6 scenario corresponds to a sustainable route, in which zero CO_2 emissions for the second half of this century are achieved. The SSP3-7.0 scenario considers doubled CO_2 emissions by 2100 due to regional rivalries and the absence of additional mitigation measures. The SSP5-8.5 scenario considers doubled CO_2 emissions by 2050 and the heavy use of fossil fuels to achieve human development. No further political measures are taken [14]. ETo was calculated using the Hargreaves method in the EToCalculator based on maximum and minimum projected temperature data for the selected model's future scenarios. To simulate ulluco yield for 2024–2100, rainfall, temperature, and ETo results were used as inputs for the climate module of the previously validated AquaCrop model, and atmospheric CO_2 concentration corresponded to the software database.

Table 4. Used CMIP6 models.

Model	Member	Historical (rf, Mintemp, Maxtemp)
IPSL-CM6A-LR	r14i1p1f1	[65]
CNRM-CM6-1	r1i1p1f2	[66]
CNRM-ESM2-1	r1i1p1f2	[67]
MIROC6	r50i1p1f1	[68]
MRI-ESM2-0	r5i1p1f1	[69]
MPI-ESM1-2-HR	r2i1p1f1	[70]
EC-Earth3	r150i1p1f1	[71]

3. Results

3.1. *Tuber Yields by Sowing Date and Type of Fertilization*

In addition to modeling, whether the tested variables influenced ulluco's yield was determined, as shown in Table 5.

At the main effects level, the sowing date (F) had no significant effect on fresh and dry weight yield, nor the number of tubers per kilogram. On the other hand, the fertilization method significantly affected the three evaluated variables for ulluco. The recommended dose values doubled under conventional management conditions in fresh weight (FW) and dry weight (DW) cases. It was found that yield decreases the later the sowing date takes place. However, mineral fertilization efficiency also decreases because less water is available as the sowing date is delayed (Table 5). Its efficiency is similar to traditional manure fertilization (Figure 4). The cumulative rainfall for the F1 growing season was 677 mm, 657 mm for F2, and 635 mm for F3.

Table 5. Ulluco tuber's yield by sowing dates and type of fertilization.

Treatment	Fresh Weight Yield ($Mg{\cdot}ha^{-1}$)	Dry Weight Yield ($Mg{\cdot}ha^{-1}$)	Number of Tubers per kg
F	ns	ns	ns
M	***	**	***
F × M	*	ns	ns

Table 5. *Cont.*

Treatment		Fresh Weight Yield (Mg·ha^{-1})	Dry Weight Yield (Mg·ha^{-1})	Number of Tubers per kg
		Sowing Date (F)		
F1		10.50 ± 5.0	1.28 ± 0.6	117.7 ± 24
F2		9.53 ± 6.8	1.10 ± 0.8	102.6 ± 27
F3		8.41 ± 1.6	1.16 ± 0.3	108.2 ± 17
		Fertilization Method (M)		
Traditional (M1)		6.4 ± 2.3 ^{b}	0.77 ± 0.3 ^{b}	123.0 ± 22 ^{a}
Recommended (M2)		12.56 ± 4.7 ^{a}	1.59 ± 0.5 ^{a}	96.0 ± 15 ^{b}
		F × M		
F1	M1	6.8 ± 2.8 ^{c}	0.79 ± 0.3	135.38 ± 20.9
	M2	14.2 ± 3.7 ^{a}	1.77 ± 0.5	100.06 ± 5.7
F2	M1	5.0 ± 2.7 ^{c}	0.57 ± 0.3	111.96 ± 30.1
	M2	14.1 ± 6.8 ^{ab}	1.64 ± 0.8	93.14 ± 24.2
F3	M1	7.4 ± 0.5 ^{bc}	0.96 ± 0.1	121.7 ± 7
	M2	9.4 ± 1.9 ^{abc}	1.36 ± 0.3	94.7 ± 12

*** Implies statistical significance at 0.1%; ** implies statistical significance at 1%; * implies statistical significance at 5%; ns: indicates not significant ($p > 0.05$); different letters in each testing parameter represent statistical significance among groups at $p < 0.05$ for Tukey Test.

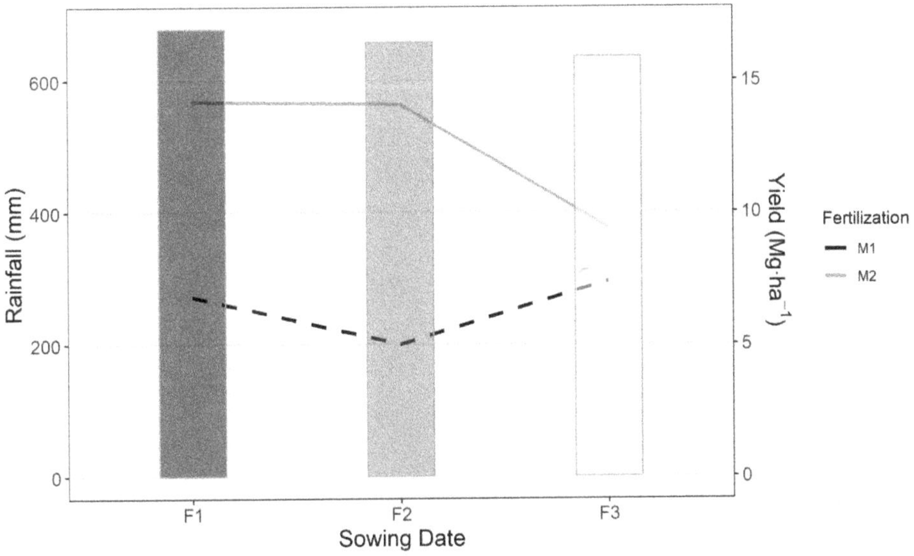

Figure 4. Bar graph showing rainfall (mm); and line graph showing marginal means of fresh weight yield of ulluco by sowing date (F1: 13 October 2022, F2: 28 October 2022, F3: 12 November 2022) response, and fertilization management (M1: traditional, M2: recommended).

3.2. *Ullucus Tuberosus Cultivation Development Model under Dryland Conditions*

The 2022–2023 season was characterized by wetter rainy season, maximum temperatures below the historical average, and minimum temperatures similar to the historical average, even reaching below 0 °C in the last months of the crop's phenological development (Figure 5). The F1M2 and F2M2 field data allowed for the model's calibration of conservative parameters in AquaCrop. Efficiency values for this phase of r: 0.99, NSE: 0.98, and r: 0.91—NSE: 0.82 were obtained for F1 and F2 plant coverage, respectively. Generated canopy cover curves from the parameterization are shown in Figures 6 and 7, showing the model's slight difficulty in reaching the maximum plant coverage of F2 (simulated canopy cover [CCsim] vs. observed canopy cover [CCobs]). In the next step, based on the obtained parameters, the model was validated with F3M2 data, obtaining efficiency indicators of r: 0.94 and NSE: 0.88 (Figure 8). The AquaCrop model calibrated parameters are shown in Table 6.

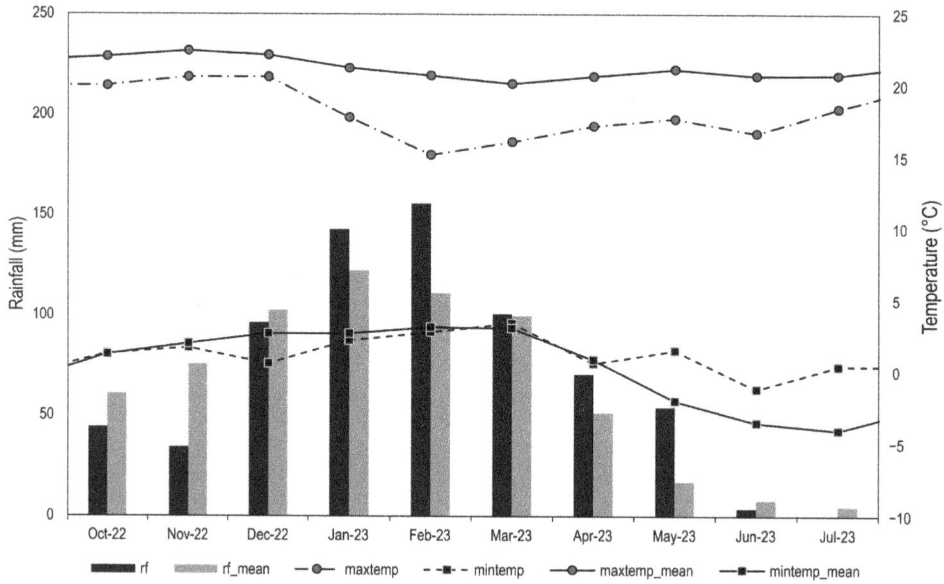

Figure 5. Apata station data (2022–2023) and historical averages on a monthly scale. Rainfall (rf), maximum temperature (maxtemp), and minimum temperature (mintemp).

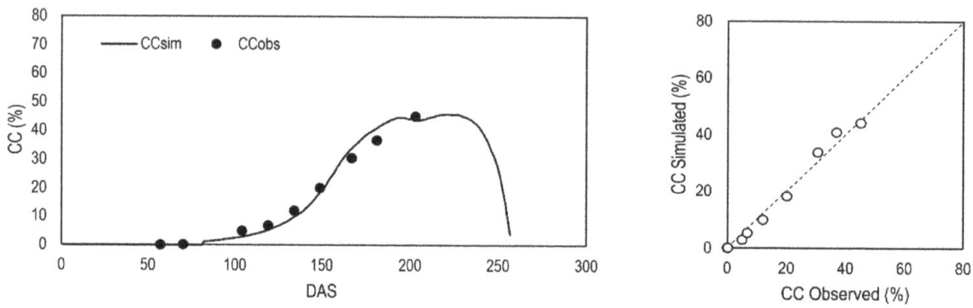

Figure 6. F1M2 calibration. DAS: days after sowing. CC: canopy cover. CCsim: simulated canopy cover. CCobs: observed canopy cover.

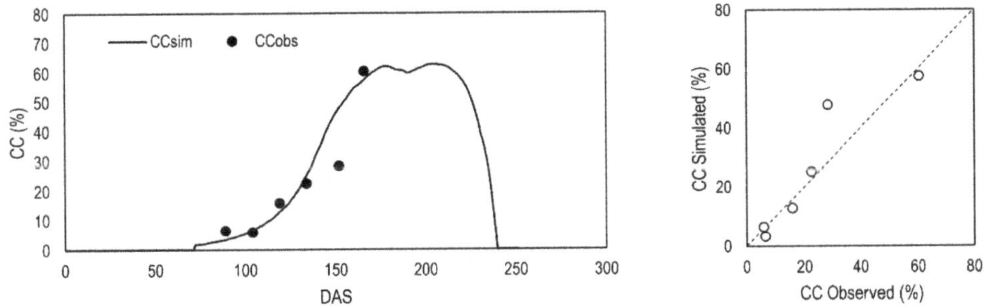

Figure 7. F2M2 calibration. DAS: days after sowing. CC: canopy cover. CCsim: simulated canopy cover. CCobs: observed canopy cover.

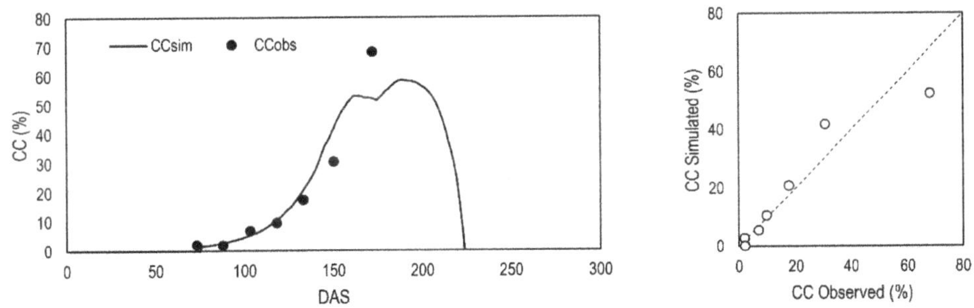

Figure 8. F3M2 validation. DAS: days after sowing. CC: canopy cover. CCsim: simulated canopy cover. CCobs: observed canopy cover.

Table 6. AquaCrop model parameter calibration for ulluco dryland crop.

Parameter	Initials	Value	Unit	Method of Determination	Base Literature
Development					
Plant density		41,667	plant ha^{-1}	M	[1]
Canopy development					
Canopy growth coefficient	CGC	0.677	% GDD^{-1}	C	[9]
Canopy decline coefficient	CDC	1.337	% GDD^{-1}	C	[9]
Root deepening					
Max effective rooting depth		0.25	m	M	
Average root zone expansion		0.1	cm day^{-1}	D	
Minimum effective rooting depth		0.2	m	E	
Shape factor of root deepening		1.5		D	
Evapotranspiration					
Soil evaporation					
Effect of canopy shelter in late season		60	%	D	
Crop transpiration	Kc tr,x	1.15		D	
Aging		0.15	% day^{-1}	D	
Water extraction pattern		40–30–20–10	%	D	
Production					
Normalized Water Productivity	(WP*)	13	g m^{-2}	L	[16]
Adjustment for yield formation		77	% of (WP*)	L	[16]
Performance under elevated CO_2					
Sink strength		50	%	D	
Reference Harvest Index	HIo	60	%	C	[9]
Composition of fresh yield					
%water		85	%	C	[2,9]
% dry matter		15	%	C	[2,9]

Table 6. *Cont.*

Parameter		Initials	Value	Unit	Method of Determination	Base Literature
Water stress						
Canopy expansion						
	p (upper)		0.3		E	[1,72]
	p (lower)		0.65		E	[1,72]
	shape factor		3		D	
Stomatal closure						
	p (upper)		0.6		D	
	shape factor		3		D	
Early canopy senescence						
	p (upper)		0.7		D	
	shape factor		3		D	
Temperature stress						
Base temperature			2	°C	E	[1,43]
Upper temperature			26	°C	E	[1,43]
Crop transpiration affected by cold stress						
	Full stress		0	°C		
	No stress		7	°C	D	[1,43]
Fertility stress						
CCx reduction			47	%	C	
CGC reduction			16	%	C	
Average decline canopy cover			0.2	% day^{-1}	C	
WP* reduction			25	%	C	
Salinity stress					NA	

C: calibration; D: AquaCrop default for potato; E: estimation; L: literature; M: measured; NA: not applicable. GDD: growing degree days; CCx: maximum canopy; Kc tr,x: coefficient for maximum crop transpiration, WP*: normalized water productivity. Note: gray hatched cells are conservative parameters.

The modelled/measured fresh weight yield was F1: 13.8 $t{\cdot}ha^{-1}$/15.28 $t{\cdot}ha^{-1}$, F2: 14.9 $t{\cdot}ha^{-1}$/17.36 $t{\cdot}ha^{-1}$, and F3: 9.19 $t{\cdot}ha^{-1}$/10.23 $t{\cdot}ha^{-1}$, while the modelled/measured dry weight yield was F1: 2.07 $t{\cdot}ha^{-1}$/1.86 $t{\cdot}ha^{-1}$, F2: 2.23 $t{\cdot}ha^{-1}$/2.02 $t{\cdot}ha^{-1}$, and F3: 1.38 $t{\cdot}ha^{-1}$/1.49 $t{\cdot}ha^{-1}$.

3.3. Climate Change Scenarios

Bias correction was performed for the three variables of interest in the historical period per station, and extreme values were represented (Figure 9). The EC-Earth 3 model was selected from which projected data bias for the 2015–2100 period was extracted and corrected for the SSP1-2.6 [73], SSP3-7.0 [74], and SSP5-8.5 [75] scenarios (Figure 10).

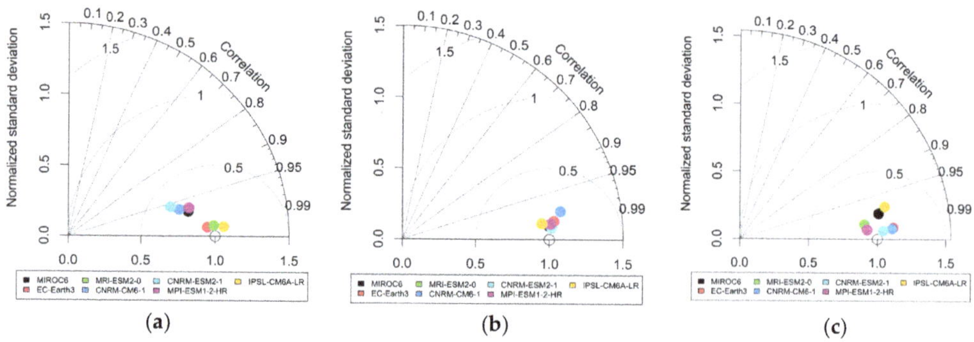

Figure 9. Taylor diagrams for extreme values of GCM historical corrected data; (**a**) rainfall above the 95th percentile of analyzed values; (**b**) under 5th minimum temperature percentile of analyzed values; and (**c**) above 95th maximum temperature percentile, evaluating monthly aggregates.

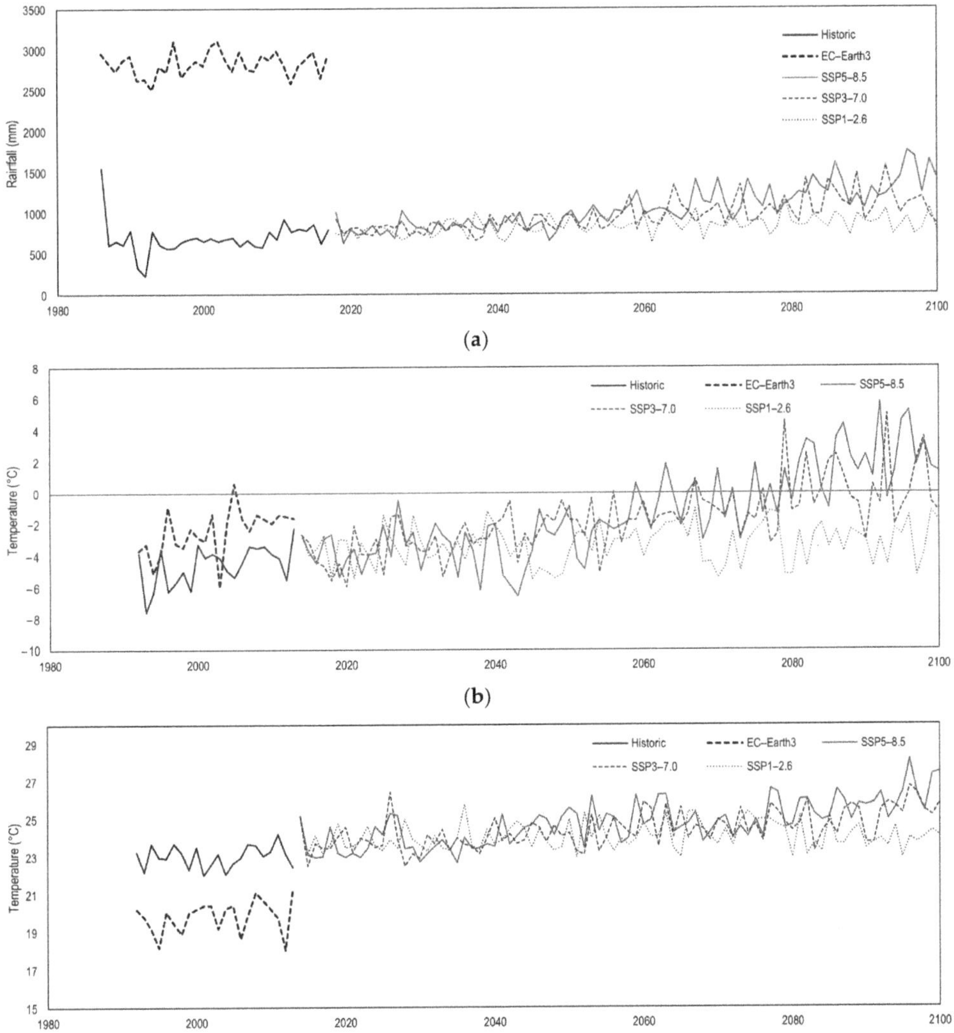

Figure 10. Climate variables projection for SSP126, SSP370, and SSP585 from corrected data of EC-Earth3; (**a**) rainfall; (**b**) minimum temperature; (**c**) maximum temperature. Annual scale grouping for the three variables.

Regarding the three analyzed scenarios (SSP1-2.6, SSP3-7.0, and SSP5-8.5) it is possible to observe that the EC-Earth 3 model projects an increasing trend in cumulative annual precipitation. In the last projected third (2075–2100), mean monthly value differences between the dry and wet seasons are accentuated (Figure 11a). On the other hand, increases in minimum and maximum temperature are observable for the three analyzed scenarios with an emphasis on the dry season (June–October) and for the minimum temperature.

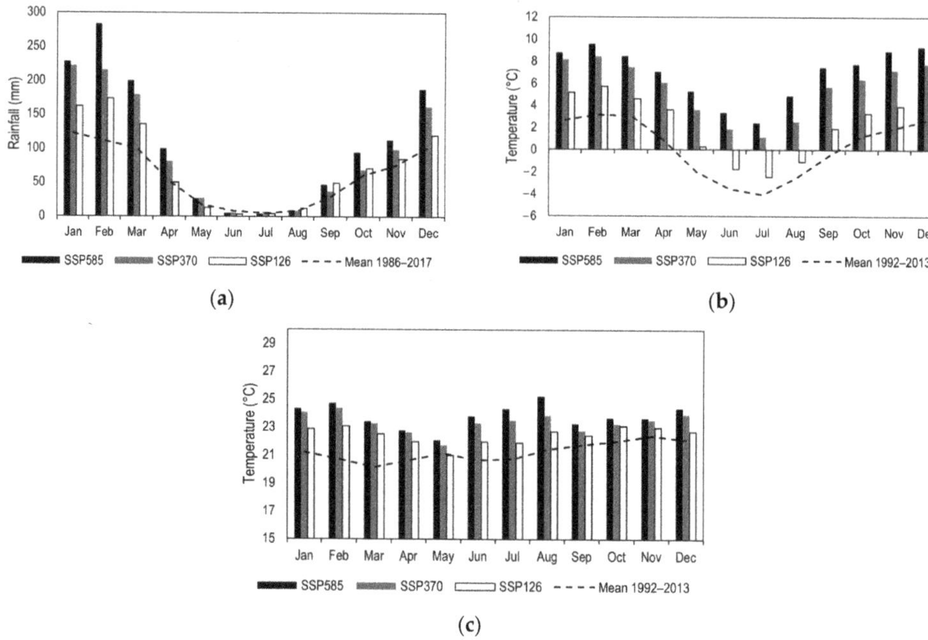

Figure 11. Monthly average values for the 2075–2100 period for SSP126, SSP370, and SSP585 from the EC-Earth3 model, in contrast to the observed historical average; (**a**) rainfall; (**b**) minimum temperature; (**c**) maximum temperature.

3.4. Yield Projections under Climate Change Conditions

The assessment results show that climate change significantly impacts ulluco yields under dryland conditions. Projections suggest a rainfall increase, accentuating differences between dry and rainy seasons, and temperature increase, mainly in the minimum temperatures. Regarding F1 and F2 conditions, a decrease in variability is observed in yield towards the last projected third in SSP1-2.6, while yield variability persists in SSP3-7.0. By contrast, a decreasing tendency toward projected yield is observed from the middle of the analysis period in SSP5-8.5. Regarding F3, it is observed that for the SSP1-2.6 scenarios, there is higher yield variability compared to SSP3-7.0 and SSP5-8.5. However, for the second half of the analysis period, this changes. In SSP1-2.6 and SSP3-7.0, a trend towards more constant yields is noted, while in SSP5-8.5, a decreasing trend is observed from the middle of the 2024–2100 period.

In all evaluated scenarios, there is a prevalence of F3 achieving higher yields compared to F1 and F2, as well as a lower failed campaign percentage (F1: 18%; F2: 13%; F3: 6%). However, an initial trend of slightly increasing yields is identified followed by a negative trend from the projection period's latter half.

4. Discussion

This study aimed to assess three climate change scenarios and their impact on the productivity of ulluco crops. To achieve this, historical weather data on rainfall and temperature were used to simulate changes from 2024 to 2100. Results from these simulations and data collected from the experimental plots were used as inputs for the required modules of the AquaCrop model. It is important to mention that the chosen area very well represents the conditions of a typical ulluco-producing area in a rainy, cold climate with dry autumns and winters [76]. Therefore, the results of this study are relevant for growing areas at the same altitude and similar soil conditions in the Andean region.

The results of the climate change simulations indicate an increasing trend in accumulated rainfall for the period 2024–2100 for three scenarios (Figure 10). Our findings contrast with those of Wongchuig et al. [77], who forecasted a reduction in annual precipitation for the same period in the Suni and Quechua regions of the Mantaro River basin. Arana et al. [78] found that historical data from the Jauja, Ingenio, and Ricrán weather stations also showed an ascending pattern. Seeing such discrepancies in the forecasts is expected because the complexity of the Andean topography largely influences bias correction processes [62] as a result of constantly changing thermal and moisture fluxes [79]. An important finding in our results was that in the last 25 years of this century, the distribution of rain throughout the year will remain similar in the sense that June, July, and August will remain predominantly dry, and the increase in cumulative precipitation will be allocated during the wet season. In other words, the rainy season will be rainier, especially in December, January, February, and March (Figure 11). This finding is in line with that of Almazroui et al. [61] who indicate stronger seasonality for this century's last period. The simulations suggest a sharp increase in minimum temperatures (Figure 11), especially for the SSP585 and SSP370 scenarios, which are the most concerning regarding climate change mitigation. Maximum temperatures seem to remain similar. All this is in accordance with the historical trends analyzed by Arana-Ruedas et al. [78] and Sanabria et al. [38], who explained that temperature increases in the studied area will be more pronounced for the minimum rather than maximum values but without concluding that frost damages in crops will necessarily be more harmful due to the complexity of this process.

The results of the AquaCrop simulations indicate that the latest sowing date, F3 (Table 3), will have a better fresh weight yield (Figure 12) in all three scenarios, and, on average, would have a greater increase in fresh weight yield concerning the 2022–2023 season's base yield (Figure 13). Given that the maximum temperatures will most likely remain constant, the underlying explanation can be found in minimum temperature and precipitation changes, specifically from 2075 to 2100 (Figure 10). The optimum temperature for ulluco has been studied as ranging between 11 and 13 °C [1]. In this sense, warmer minimum temperatures will affect the daily temperature average, impacting the duration of phenological transitions [80]. This indicates that the cropping period can be shortened, which could detrimentally affect yields. On the other hand, this species needs a low-temperature stimulus to start forming tubers [41]. If late sowing occurs in mid-November, tuberization induction is expected to happen by the end of February, when minimum temperatures reach 2°C under current conditions (Figure 11). All three scenarios forecast that minimum temperatures will be between 5 and 10 °C, which could affect tuber formation. In addition, maximum temperatures are expected to become 2 degrees higher during the tuber filling stage (March and April) (Figure 11), adding extra physiological problems, as it could even exceed the optimum range of 11–13 °C. Wahid et al. [81] define heat stress as plant growth and development limitations, caused by an air temperature increase above the optimal crop temperature, for a sufficient period to cause damage. Sage et al. [82] mention that between 5 and 6 °C increases in the plant's optimal temperature generate dry matter accumulation inhibition in reserve organs and greater infection susceptibility to pathogens. Likewise, high temperatures in heat-sensitive plants cause stomatal conductance decrease, CO_2 fixation inhibition, and oxidative damage in photosystem II [83,84]. It is important to consider how temperature increase involves an increase in incidences of pests and diseases [72]. Unfortunately, AquaCrop does not incorporate these effects in its calculations [29], so we should be critical regarding this factor.

On the other hand, the fact that wet seasons are expected to become rainier (Figure 11) might positively affect the early stages of plant development. This can be associated with the results of the experimental plot (Figures 6–8), in which plant emergence at early sowing treatments (Figure 6) took place 53 days after sowing, in comparison with normal sowing (Figure 7) and late sowing (Figure 8), which took place at 90 and 75 days, respectively. By relating this phenological transition with rainfall registered at the weather station (Figure 3), we can see that an intense precipitation event before early sowing (13 October 2022) was the single different weather variable compared to the other two sowing dates. On the other hand, during October and November, when sowing of the two latest treatments took place, only 34–44 mm of accumulated rainfall and around 8.6 °C average temperatures were recorded, which may have influenced the emergence delay of the last two sowing date treatments [49]. This suggests that more significant water availability may be beneficial for accelerating canopy emergence, which is a crucial factor in guaranteeing crop success. Despite this positive effect of water availability at the early stages of the crop, it is important to mention that an even distribution of precipitation is crucial, particularly after tuber induction, which usually takes place 90 days after sowing. Unfortunately, during April and May 2023, when tubers fill in, precipitation decreased drastically compared to February and March (Figure 3). This might be why the late sowing treatment at optimum fertilization rates performed worse than the other two treatments. This indicates the importance of having good soil moisture content during the filling stage. Fertilization proved to play a differentiating role when water was available (Figure 4). On its own, proper fertilization doubled yields compared to the traditional way, which far exceeds that reported by Nieto-Cabrera et al. [85] who indicated an 18.5% increase due to fertilization effects. However, if there is a water shortage during the filling stage, fertilization becomes an irrelevant factor, as the yields will be similar to plots that are not fertilized (Figure 4). The experimental plots yielded results comparable to those described by Sánchez-Portillo et al. [1], in which 2–10 $t \cdot ha^{-1}$ were obtained under Ecuadorian conditions. In addition to this, fresh and dry weight relations coincided with the high-water content expected for this tuber (72–87%) [2,9]. However, our fresh yield results were far from the optimal of 26 $t\ ha^{-1}$, as reported by Condori et al. [9] for irrigated ulluco crops, under Bolivian highland conditions. This yield gap between irrigated ulluco crops and dryland ones indicates how susceptible this crop is to the lack of a proper water supply. Although the models predict an increasing trend of accumulated precipitation over the years (Figure 10), it is not certain that such volumes will be evenly distributed during the growing season, or most importantly, that they will take place during periods of higher water demand for the crop (emergence and tuber filling).

All in all, this exploratory study alerts stakeholders and decision-makers to the necessity of strategically planning how to address the previously mentioned problems (i.e., temperature increases and excess water in the soil during future rainy seasons). Water management practices such as rain harvesting can be considered a way of storing excessive rainwater, and complimentary irrigation systems can prevent water shortage during critical phenological stages. On the other hand, excessive soil moisture could damage the crop; therefore, drainage systems can be included in plots that become easily flooded. Furthermore, this is an opportunity to suggest future plant breeding programs that allow for tuberization at higher temperatures and achieve heat stress-resistant cultivars.

The findings of this study highlight the susceptibility of ulluco, a native tuber, to climate change. Extrapolation of this potential effect to other minor native tubers due to their characteristic long growth cycles [86] can predict a shift in cultivated crops in highlands. It could increase the production costs, the loss of traditional field management, and the disruption of native crop roles in agroecosystems [86]. Losing traditions of native crop management such as crop rotation could cause soil health degradation, jeopardize agrobiodiversity conservation, and emphasize poverty patterns due to the settled subsistence of the agriculture–poverty relation [87]. Therefore, building the resilience of the local food system and investing in research on smallholder agriculture will be critical for reaching Sustainable Development Goal (SDG) 2 (Zero hunger) and SDG 1 (No poverty) [87]. In addition, the good performance of AquaCrop to model tubers allows us to perform climate change impact predictions to improve national climate change adaptation policies toward SDG 13 (Take urgent action to combat climate change and its impacts).

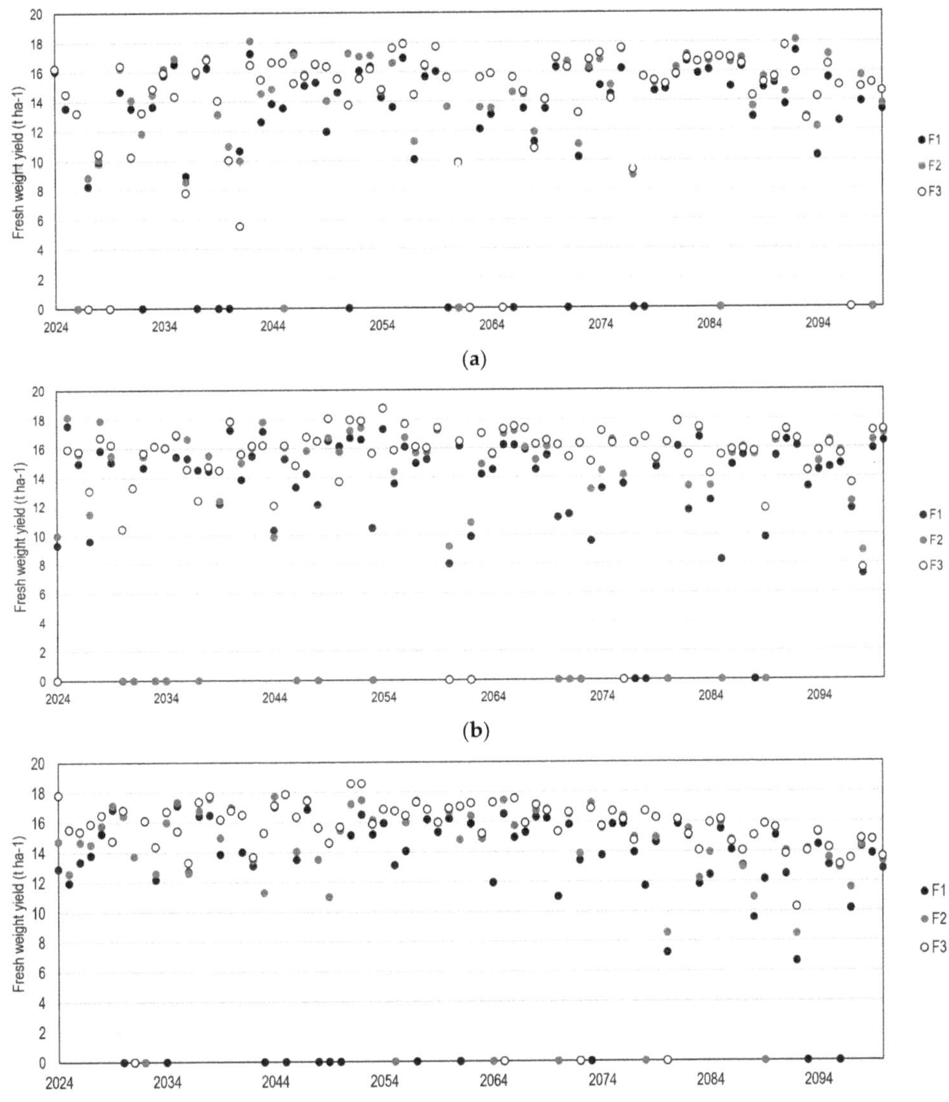

Figure 12. Ulluco's fresh weight yields (t·ha^{-1}) projected for (**a**) SSP1-2.6; (**b**) SSP3-7.0; and (**c**) SSP5-8.5 considering the three sowing dates (F1: 13-oct, F2: 28-oct, F3: 12-nov).

In addition, some policy suggestions are presented comparing this study and the strategic actions proposed in Peru's National Adaptation Plan Against Climate Change [88]. Modeling studies should be increased to broaden the prediction of possible climate change effects on critical food security crops and regional agrobiodiversity conservation. This measure can enhance the performance of early warning services and the scope of agro-climatic risk forecasts. However, in Peru, the results of the forecasting service developed by SENAMHI have restricted scope due to the limited Internet connection in the highlands [89] and the short list of commercial crops analyzed [90]. Furthermore, the water resources management policies for highlands should focus on Natural Infrastructure and

Water Sowing and Harvesting (WSH) due to their more straightforward and lower-cost implementation process and easier adaptation to local traditional practices. Therefore, the more democratic the construction of knowledge is, the easier it will be for farmers to adopt proposed policies and technologies.

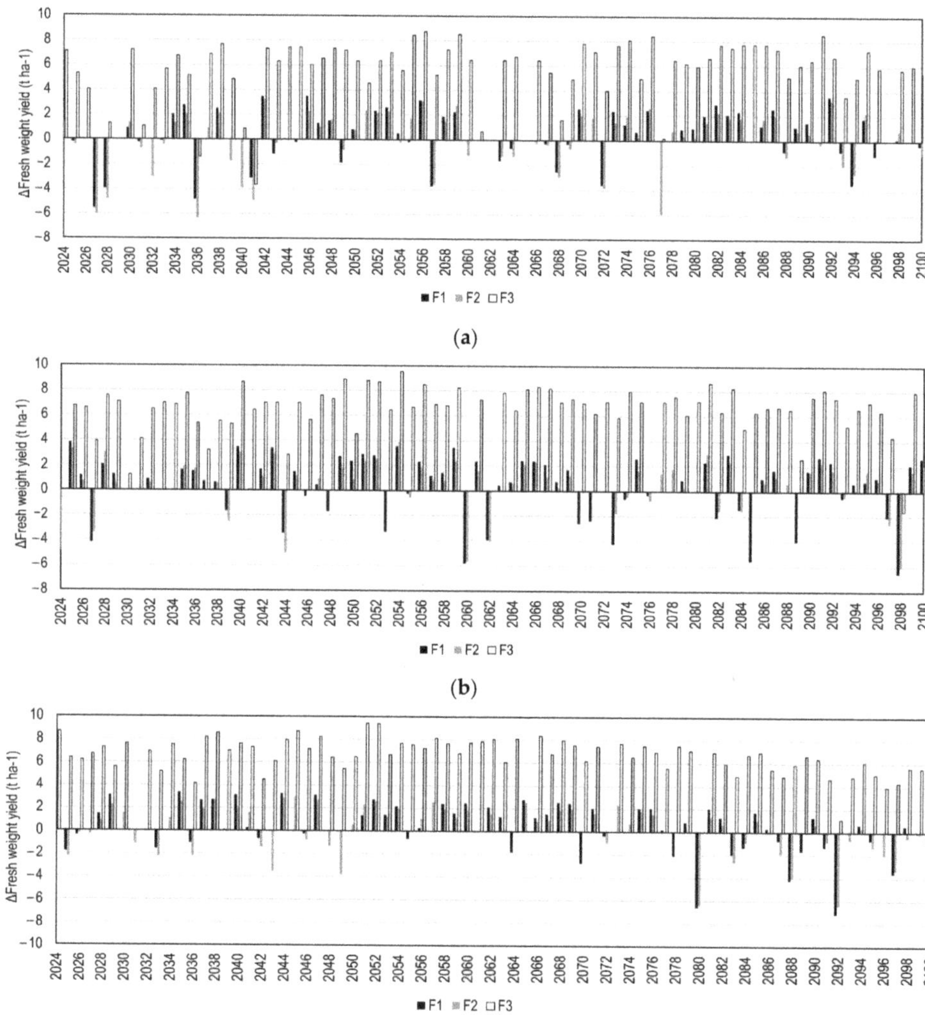

Figure 13. Ulluco's fresh weight yield (t·ha^{-1}) variations projected concerning the 2022–2023 season's base yield for (**a**) SSP1-2.6; (**b**) SSP3-7.0; and (**c**) SSP5-8.5 considering the three sowing dates (F1: 13-oct, F2: 28-oct, F3: 12-nov).

5. Conclusions

The AquaCrop model adequately represents ulluco cultivation development under dryland conditions for Peru's central highlands, allowing for yield projection analysis and crop management for changing environmental conditions. In this context, GCM analysis corresponding to CMIP 6 outlined how the more substantial mitigation and adaptation measures scenario (SSP1-2.6) projects an increase of up to 50% in precipitation for the rainy season in the period 2075–2100 and an average increase of 2 °C for minimum temperature. The maximum temperature in the rainy season (January–March) will increase regardless of mitigation and adaptation measures. By contrast, extreme values in the dry season (June–August) will be directly related to adopted measures on climate change—more intense for SSP5-8.5 and SSP3-7.0. Thus, for the intensive fossil fuel scenario (SSP5-8.5), maximum temperatures will reach 24–25 °C peaks, and minimum temperatures will increase sharply by 2050 with increases of up to 6 °C and exceeding 0 °C. These changes in maximum and minimum temperatures will impact ulluco's yield. In addition, for pessimistic scenarios (SSP5-8.5 and SSP3-7.0), rainfall in the rainy season can double the historical monthly precipitation volume, which means more intense events with greater erosive potential.

Increases in rainfall and environmental CO_2 concentrations show an opportunity window for yield increment in the early stage of the 2024–2100 period. However, in very high greenhouse gas concentration scenarios, a negative trend is observed for 2050–2100 mainly due to crop temperature stress, increased soil evaporation, and crop transpiration. A trend toward yield stability is achieved throughout the projected period for more optimistic scenarios. In summary, evaluating dryland ulluco cultivation yield under climate change scenarios using the AquaCrop model highlights the urgency of mitigating potential negative impacts on this agricultural system. This study remarks on the importance of carrying out adaptive sustainable practices concerning climate change such as rain harvesting for storing the excessive amount of rainwater and complimentary irrigation systems to prevent water shortage during critical phenological stages; the continuity of modeling future climate change effects to improve the information for agroclimatic risk forecast; the recommendation to plan plant breeding programs looking for tuberization at higher temperatures and heat stress-resistant cultivars; or investing in research on smallholder agriculture to reach SDG 1 and SDG 2.

Author Contributions: R.F.-M., conceptualization, methodology, validation, writing—original draft, visualization, investigation; J.V.-V., conceptualization, methodology; P.V.-M., methodology; S.L., writing—review and editing; R.S.-A., supervision, writing—review and editing. All authors have read and agreed to the published version of the manuscript.

Funding: This research was funded by the CUI 2487112 INIA project "Research and technology transfer services improvement in the management and recovery of degraded agricultural soils and irrigation water for small and medium-scale agriculture in the departments of Lima, Ancash, San Martín, Cajamarca, Lambayeque, Junín, Ayacucho, Arequipa, Puno, and Ucayali".

Institutional Review Board Statement: Not applicable.

Informed Consent Statement: Not applicable.

Data Availability Statement: The data presented in this study are available on request from the corresponding author.

Conflicts of Interest: The authors have no conflicts of interest to declare.

References

1. Sanchez-Portillo, S.; del Sánchez, M.R.S.; Solanilla-Duque, J.F.; Herrera, R.R. Andean Tubers, Morphological Diversity, and Agronomic Management: A Review. *Plant Sci. Today* **2023**, *10*, 98–105. [CrossRef]
2. Manrique, I.; Arbizu, C.; Vivanco, F.; Gonzales, R.; Ramírez, C.; Chávez, O.; Tay, D.; Ellis, D. *Ullucus Tuberosus Caldas: Colección de Germoplasma de ulluco Conservada en el Centro Internacional de la Papa (CIP)*, 1st ed.; International Potato Center: Lima, Peru, 2017; ISBN 978-92-9060-482-2. [CrossRef]
3. Tapia Núñez, M.E.; Fries, A.M.; Mazar, I. *Guía de Campo de los Cultivos Andinos*, 1st ed.; Rosell, C., Ed.; Asociación Nacional de Productores Ecológicos del Perú: Lima, Roma; Organización de las Naciones Unidas para la Agricultura y la Alimentación: Lima, Roma, 2007; ISBN 978-92-5-305682-8.
4. Busch, J.M.; Sangketkit, C.; Savage, G.P.; Martin, R.J.; Halloy, S.; Deo, B. Nutritional Analysis and Sensory Evaluation of Ulluco (*Ullucus tuberosus* Loz) Grown in New Zealand. *J. Sci. Food Agric.* **2000**, *80*, 2232–2240. [CrossRef]
5. Campos, D.; Chirinos, R.; Gálvez Ranilla, L.; Pedreschi, R. Bioactive Potential of Andean Fruits, Seeds, and Tubers. In *Advances in Food and Nutrition Research*; Elsevier: Amsterdam, The Netherlands, 2018; Volume 84, pp. 287–343. ISBN 978-0-12-814990-4. [CrossRef]

6. O'Hair, S.K.; Maynard, D.N. VEGETABLES OF TROPICAL CLIMATES | Root Crops of Uplands. In *Encyclopedia of Food Sciences and Nutrition*, 2nd ed.; Caballero, B., Ed.; Academic Press: Oxford, UK, 2003; pp. 5962–5965. Available online: https://www.sciencedirect.com/science/article/pii/B012227055X012438 (accessed on 10 March 2024)ISBN 978-0-12-227055-0.
7. MIDAGRI Perfil Productivo y Competitivo de Los Principales Cultivos Del Sector. Available online: https://app.powerbi.com/view?r=eyJrIjoiYjYwYTk5MDgtM2M0MS00NDMyLTgzNDEtMjNhNjEzYWQyOTNlIiwidCI6IjdmMDg0NjI3LTdmNDAtNDg3OS04OTE3LTk0Yjg2ZmQzNWYzZiJ9 (accessed on 29 February 2024).
8. INEI IV Censo Nacional Agropecuario 2012—Base de Datos REDATAM. Available online: http://censos1.inei.gob.pe/Cenagro/redatam/ (accessed on 21 May 2024).
9. Condori, B.; Mamani, P.; Botello, R.; Patiño, F.; Devaux, A.; Ledent, J.F. Agrophysiological Characterisation and Parametrisation of Andean Tubers: Potato (*Solanum* Sp.), Oca (*Oxalis tuberosa*), Isaño (*Tropaeolum tuberosum*) and Papalisa (*Ullucus tuberosus*). *Eur. J. Agron.* **2008**, *28*, 526–540. [CrossRef]
10. Muñoz, A.M.; Jimenez-Champi, D.; Contreras-López, E.; Fernández-Jerí, Y.; Best, I.; Aguilar, L.; Ramos-Escudero, F. Valorization of Extracts of Andean Roots and Tubers and Its Byproducts: Bioactive Components and Antioxidant Activity in Vitro. *Food Res.* **2023**, *7*, 55–63. [CrossRef]
11. INEI Perú: Consumo Per Cápita de los Principales Alimentos 2008–2009. In *Encuesta Nacional de Presupuestos Familiares (ENAPREF)*; Dirección Técnica de Demografía e Indicadores Sociales-INEI: Lima, Perú, 2012.
12. Abbas, A.; Waseem, M.; Ahmad, R.; Khan, K.A.; Zhao, C.; Zhu, J. Sensitivity Analysis of Greenhouse Gas Emissions at Farm Level: Case Study of Grain and Cash Crops. *Environ. Sci. Pollut. Res.* **2022**, *29*, 82559–82573. [CrossRef]
13. Elahi, E.; Li, G.; Han, X.; Zhu, W.; Liu, Y.; Cheng, A.; Yang, Y. Decoupling Livestock and Poultry Pollution Emissions from Industrial Development: A Step towards Reducing Environmental Emissions. *J. Environ. Manag.* **2024**, *350*, 119654. [CrossRef] [PubMed]
14. IPCC. *Climate Change 2021—The Physical Science Basis: Working Group I Contribution to the Sixth Assessment Report of the Intergovernmental Panel on Climate Change*, 1st ed.; Cambridge University Press: Cambridge, UK, 2023; ISBN 978-1-00-915789-6. [CrossRef]
15. O'Neill, B.C.; Tebaldi, C.; van Vuuren, D.P.; Eyring, V.; Friedlingstein, P.; Hurtt, G.; Knutti, R.; Kriegler, E.; Lamarque, J.-F.; Lowe, J.; et al. The Scenario Model Intercomparison Project (ScenarioMIP) for CMIP6. *Geosci. Model Dev.* **2016**, *9*, 3461–3482. [CrossRef]
16. Ahmadi, S.H.; Reis Ghorra, M.R.; Sepaskhah, A.R. Parameterizing the AquaCrop Model for Potato Growth Modeling in a Semi-Arid Region. *Field Crops Res.* **2022**, *288*, 108680. [CrossRef]
17. Wang, H.; Cheng, M.; Liao, Z.; Guo, J.; Zhang, F.; Fan, J.; Feng, H.; Yang, Q.; Wu, L.; Wang, X. Performance Evaluation of AquaCrop and DSSAT-SUBSTOR-Potato Models in Simulating Potato Growth, Yield and Water Productivity under Various Drip Fertigation Regimes. *Agric. Water Manag.* **2023**, *276*, 108076. [CrossRef]
18. Wellens, J.; Raes, D.; Fereres, E.; Diels, J.; Coppye, C.; Adiele, J.G.; Ezui, K.S.G.; Becerra, L.-A.; Selvaraj, M.G.; Dercon, G.; et al. Calibration and Validation of the FAO AquaCrop Water Productivity Model for Cassava (*Manihot esculenta* Crantz). *Agric. Water Manag.* **2022**, *263*, 107491. [CrossRef]
19. Rankine, D.R.; Cohen, J.E.; Taylor, M.A.; Coy, A.D.; Simpson, L.A.; Stephenson, T.; Lawrence, J.L. Parameterizing the FAO AquaCrop Model for Rainfed and Irrigated Field-Grown Sweet Potato. *Agron. J.* **2015**, *107*, 375–387. [CrossRef]
20. Puma-Cahua, J.; Belizario, G.; Laqui, W.; Alfaro, R.; Huaquisto, E.; Calizaya, E. Evaluating the Yields of the Rainfed Potato Crop under Climate Change Scenarios Using the AquaCrop Model in the Peruvian Altiplano. *Sustainability* **2023**, *16*, 71. [CrossRef]
21. Chumbe, R.; Silva, S.; Garcia, Y. Comparison of the Machine Learning and AquaCrop Models for Quinoa Crops. *Res. Agric. Eng.* **2023**, *69*, 65–75. [CrossRef]
22. Geerts, S.; Raes, D.; Garcia, M.; Taboada, C.; Miranda, R.; Cusicanqui, J.; Mhizha, T.; Vacher, J. Modeling the Potential for Closing Quinoa Yield Gaps under Varying Water Availability in the Bolivian Altiplano. *Agric. Water Manag.* **2009**, *96*, 1652–1658. [CrossRef]
23. Alavi, G.; Diels, J.; Willems, P.; García, M. Simulación de la producción de Quinua en el Altiplano Boliviano con el modelo de Aquacrop con escenarios futuros generados por LARS-WG y QMP. *Rev. Investig. E Innov. Agropecu. Recur. Nat.* **2015**, *2*, 7–13.
24. Bwalya, A. Validation of the Aquacrop Model for Irrigated African Eggplant (*Solanum macrocarpon*) at the Unza Field Station. Ph.D. Thesis, University of Zambia, Lusaka, Zambia, 2012. Available online: https://library.adhl.africa/handle/123456789/12619 (accessed on 10 March 2024).
25. Cobeña, O.; Pamela, C. Calibración, validación y utilización del modelo AquaCrop para el cultivo de la chufa (*Cyperus esculentus* L. var. sativus Boeck.) en Valencia. Master's Thesis, Universitat Politècnica de Valencia, Valencia, Spain, 2020. Available online: https://riunet.upv.es/handle/10251/134097 (accessed on 10 March 2024).
26. Mabhaudhi, T.; Modi, A.T.; Beletse, Y.G. Parameterization and Evaluation of the FAO-AquaCrop Model for a South African Taro (*Colocasia esculenta* L. Schott) Landrace. *Agric. For. Meteorol.* **2014**, *192*, 132–139. [CrossRef]
27. Sharma, V.; Singh, P.K. Performance of AquaCrop Model for Predicting Yield and Biomass of Okra (*Abelmoschus esculentus*) Crop. *Indian J. Agric. Sci.* **2023**, *93*, 899–905. [CrossRef]
28. Walker, S.; Bello, Z.A.; Mabhaudhi, T.; Modi, A.T.; Beletse, Y.G.; Zuma-Netshiukhwi, G. Calibration of AquaCrop Model to Predict Water Requirements of Traditional African Vegetables. *Acta Hortic.* **2013**, *1007*, 943–949. [CrossRef]

29. Steduto, P.C.; Hsiao, T.; Fereres, E.; Raes, D. Land and Water Division. In *Crop Yield Response to Water*; FAO Irrigation and Drainage Paper; FAO: Rome, Italy, 2012; ISBN 978-92-5-107274-5. Available online: https://www.fao.org/documents/card/en/c/c355da16-217c-555b-acbc-505d87bade00 (accessed on 10 March 2024).
30. Vanuytrecht, E.; Raes, D.; Steduto, P.; Hsiao, T.C.; Fereres, E.; Heng, L.K.; Garcia Vila, M.; Mejias Moreno, P. AquaCrop: FAO's Crop Water Productivity and Yield Response Model. *Environ. Model. Softw.* **2014**, *62*, 351–360. [CrossRef]
31. Dewedar, O.; Plauborg, F.; El-Shafie, A.; Marwa, A. Response of Potato Biomass and Tuber Yield under Future Climate Change Scenarios in Egypt. *J. Water Land Dev.* **2021**, *49*, 139–150. [CrossRef]
32. Yaghoobzadeh, M.; Azarmi Atajan, F.; Arabi Ayask, M.; Ghadirian, A.H. Investigating the Effect of Drip and Furrow Irrigation Methods on the Simulation of Sugar Beet Yield by Using the AquaCrop Model. *J. Water Res. Agric.* **2023**, *37*, 323–335. [CrossRef]
33. FAO Limitations | AquaCrop | Food and Agriculture Organization of the United Nations. Available online: https://www.fao.org/aquacrop/overview/limitations/en/ (accessed on 12 May 2024).
34. IPCC. *Climate Change 2022—Impacts, Adaptation and Vulnerability: Working Group II Contribution to the Sixth Assessment Report of the Intergovernmental Panel on Climate Change*, 1st ed.; Cambridge University Press: Cambridge, UK, 2023; ISBN 978-1-00-932584-4.
35. Tito, R.; Vasconcelos, H.L.; Feeley, K.J. Global Climate Change Increases Risk of Crop Yield Losses and Food Insecurity in the Tropical Andes. *Glob. Chang. Biol.* **2018**, *24*, e592–e602. [CrossRef]
36. Yzarra, W.; Trebejo, I.; Kanamura, H.; Fujisawa, M.; Guerrero, J.; Bello, D.; Villanueva, J.; Soriano, V. *Componente 1: Evaluación de Los Impactos del Cambio Climático Sobre el Rendimiento de Los Cultivos en el Perú*; Servicio Nacional de Meteorologia e Hidrologia: Lima, Peru, 2015.
37. Tarmizi, A.H.A. Climate Change and Its Impact on Rainfall. *Int. J. Integr. Eng.* **2019**, *11*, 170–177.
38. Sanabria, J.; Calanca, P.; Alarcón, C.; Canchari, G. Potential Impacts of Early Twenty-First Century Changes in Temperature and Precipitation on Rainfed Annual Crops in the Central Andes of Peru. *Reg. Environ. Chang.* **2014**, *14*, 1533–1548. [CrossRef]
39. Bisbis, M.B.; Gruda, N.; Blanke, M. Potential Impacts of Climate Change on Vegetable Production and Product Quality—A Review. *J. Clean. Prod.* **2018**, *170*, 1602–1620. [CrossRef]
40. Kalliola, R.; Jokela, P.; Pietila, L.; Rousi, A.; Salo, J.; Yli-Rekola, M. Influencia del fotoperíodo en el crecimiento y formación de tubérculos de ulluco (*Ullucus tuberosus*, Basellaceae), Oca (*Oxalis tuberosa*, Oxalidaceae) y Añu (*Tropaeolum tuberosum*, Tropaeolaceae). In *Turrialba*; IICA: San José, Costa Rica, 1990; Available online: https://repositorio.catie.ac.cr/handle/11554/10460 (accessed on 10 March 2024).
41. Markarov, A.M. Causes of Flowering of Long-Day Potato Species under Short-Day and Cold-Night Conditions. *Russ. J. Plant Physiol.* **2002**, *49*, 465–469. [CrossRef]
42. Martin, R.J.; Scheffer, J.J.C.; Deo, B.; Halloy, S.R.P. Effect of Planting Date on Ulluco Yield. *Acta Hortic.* **2005**, *670*, 181–187. [CrossRef]
43. Scheffer, J.J.C.; Douglas, J.A.; Martin, R.J.; Halloy, S.R.P.; Deo, B.; Triggs, C.M. Agronomic Requirements of Ulluco (*Ullucus Tuberosus*)—A South American Tuber. *Agronomy* **2002**, *32*, 41–47.
44. Acurio, L.; Salazar, D.; Castillo, B.; Santiana, C.; Martínez-Monzó, J.; Igual, M. Characterization of Second-Generation Snacks Manufactured from Andean Tubers and Tuberous Root Flours. *Foods* **2024**, *13*, 51. [CrossRef] [PubMed]
45. Mosquera, N.; Cejudo-Bastante, M.J.; Heredia, F.J.; Hurtado, N. Identification of New Betalains in Separated Betacyanin and Betaxanthin Fractions from Ulluco (*Ullucus Tuberosus* Caldas) by HPLC-DAD-ESI-MS. *Plant Foods Hum. Nutr. Dordr. Neth.* **2020**, *75*, 434–440. [CrossRef] [PubMed]
46. Velasquez Barreto, F.F.; Bello-Pérez, L.A. Chemical, Structural, Technological Properties and Applications of Andean Tuber Starches: A Review. *Food Rev. Int.* **2023**, *39*, 1293–1308. [CrossRef]
47. Keleman Saxena, A.; Cadima Fuentes, X.; Gonzales Herbas, R.; Humphries, D.L. Indigenous Food Systems and Climate Change: Impacts of Climatic Shifts on the Production and Processing of Native and Traditional Crops in the Bolivian Andes. *Front. Public Health* **2016**, *4*, 175055. [CrossRef] [PubMed]
48. Ponce, N.L.C.; Martínez, M.E.P. Tubérculos andinos y conocimiento agrícola local en comunidades rurales de Ecuador y Colombia. *Cuad. Desarro. Rural* **2014**, *11*, 149–166. [CrossRef]
49. López, G.; Yupanqui, A.T.; Fierro, R.E. Fenología y Agronomía del Cultivo. In *El Cultivo del Ulluco en la Sierra Central del Perú*; López, G., Hermann, M., Eds.; Conservación y uso de la biodiversidad de raíces y tubérculos andinos: Una década de investigación para el desarrollo (1993–2003); Centro Internacional de la Papa (CIP): Lima, Peru, 2004.
50. SENAMHI. SENAMHI—Descarga de Datos. Available online: https://www.senamhi.gob.pe/site/descarga-datos/ (accessed on 29 February 2024).
51. López, G. Tubérculos-Semilla. In *El Cultivo del Ulluco en la Sierra Central del Perú*; López, G., Hermann, M., Eds.; Conservación y uso de la biodiversidad de raíces y tubérculos andinos: Una década de investigación para el desarrollo (1993–2003); Centro Internacional de la Papa (CIP): Lima, Peru, 2004.
52. *ISO 11272:2017*; Soil Quality—Determination of Dry Bulk Density. ISO (International Organization for Standardization): Geneva, Switzerland, 2017. Available online: https://www.iso.org/standard/68255.html (accessed on 25 March 2024).
53. *NOM-021-RECNAT-2000*; Norma Oficial Mexicana Que Establece Las Especificaciones de Fertilidad, Salinidad y Clasificación de Suelos. Estudios, Muestreo y Análisis. 2002. Available online: https://faolex.fao.org/docs/pdf/mex50674.pdf (accessed on 10 March 2024).

54. USEPA. *METHOD 9045D. SOIL AND WASTE pH 2004*; USEPA: Washington, DC, USA, 2004. Available online: https://www.epa.gov/sites/default/files/2015-12/documents/9045d.pdf (accessed on 25 March 2024).
55. *ISO 11265:1994*; Soil Quality—Determination of the Specific Electrical Conductivity. ISO (International Organization for Standardization): Geneva, Switzerland, 1994. Available online: https://www.iso.org/standard/19243.html (accessed on 25 March 2024).
56. *ISO 11461:2001*; Soil Quality—Determination of Soil Water Content as a Volume Fraction Using Coring Sleeves—Gravimetric Method. ISO (International Organization for Standardization): Geneva, Switzerland, 2001. Available online: https://www.iso.org/standard/33031.html (accessed on 25 March 2024).
57. Horikoshi, M.; Tang, Y.; Dickey, A.; Grenié, M.; Thompson, R.; Selzer, L.; Strbenac, D.; Voronin, K.; Pulatov, D. Ggfortify: Data Visualization Tools for Statistical Analysis Results 2024. Available online: https://cran.r-project.org/web/packages/ggfortify/index.html (accessed on 26 January 2024).
58. Mendiburu, F. de Agricolae: Statistical Procedures for Agricultural Research, version 1.3-7. 2023. Available online: https://cran.r-project.org/web/packages/agricolae/index.html (accessed on 26 January 2024).
59. Paulhus, J.L.H.; Kohler, M.A. Interpolation of Missing Precipitation Records. *Mon. Weather Rev.* **1952**, *80*, 129–133. [CrossRef]
60. Guijarro, J.A. Climatol: Climate Tools (Series Homogenization and Derived Products), version 4.1.0. 2023. Available online: https://cran.r-project.org/web/packages/climatol/index.html (accessed on 26 January 2024).
61. Almazroui, M.; Ashfaq, M.; Islam, M.N.; Rashid, I.U.; Kamil, S.; Abid, M.A.; O'Brien, E.; Ismail, M.; Reboita, M.S.; Sörensson, A.A.; et al. Assessment of CMIP6 Performance and Projected Temperature and Precipitation Changes Over South America. *Earth Syst. Environ.* **2021**, *5*, 155–183. [CrossRef]
62. Fernandez-Palomino, C.A.; Hattermann, F.F.; Krysanova, V.; Vega-Jácome, F.; Menz, C.; Gleixner, S.; Bronstert, A. High-Resolution Climate Projection Dataset Based on CMIP6 for Peru and Ecuador: BASD-CMIP6-PE. *Sci. Data* **2024**, *11*, 34. [CrossRef] [PubMed]
63. Hijmans, R.J.; Bivand, R.; Pebesma, E.; Sumner, M.D. Terra: Spatial Data Analysis, version 1.7-71. 2023. Available online: https://cran.r-project.org/web/packages/terra/index.html (accessed on 26 January 2024).
64. Gudmundsson, L. Qmap: Statistical Transformations for Post-Processing Climate Model Output, version 1.0-4. 2016. Available online: https://cran.r-project.org/web/packages/qmap/index.html (accessed on 26 January 2024).
65. Boucher, O.; Denvil, S.; Levavasseur, G.; Cozic, A.; Caubel, A.; Foujols, M.-A.; Meurdesoif, Y.; Cadule, P.; Devilliers, M.; Ghattas, J.; et al. IPSL IPSL-CM6A-LR Model Output Prepared for CMIP6 CMIP Historical. 2018. [CrossRef]
66. Voldoire, A. *CMIP6 Simulations of the CNRM-CERFACS Based on CNRM-CM6-1 Model for CMIP Experiment Historical*; Earth System Grid Federation: Greenbelt, MD, USA, 2018; Available online: http://cera-www.dkrz.de/WDCC/meta/CMIP6/CMIP6.CMIP.CNRM-CERFACS.CNRM-CM6-1.historical (accessed on 28 February 2024).
67. Seferian, R. *CNRM-CERFACS CNRM-ESM2-1 Model Output Prepared for CMIP6 CMIP Historical*; Earth System Grid Federation: Greenbelt, MD, USA, 2018; Available online: http://cera-www.dkrz.de/WDCC/meta/CMIP6/CMIP6.CMIP.CNRM-CERFACS.CNRM-ESM2-1.historical (accessed on 28 February 2024).
68. Tatebe, H.; Watanabe, M. *MIROC MIROC6 Model Output Prepared for CMIP6 CMIP Historical*; Earth System Grid Federation: Greenbelt, MD, USA, 2018; Available online: http://cera-www.dkrz.de/WDCC/meta/CMIP6/CMIP6.CMIP.MIROC.MIROC6.historical (accessed on 28 February 2024).
69. Yukimoto, S.; Koshiro, T.; Kawai, H.; Oshima, N.; Yoshida, K.; Urakawa, S.; Tsujino, H.; Deushi, M.; Tanaka, T.; Hosaka, M.; et al. *MRI MRI-ESM2.0 Model Output Prepared for CMIP6 CMIP Historical*; WCRP: Geneva, Switzerland, 2019; Available online: http://cera-www.dkrz.de/WDCC/meta/CMIP6/CMIP6.CMIP.MRI.MRI-ESM2-0.historical (accessed on 28 February 2024).
70. Jungclaus, J.; Bittner, M.; Wieners, K.-H.; Wachsmann, F.; Schupfner, M.; Legutke, S.; Giorgetta, M.; Reick, C.; Gayler, V.; Haak, H.; et al. *MPI-M MPI-ESM1.2-HR Model Output Prepared for CMIP6 CMIP Historical*; WCRP: Geneva, Switzerland, 2019.
71. EC-Earth Consortium (EC-Earth). *EC-Earth-Consortium EC-Earth3 Model Output Prepared for CMIP6 CMIP Historical*; Earth System Grid Federation: Greenbelt, MD, USA, 2019. [CrossRef]
72. EFSA Panel on Plant Health (PLH); Bragard, C.; Dehnen-Schmutz, K.; Di Serio, F.; Gonthier, P.; Jacques, M.; Jaques Miret, J.A.; Justesen, A.F.; MacLeod, A.; Magnusson, C.S.; et al. Commodity Risk Assessment of *Ullucus Tuberosus* Tubers from Peru. *EFSA J.* **2021**, *19*, e06428. [CrossRef] [PubMed]
73. EC-Earth Consortium (EC-Earth). *EC-Earth-Consortium EC-Earth3 model output prepared for CMIP6 ScenarioMIP ssp126*; Earth System Grid Federation: Greenbelt, MD, USA, 2019. [CrossRef]
74. EC-Earth Consortium (EC-Earth). *EC-Earth-Consortium EC-Earth3 model output prepared for CMIP6 ScenarioMIP ssp370*; Earth System Grid Federation: Greenbelt, MD, USA, 2019. [CrossRef]
75. EC-Earth Consortium (EC-Earth). *EC-Earth-Consortium EC-Earth3 model output prepared for CMIP6 ScenarioMIP ssp585*; Earth System Grid Federation: Greenbelt, MD, USA, 2019. [CrossRef]
76. Castro, A.; Davila, C.; Laura, W.; Cubas, F.; Avalos, G.; López-Ocaña, C.; Villena, D.; Valdez, M.; Urbiola, J.; Trebejo, I.; et al. CLIMAS DEL PERÚ—Mapa de Clasificación Climática Nacional, 1st ed. 2021. Available online: https://www.senamhi.gob.pe/load/file/01404SENA-4.pdf (accessed on 22 May 2024).
77. Wongchuig, S.C.; Mello, C.R.; Chou, S.C. Projections of the Impacts of Climate Change on the Water Deficit and on the Precipitation Erosive Indexes in Mantaro River Basin, Peru. *J. Mt. Sci.* **2018**, *15*, 264–279. [CrossRef]

78. Arana Ruedas, D.P.R.; Soto Guerra, L.; Popli, K.; Madaki, S.G.; Arana Ruedas, D.P.R.; Soto Guerra, L.; Popli, K.; Madaki, S.G. Evaluación Espacio-Temporal de Sequías Usando El Índice Estandarizado de Precipitación y Evapotranspiración (SPEI) En El Valle Del Mantaro, Perú. *Rev. Investig. Altoandinas* **2023**, *25*, 159–170. [CrossRef]
79. Saavedra, M.; Junquas, C.; Espinoza, J.-C.; Silva, Y. Impacts of Topography and Land Use Changes on the Air Surface Temperature and Precipitation over the Central Peruvian Andes. *Atmos. Res.* **2020**, *234*, 104711. [CrossRef]
80. Yin, Y.; Deng, H.; Wu, S. Spatial-Temporal Variations in the Thermal Growing Degree-Days and Season under Climate Warming in China during 1960–2011. *Int. J. Biometeorol.* **2019**, *63*, 649–658. [CrossRef]
81. Wahid, A.; Gelani, S.; Ashraf, M.; Foolad, M.R. Heat Tolerance in Plants: An Overview. *Environ. Exp. Bot.* **2007**, *61*, 199–223. [CrossRef]
82. Sage, T.L.; Bagha, S.; Lundsgaard-Nielsen, V.; Branch, H.A.; Sultmanis, S.; Sage, R.F. The Effect of High Temperature Stress on Male and Female Reproduction in Plants. *Field Crops Res.* **2015**, *182*, 30–42. [CrossRef]
83. Chaves-Barrantes, N.F.; Gutiérrez-Soto, M.V. Respuestas al estrés por calor en los cultivos. I. aspectos moleculares, bioquímicos y fisiológicos. *Agron. Mesoam.* **2017**, *28*, 237–253. [CrossRef]
84. Sita, K.; Sehgal, A.; HanumanthaRao, B.; Nair, R.M.; Vara Prasad, P.V.; Kumar, S.; Gaur, P.M.; Farooq, M.; Siddique, K.H.M.; Varshney, R.K.; et al. Food Legumes and Rising Temperatures: Effects, Adaptive Functional Mechanisms Specific to Reproductive Growth Stage and Strategies to Improve Heat Tolerance. *Front. Plant Sci.* **2017**, *8*, 298433. [CrossRef] [PubMed]
85. Nieto-Cabrera, C.; Francis, C.; Caicedo, C.; Gutierrez, P.F.; Rivera, M. Response of Four Andean Crops to Rotation and Fertilization. *Mt. Res. Dev.* **1997**, *17*, 273. [CrossRef]
86. Meldrum, G.; Mijatović, D.; Rojas, W.; Flores, J.; Pinto, M.; Mamani, G.; Condori, E.; Hilaquita, D.; Gruberg, H.; Padulosi, S. Climate Change and Crop Diversity: Farmers' Perceptions and Adaptation on the Bolivian Altiplano. *Environ. Dev. Sustain.* **2018**, *20*, 703–730. [CrossRef]
87. FAO. *Climate Change and Food Security: Risks and Responses*; FAO: Rome, Italy, 2015; ISBN 978-92-5-108998-9. Available online: https://openknowledge.fao.org/server/api/core/bitstreams/a4fd8ac5-4582-4a66-91b0-55abf642a400/content (accessed on 22 May 2024).
88. MINAM. *Plan Nacional de Adaptación al Cambio Climático Del Perú: Un Insumo Para La Actualización de La Estrategia Nacional Ante El Cambio Climático—Resumen Ejecutivo*, 1st ed.; Ministerio del Ambiente (MINAM): Lima, Peru, 2022. Available online: https://cdn.www.gob.pe/uploads/document/file/2827898/220214_Resumen%20Ejecutivo%20del%20Plan%20Nacional%20de%20Adaptaci%C3%B3n_compressed.pdf.pdf?v=1664915422 (accessed on 22 May 2024).
89. Calderón, R.R. *Estadisticas de las Tecnologías de Información y Comunicación en los Hogares*; Instituto Nacional de Estadística e Informática (INEI): Lima, Peru, 2021. Available online: https://cdn.www.gob.pe/uploads/document/file/3093592/Las%20Tecnolog%C3%ADas%20de%20Informaci%C3%B3n%20y%20Comunicaci%C3%B3n%20en%20los%20Hogares:%20Abr-May-Jun%202021.pdf?v=1652377629 (accessed on 5 June 2024).
90. SENAMHI. Pronóstico de Riego Agroclimático Nacional. Available online: https://www.senamhi.gob.pe/?&p=impacto-en-la-agricultura (accessed on 5 June 2024).

Disclaimer/Publisher's Note: The statements, opinions and data contained in all publications are solely those of the individual author(s) and contributor(s) and not of MDPI and/or the editor(s). MDPI and/or the editor(s) disclaim responsibility for any injury to people or property resulting from any ideas, methods, instructions or products referred to in the content.

sustainability

MDPI

Article

Evaluation of Climate Suitability for Maize Production in Poland under Climate Change

Aleksandra Król-Badziak [1,*], Jerzy Kozyra [1] and Stelios Rozakis [2]

1 Institute of Soil Science and Plant Cultivation—State Research Institute, 24-100 Puławy, Poland; kozyr@iung.pulawy.pl
2 Bioeconomy and Biosystems Economics Lab, School of Chemical and Environmental Engineering, Technical University of Crete, 73100 Chania, Greece; srozakis@tuc.gr
* Correspondence: akrol@iung.pulawy.pl

Abstract: Climatic conditions are the main factor influencing the suitability of agricultural land for crop production. Therefore, the evaluation of climate change impact on crop suitability using the best possible methods and data is needed for successful agricultural climate change adaptation. This study presents the application of a multi-criteria evaluation approach to assess climate suitability for maize production in Poland, for a baseline period (BL, 1981–2010) and two future periods 2041–2070 (2050s) and 2071–2100 (2080s) under two RCP (Representative Concentration Pathways) scenarios: RCP4.5 and RCP8.5. The analyses incorporated expert knowledge using the Analytical Hierarchy Process (AHP) into the evaluation of criteria weights. The results showed that maturity and frost stress were the most limiting factors in assessing the climatic suitability of maize cultivation in Poland, with 30% and 11% of Poland classified as marginally suitable or not suitable for maize cultivation, respectively. In the future climate, the area limited by maturity and frost stress factors is projected to decrease, while the area of water stress and heat stress is projected to increase. For 2050 climate projections, water stress limitation areas occupy 7% and 8% of Poland for RCP4.5 and RCP8.5, respectively, while for 2080 projections, the same areas occupy 12% and 32% of the country, respectively. By 2080, heat stress will become a limiting factor for maize cultivation; according to our analysis, 3% of the Polish area under RCP8.5 will be marginally suitable for maize cultivation because of heat stress. The overall analyses showed that most of Poland in the BL climate is in the high suitability class (62%) and 38% is moderately suitable for maize cultivation. This situation will improves until 2050, but will worsen in the 2080s under the RCP8.5 scenario. Under RCP8.5, by the end of the century (2080s), the highly suitable area will decrease to 47% and the moderately suitable area will increase to 53%.

Keywords: multi-criteria; analytic hierarchy process; EURO-CORDEX; climate risk

1. Introduction

Observed climate change is seen as one of the major concerns for agriculture, which poses a challenge for the adaptation of agriculture to ongoing climate change [1]. At the same time, scientific progress has achieved considerable control over various parameters affecting agricultural production, such as biophysical soil characteristics. But when climate is added to the equation of shaping pedoclimatic conditions, things become more complicated, especially with current frequent weather perturbations that pose great challenges to agricultural science. Various stresses, such as water shortage, heat, frost, pests and diseases determine both the quantity and quality of crop yields [2]. To minimise risks to crop production in the context of climate change [3], it is crucial to evaluate changes in land suitability with a multivariable approach [4], and thus more and more detailed analyses can be carried out to support the preparation of improved adaptation strategies [3,5]. Refining land suitability assessments may help decision-making on agricultural land use, which can then be used by governments and institutions to improve the resilience of farming systems to climate change [6].

Maize, as a thermophilic plant, is used in many studies as a model crop to show the effects of climate change on land suitability [4,6–8]. In previous studies it has been shown that the expected temperature increase would improve conditions for maize production in Poland [9], which seems to be already being realised and is shown by statistics. In Poland, the area under grain maize cultivation has increased significantly, from 152,000 ha in 2000 to 1,196,000 ha in 2022 [10]. The yield over the last 24 years (2000–2023) shows an increasing trend from 5.6 to 7.0 t/ha. However, this has been accompanied by relatively high year-to-year yield variability, ranging from 4.2 t/ha in 2006 to 7.5 t/ha in 2021 [11]. It is mainly due to extreme weather events, with droughts being the most important factor [12,13]. These conditions require not only a review of the suitability of land for maize production, but also a different approach to analysis, not just one that is variable-based, but which is also based on multivariables. An additional condition is the need to rebalance production systems for proper adaptation actions for a more sustainable model, so-called short-term and long-term strategies, which should also be taken into account in land suitability analyses [14] and farmers' perception [15]. In addition, agriculture should support climate change mitigation, with a particular focus on reducing emissions from agriculture. This can be achieved through fertilisation adapted to crop needs and soil conditions, care of the soil environment and rational management of resources [14].

Through a multicriteria approach in suitability areas assessment for agricultural production, many factors can be assessed [16,17], involving expert opinions [18]. In the view of the fact that these criteria have different levels of importance for land suitability, there are many methods used for criteria weight evaluation [19]. The AHP has been widely used to determine the weights of the parameters in the land suitability assessment for maize [20–23] and other crops such as tea [24,25], rice and soybean [23]. The AHP has the advantage of incorporating experts' opinion in order to prioritise the criteria [21]. Another advantage over other methods is the measurement and control of judgment inconsistency. One drawback of the AHP methodology is that it involves a subjective judgement of the relative importance of two criteria being compared, while there is a possibility of omitting interrelationships between different criteria [20].

In spite of many studies on the impact of climate change on the conditions for growing maize in Poland, it is difficult to obtain a clear answer as to whether these changes are beneficial or not [1]. Therefore, the overall objective of this study is to assess the climatic conditions for maize cultivation, considering both the positive effects of climate change, such as the increase in accumulated temperature leading to crop maturity, and the potential risks, such as those of water deficiency, frost and heat stress. These factors were integrated using a multicriteria assessment approach. Additionally, the experts' opinion were also integrated during the evaluation of selected criteria importance. This study expands on earlier work [26] in which the climate suitability of areas for maize cultivation was assessed, but only in relation to water deficit stress. Using the same methodology, and the developed water stress, and adding further factors: heat stress, frost stress and reaching maturity, we re-evaluated climate suitability for maize cultivation in Poland. Furthermore, in order to adapt the crop to the climatic conditions and provide more realistic results, the climate risk indicators were assessed using possible sowing and harvesting dates based on air temperature.

2. Materials and Methods

2.1. Study Area

Poland is located in Central Europe, from 14°07′ E to 24°09′ E and from 49° N to 54°50′ N, and covers an area of 311,888 km^2. It is mainly a lowland area with an average altitude of 173 m above sea level, but it has mountains in the south (the Carpathians and the Sudetes). Poland's climate is influenced by various air masses: maritime from Western Europe, continental from the East, polar from the Arctic and subtropical from the Atlantic, Mediterranean and Black Seas [27]. There are six distinct seasons in Poland: early spring, spring, summer, autumn, early winter and winter. In Poland the seasons vary from year to year due to significant natural variations in the climate [27]. In 2022, the average annual temperature in Poland was 9.5 °C, and it varied between 8.9 and 9.9 °C in different areas. In 2022, the average annual precipitation sum in Poland was 534.4 mm and it was characterised by spatial variability, ranging from over 350 mm to almost 950 mm. The 2022 year should be classified as a very warm and dry year [28]. Based on climatological norms for the period 1991–2020, the average annual temperature in Poland is 8.4 °C, while area-averaged precipitation total in Poland is 678.6 mm [29]. The growing season also varies between areas; in the northeast of Poland it is about 40 days shorter than in the southwest [27]. The average duration of the growing season in Poland for the period 1991–2020 was 224 days [29].

Natural conditions in Poland are fairly favourable for agriculture, but climate change is making droughts and floods more frequent and intense [30,31]. Decreases in yields of traditional crops are also predicted for most of the country, with their inter-annual variability. However, despite the negative impacts, there are also positive changes observed in agriculture: the length of the vegetative period will increase, and more favourable conditions for growing crops typical of warmer areas (maize, soya bean, etc.) are predicted [27].

2.2. Climate Data Dataset

Regional climate model (RCM) projections are used for regional climate change studies and impact assessments. Compared to global climate models (GCMs), they have the advantage of superior resolution and better representation of the local climate. However, relying solely on the output of RCMs can be subject to uncertainties and biases that may affect the accuracy of the results. As part of the Coordinated Regional Downscaling Experiment Project (CORDEX) project, the EURO-CORDEX initiative is providing high resolution RCM projections for the European domain that are openly available through the Earth System Grid Federation (ESGF) [32,33]. The climate projections data used in this work were obtained from the ESGF server (https://esgf-data.dkrz.de/search/cordex-dkrz, accessed on 6 July 2021). The bias-adjusted daily mean/max/min temperature and precipitation data are available from the CORDEX-Adjust project. These data are bias-corrected using a number of different methods [33]. For analysis, the daily precipitation, minimum, maximum and mean temperature bias-adjusted outputs of RCMs were obtained from the CORDEX Adjust database. These outputs were generated by the Swedish Meteorological and Hydrological Institute (SMHI) with Distribution-Based Scaling (DBS45) method [34] and were calibrated by SMHI according to the MESoscale ANalysis system (MESAN) reanalysis data [35]. These datasets had an original spatial resolution of 0.11°. CDO (Climate Data Operators; v. 1.7.0) software [36] was used to bilinearly interpolate the original gridded data to a regular grid and for further calculations. Due to the fact that there are considerable uncertainties in the projections of future climate by climate models, previous studies suggest that an averaged ensemble of several climate models, rather than a single one whose output is unreliable, should be used to assess future climate [30,32]. Therefore, in this and our previous study [26], we decided to use a mean ensemble of six GCM-RCM chain simulations (combination of 4 GCMs and 3 RCMs) to obtain a reliable projection of future climate variables (Table 1). In this research, the change in climate suitability and climatic conditions for maize cultivation in Poland was analysed under two Representative Concentration Pathways (RCP) scenarios: RCP4.5 (low emissions, LE) and RCP8.5 (high emissions, HE). RCP scenarios correspond to radiative forcing by 2100 relative to pre-industrial values of +4.5 Wm^{-2} (RCP 4.5), and to +8.5 Wm^{-2} (RCP 8.5) [37,38]. We selected equal 30-year periods for the reference period (1981–2010) and two future periods: the 2050s (2041–2070) and the 2080s (2071–2100) for our analyses.

Table 2 shows the area-averaged annual temperature and annual precipitation sum in Poland and their projected changes in future climate conditions. According to the ensemble mean of six analysed RCMs, the mean annual air temperature in Poland was 8.2 °C for BL, which increased by 2.1 °C and 3.8 °C for LE and HE scenarios, respectively, by the 2080s. Furthermore, the mean annual precipitation sum was about 622 mm in BL, with projected increase of 9–14% and 11–19% in the 2050s and the 2080s, respectively. When using climate change scenarios, it is important to note that the results of numerical simulations may differ from observations. It is therefore essential to make comparisons between the two in order to assess the performance and capability of the climate models used [30,39]. The detailed information related to the comparison between observations and climate scenarios data used in this study can be found in our previous study [26]. The RMSE (root mean square error) of the annual mean values was 0.3 °C for the temperature and 36 mm/year (6%) for the precipitation [26].

Table 1. Combinations of GCMs and RCMs that were used in this study.

Global Climate Models	Regional Climate Models		
	SMHI-RCA4	KNMI-RACMO22E	DMI-HIRHAM5
CNRM-CERFACS-CNRM-CM5	X		
ICHEC-EC-EARTH	X	X	X
IPSL-IPSL-CM5A-MR	X		
MPI-M-MPI-ESM-LR	X		

Table 2. The area-averaged annual temperature, annual precipitation sum and their changes (in brackets) in Poland [26].

Period [1]	BL	2050s		2080s	
Emission Scenario [2]		LE	HE	LE	HE
Mean annual temperature [°C]	8.2	9.7 (+1.5)	10.3 (+2.1)	10.3 (+2.1)	12.0 (+3.8)
Mean annual precipitation [mm]	622	676 (+9%)	710 (+14%)	692 (+11%)	741 (+19%)

[1] BL—baseline; [2] LE—low emissions scenario; HE—high emissions scenario.

2.3. Climate Suitability Assessment

The selection of different criteria for climate suitability assessment was based on a literature review in respect of their relation to maize growth conditions. The weights for the selected criteria were calculated using the AHP. Then, the suitability map was generated by application of a weighted overlay. The methodology employed is presented in the form of a flowchart in Figure 1.

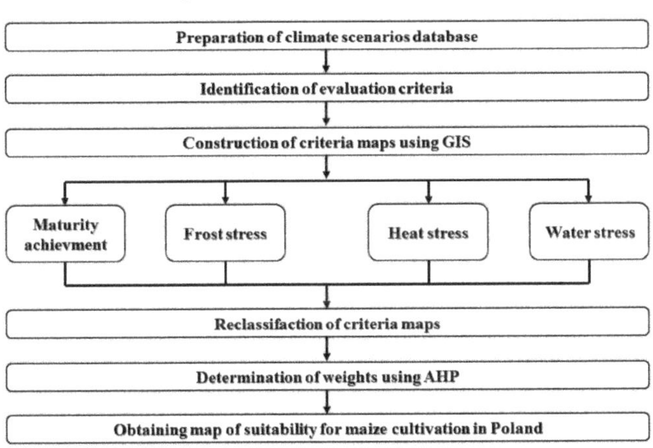

Figure 1. Methodology flowchart.

2.4. Identification of Evaluation Criteria

The planting and growth of crops are affected by many factors, and therefore it is impossible to take all of them into account [23]. In accordance with maize growth conditions [4,40,41], and their potential changes due to climate change, the following criteria were selected for the evaluation of land suitability for maize cultivation in Poland: maturity, spring frost stress, heat stress and water stress. By means of FAO [42] and the literature review [20,21,23,43,44] the suitability classes are classified as follows: highly suitable (S1), moderately suitable (S2), marginally suitable (S3) and not suitable (N). Subsequently, a specific level of suitability for each criterion was defined and presented in Table 3.

Table 3. Suitability level per selected criteria.

Characteristics		**Suitability Rating Score ***				**References**
Criteria	Sub-criteria	S1 100–85	S2 85–60	S3 60–40	N 40–0	[23]
Maturity of maize	FAO210 [number of years]	>27	24–27	18–24	<18	[45,46]
	FAO240 [number of years]	>27	24–27	18–24	<18	[45,46]
	FAO290 [number of years]	>27	24–27	18–24	<18	[45,46]
Frost stress	Tmin < 0 °C [number of years]	<3	3–6	6–12	>12	[46]
Heat stress	Tmax > 35 °C [number of years]	<3	3–6	6–12	>12	[46]
	Tmax > 40 °C [number of years]	<3	3–6	6–12	>12	[46]
Water stress	CWB on very light soils [mm]	>−150	−200–−150	−250–−200	<−250	[47]
	CWB on light soils [mm]	>−210	−260–−210	−310–−260	<−310	[47]
	CWB on medium soils [mm]	>−250	−300–−250	−350–−300	<−350	[47]
	CWB on heavy soils [mm]	>−290	−340–−290	−390–−340	<−390	[47]

* S1—highly suitable, S2—moderately suitable, S3—marginally suitable and N—not suitable.

It is evident that climate exerts a significant influence on agricultural production potential [23,40]. Indeed, the necessity of agricultural management adjustment to local climatic conditions has been pointed out for the best use of these potentials. Due to changes in these conditions over time and space, it is important to consider them in the planning of adaptation measures [40]. Climatic suitability is largely influenced by heat, moisture and sunshine [23]. In agricultural practice, the temperature is typically regarded as the primary indicator that determines the sowing date [48]. The yield-forming effect of temperature is reflected in both the heat sum necessary to ensure all phenological phases occur and the temperature thresholds for critical phenological phases [49]. It is anticipated that extreme temperature events resulting from global warming will occur with greater frequency and greater intensity, which will have an impact on the growth and development of crops [50]. Therefore, frost and heat stress parameters are considered to be limiting factors for grain maize [40].

2.4.1. Maturity of Maize (GDD)

Maize is a thermophilic plant, and therefore the vegetation period for this crop should be relatively long with high values of degree-days [41]. The effective temperature sum that is needed for maize maturity depends on the type of earliness that is determined by the FAO number and the utilisation mode (for grain, CCM and silage). The choice of a variety with a desired length of the growing season ensures that maize can be cultivated on the entire area of Poland, with the exception of mountain regions [51]. It was assumed that the recommended average date of maize sowing is the first day when the average temperature over the last five days exceeds 11 °C. As the end of the maize growing season, the first day when the average temperature drops below 10 °C for five consecutive days was determined. A 6 °C threshold was used to calculate Growing Degree Days (GDD) for maize. By comparing its thermal requirements with GDD, it is possible to determine the possibility of reaching maize maturity [41]. In the following study, the number of years in which maize reaches maturity within a 30-year period is used as a suitable criterion for maize cultivation for three different maize maturity types: (1) FAO210, (2) FAO240 and (3) FAO290 as examples of early, medium-early and medium-late varieties.

2.4.2. Late Spring Frost Stress (LSF)

Maize is tolerant of a wide range of environmental conditions, although it is highly susceptible to frost [50,52,53]. Frost is a phenomenon that is difficult to predict and which can cause partial or complete damage to maize [4,54]. Frost stress occurs when the daily minimum temperature drops below 0 °C [40,54]. One of the most critical extreme events occurring in temperate regions, among others, is late spring frost (LSF), which affects plant performance. When the leaves are young, plants are most susceptible to frost, so the onset of freezing temperatures in late spring has a major impact on the growth, competitiveness and range of plants [55]. Consequently, frost occurrence was defined as one day with a minimum temperature below 0 °C from the sowing date until the end of June, and the frequency of years with at least one frost occurrence in a 30-year period was calculated.

2.4.3. Heat Stress (HS)

Maize is tolerant of high temperatures, although extremely high temperatures can reduce yield [4]. The optimum temperature differs among development processes, but is usually determined as 28–34 °C [56]. Daily maximum temperatures exceeding 35 °C can affect maize grain [40,57]. We defined heat stress as five consecutive days with a maximum temperature above 35 °C. Given that maize development can cease at temperatures of more than 40 °C, the heat stress temperature was additionally defined as one day with a temperature above 40 °C [4,56]. HS criteria were evaluated within sowing and end of growing season date.

2.4.4. Water Stress (WS)

The climatic water balance (CWB), employed in Poland as an indicator for meteorological and agricultural drought monitoring represents the difference between precipitation and potential evapotranspiration [58]. Given the observed increase in the frequency of water stress, the CWB was selected as one climate suitability criterion for maize cultivation [26,47]. Numerous methods for potential evapotranspiration (PET) calculation exist in the literature, with varying data requirements. The numerous data requirements allow for the potential for implementing simplifications in PET formulae based on specific application needs. The following study utilises the results of a previous study [26] on the calculated CWB sum for the period from April to September, based on a simplified Penman formula presented by Doroszewski and Górski [59] and Żyłowska et al. [60].

2.5. Determination of Weights (AHP)

An Analytical Hierarchy Process (AHP) method is used to determine the weights of selected main criteria and sub-criteria. The weights need to be chosen carefully as they reflect the relative importance of each criterion. Thanks to the AHP method, qualitative data, assessed by experts, can be converted into quantitative data [61].

In this study, a team of ten crop specialists (comprising five experts and five farmers) was asked to determine the relative importance of each criterion. The group of experts was selected on the basis of their extensive knowledge and experience in agriculture. It included scientists from the Institute of Soil Science and Plant Cultivation (IUNG-PIB), who conduct research in crops and agrometeorology, as well as head of crop units from the Agricultural Advisory Centre and the Regional Centre of Agricultural Advisory Services. The group of farmers comprised five directors and workers from the Agricultural Experimental Stations (AES IUNG-PIB), selected on the basis of their experience in leading long-term agricultural experiments across a range of voivodeships in Poland. A questionnaire was designed to compare the four main criteria pairwise regarding their importance in achieving maize suitability. Based on the proposed hierarchy (Figure 2), the sub-criteria are then compared pairwise in terms of their importance in achieving their upper-level criteria. Table 4 shows

Saaty's [62] linguistic scales for assessing the relative importance for evaluation of the selected criteria using AHP methodology.

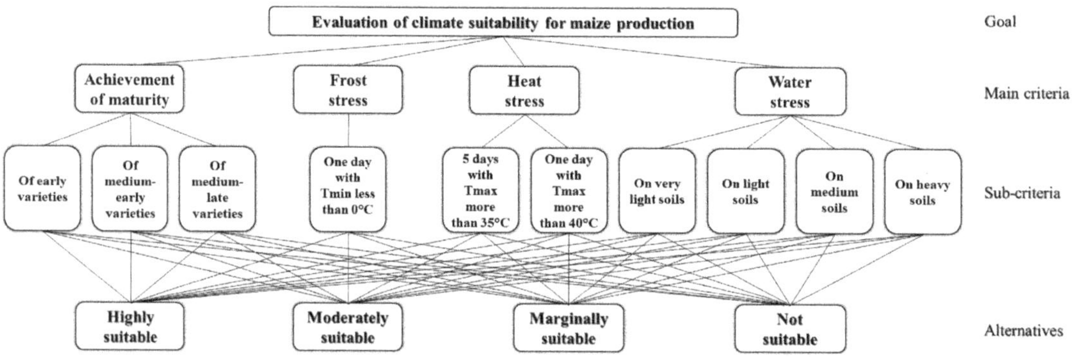

Figure 2. The AHP methodology framework for assessing climate suitability for maize production.

Table 4. Analytic Hierarchy Process linguistic pairwise comparison scale based on Saaty [62].

Degree of Importance	Definition
1	Equal importance
3	Moderate importance
5	Strong importance
7	Very strong importance
9	Extreme importance
2, 4, 6, 8	Compromise between above values
Reciprocals of above	Values for inverse comparison

Using the AHP as previously discussed, each specialist's opinion was transformed into a pairwise comparison matrix using a scale of 1–9 (Table 4). This scale was employed with 1 representing equal importance and 9 representing extreme importance between two compared criteria. Once the pairwise matrix had been generated, the weights for each criterion were computed using the eigenvector corresponding to the largest eigenvalue of the matrix (λ_{max}), and then the sum of the components was normalized to unity [23,61]. The AHP methodology involves the identification and consideration of inconsistencies in decision-maker judgment through the evaluation of consistency relationship (CR), which depend on the Consistency Index (CI) and Random Index (RI) [61]. The CI was calculated using the formula: $CI = (\lambda_{max} - n)/(n-1)$ [63]. The random consistency index (RI) related to the number of elements (n) provided in Table 5 was used to determine the CR based on the formula $CR = CI/RI$. A CR value of 0.1 or less is deemed acceptable, although in some studies a value of 0.2 is considered an acceptable consistency level [64].

Table 5. Random Consistency Index (RI) values [62].

n	1	2	3	4	5	6	7	8	9	10
RI	0.00	0.00	0.52	0.89	1.11	1.25	1.35	1.40	1.45	1.49

AHP can be used for priorities. According to Forman & Peniwati [65], there are two basic ways to aggregate individual opinions into a group opinion, depending on whether the group wants to act together as a unit (e.g., a group of department heads meeting to set corporate policy) or as separate individuals (e.g., a group consisting of representative constituencies with stakes in welfare reform, such as taxpayers, politicians, etc.). Several researchers find agreement in applying the AIJ method in case the target group is homogenous and assumed to act as one individual, while the AIP method is recommended when the group is heterogeneous and is seen as a collection of independent agents maintaining their own identities.

Given that the evaluation of criteria importance involved ten respondents with disparate opinions, the AHP group decision-making process was employed to aggregate individual judgments. There were alternative methods used to aggregate individual opinions applied in AHP methodology, depending on whether the group wanted to act together as a unit assuming consensus or as separate individuals. The Aggregation of Individual Judgments (AIJ) method is suggested in case the target group is homogenous whereas for heterogeneous groups the AIP method is recommended [65]. Calculated by the geometric mean to aggregate comparison matrices (AIJ) or to aggregate priority vectors (AIP) [66]. Srdjevic [67], who goes through a GDM process by utilizing a group of experts, split into interest subgroups, claims that AIJ is possible to apply only at the subgroup level. Among different subgroups only AIP is possible for final decision.

2.6. Weighted Overlay

There is a need to standardize the selected criteria as they are expressed in different units [61]. All selected criteria were reclassified into four suitability classes (S1, S2, S3, N) according to Table 3. Standardised criteria maps were generated using QGIS 3.28.4. Climate suitability for maize production was then assessed by combing all standardised criteria maps and weights of criteria. The final map was generated using the weighted overlay method in QGIS [44,61].

3. Results

3.1. Growing Season of Maize for Baseline and 2050s and 2080s Climates

The spatial distribution of the start (11 °C) and end (10 °C) dates of the growing season is presented in Figures 3 and 4. The beginning and end dates of the growing season are delayed in colder regions (north-east) compared to warmer parts of Poland (south-west). This is seen in both the baseline timeframe and in the future climate. The projected warmer climate leads to an acceleration in the occurrence of the beginning and to a delay of the end dates of the growing season, with the rate of change depending on the time horizon and the RCP analysed. For the 2050s, the projected rate of acceleration of the start and delay of the end dates of the growing season is about 5–10 and 10–15 days for the LE and HE scenarios, respectively. In the 2080s, a higher advancement of about 5–15 days for LE and approximately 15–25 days for HE was projected.

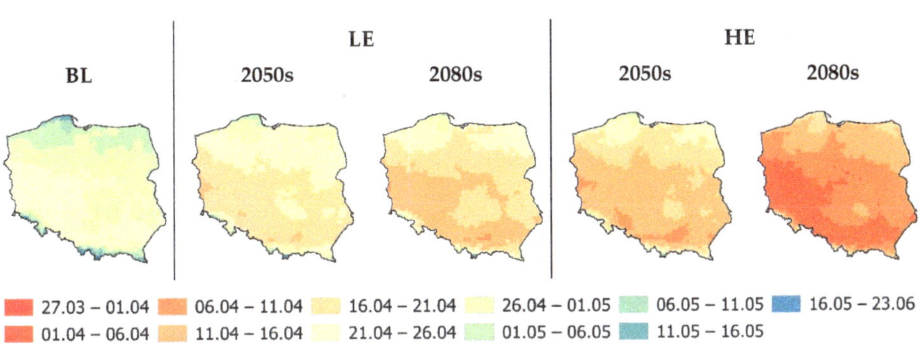

Figure 3. Multimodel ensemble mean of the date of beginning the growing season (11 °C) for the baseline (BL, 1981–2010) and projected according to the low emissions (LE, RCP4.5) and high emissions (HE, RCP8.5) scenario for the 2050s (2041–2070) and 2080s (2071–2100).

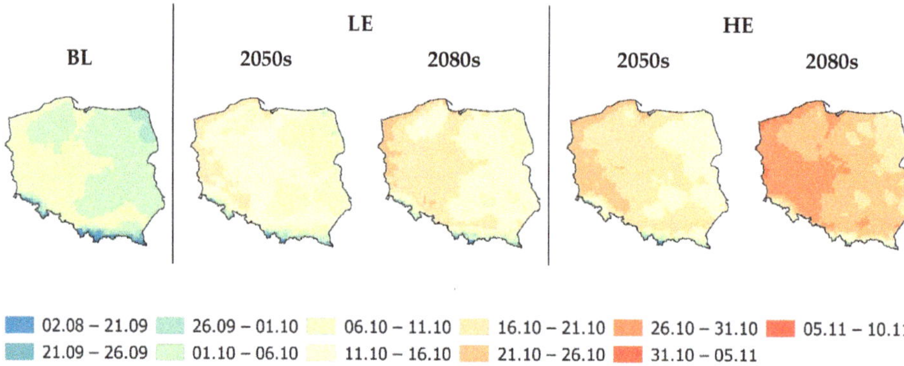

Figure 4. Multimodel ensemble mean of the date of end of growing season (10 °C) for the baseline (BL, 1981–2010) and projected according to the low emissions (LE, RCP4.5) and high emissions (HE, RCP8.5) scenario for the 2050s (2041–2070) and 2080s (2071–2100).

The mean accumulated temperature sum above 6 °C (AT) during the maize growing season was higher than 1200 °C in the baseline period, except in southern Poland where the mean AT was lower (Figure 5). Due to climate change, the mean AT is projected to increase. According to the LE scenario, the AT during the maize growing season is projected to range from 1600 to 2000 °C in the 2050s and from 1800 to 2200 °C in most areas, whereas the AT will be much lower in the northern and southern parts of Poland. According to the HE scenario, the rise in AT is much higher, and by the end of the century, the mean AT is projected to rise up to 2600 °C in the west-central part of Poland.

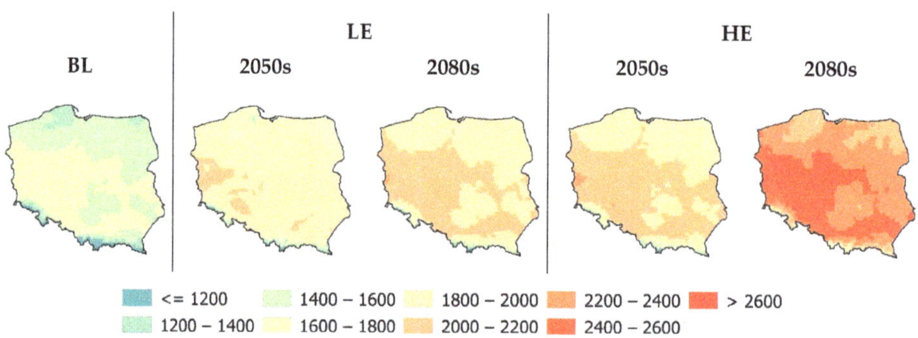

Figure 5. Multimodel ensemble mean of accumulated temperature sum of growing season [°C/yr] for the baseline (BL, 1981–2010) and projected according to the low emissions (LE, RCP4.5) and high emissions (HE, RCP8.5) scenario for the 2050s (2041–2070) and 2080s (2071–2100).

3.2. *Evaluation of Climate Criteria under Baseline and 2050s and 2080s Climates*

The probability of the maturity of the different maize varieties is shown in Figure 6 as the reference number of years in which maize reached maturity per analysed 30-year period. The base line time period (BL) is highly diverse. In the northern and southern parts of Poland, the probability of reaching maize maturity is low (<60%) even for early-maturing varieties, and this area increases for later varieties. Furthermore, for early varieties, the probability of maturation is higher than 90% in 44% of Poland, but for medium-early varieties, this area decreases to 25% of Poland, while for medium-late varieties, it is almost non-existent (2%). The projected temperature increase results in greater thermal resources, allowing the cultivation of even medium-late varieties in a large part of Poland. However, there will still be regions where choosing a particular variety should involve considering the risk of a cold year, but for most of the area (from 65 to 98%) this criterion will not limit maize cultivation.

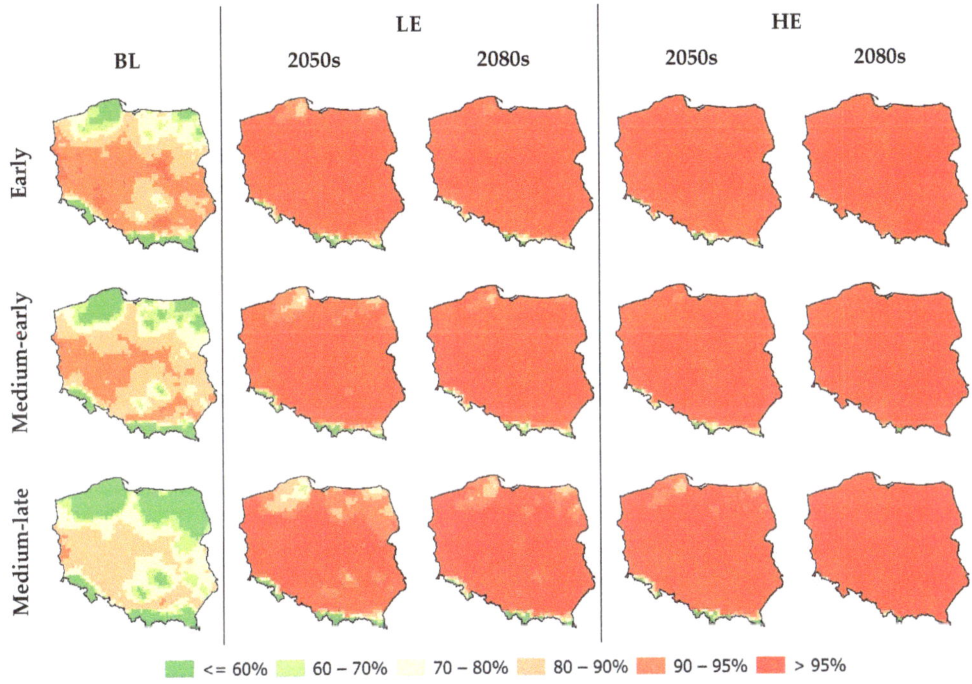

Figure 6. Multimodel ensemble-mean of the percentage of years in the 30-year period analysed in which maize can reach maturity for baseline (1981–2010) and projected according to low-emissions (LE, RCP4.5) and high-emissions (HE, RCP8.5) scenario for 2050s (2041–2070) and 2080s (2071–2100). The probability of maturity is evaluated for early, medium-early and medium-late varieties.

Frost stress (FS) was evaluated from the beginning (11 °C) of the growing season until the end of June, and its spatial distribution is presented in Figure 7. In the baseline time period, FS (minimum temperature below 0 °C) occurred mainly in 3–6 years out of the analysed 30-year time period (10–20%). The occurrence of spring FS decreases in future climate conditions. The area which is highly suitable according to spring frost risk, where the minimum temperature falls below 0 °C, will increase, and, by the 2080s, it will cover a large area of Poland (70% and 51% of Poland, according to LE and HE scenarios, respectively).

Heat stress was evaluated for the growing season and its spatial distribution is presented in Figure 8. In the baseline (BL) period, the maximum temperature did not exceed 35 °C for at least five consecutive days, while the maximum temperature was higher than 40 °C only for one day in a small area on the west part of Poland (6% of the area). In the future climate scenarios, the area with a maximum temperature higher than 35 °C for five days increased. According to RCP8.5, for the 2080s timeframe, it is going to occur up to 6 times per 30 years in a large area of Poland (97%). According to the frequency of occurrence of temperatures higher than 35 °C in Poland, there is a highly suitable area for the cultivation of maize in both current and future climate conditions, except for the 2080s for the HE scenario, where a large area of Poland (71%) is defined as moderately suitable. Furthermore, the frequency of the event when the maximum temperature exceeds 40 °C is expected to increase in the future. On the baseline, in Poland, there are highly suitable conditions for maize cultivation according to this stress. However, in the 2050s, in the western part of Poland, there is a large area defined as moderately suitable (this stress occurs in 3–6 years for 41–45% of Poland). According to the HE scenarios in the 2080s, it is projected that, in Poland, there will be marginally suitable and not suitable areas for maize cultivation in the western and central parts of Poland, with this stress occurring for more than 6 years in 62% and 1% of the area of Poland, respectively.

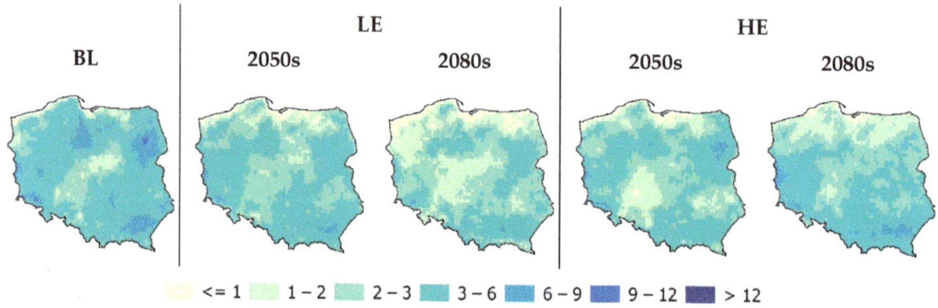

Figure 7. Multimodel ensemble-mean of number of years with frost stress occurrence for baseline (1981–2010) and projected according to low-emissions (LE, RCP4.5) and high-emissions (HE, RCP8.5) scenario for 2050s (2041–2070) and 2080s (2071–2100). Frost stress is defined as frequency of occurrence of at least one day between the beginning of growing season and the end of June with the minimum temperature falling below 0 °C during the 30 years analysed period.

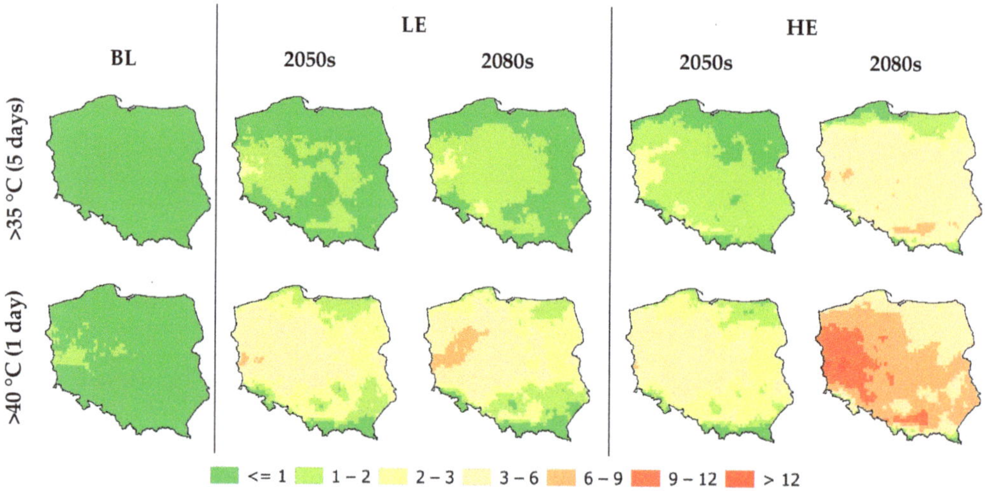

Figure 8. Multimodel ensemble-mean of number of years with heat stress occurrence for baseline (1981–2010) and projected according to low-emissions (LE, RCP4.5) and high-emissions (HE, RCP8.5) scenario for 2050s (2041–2070) and 2080s (2071–2100). Heat stress is defined as frequency of occurrence of 5 consecutive days with maximum temperature exceeding 35 °C, and at least one day with the maximum temperature greater than 40 °C during the 30 years analysed period.

Water stress (WS) was evaluated in our previous study [26] on the basis of the CWB sum from April to September and its spatial distribution is presented there [26]. The CWB values were negative when the potential evapotranspiration exceeded the precipitation sum, which occured in a large area of Poland. The CWB exceeding 100 mm was observed only in mountainous areas. The area with CWB less than −300 mm was observed in central Poland and it is projected to increase in the future.

3.3. Climate Criteria Weights for Land Suitability Assessment

The opinions of experts (5) and farmers (5) were used to weigh the four main climate suitability criteria for maize (maturity, frost risk, heat risk and water stress) and sub-criteria using an AHP pairwise comparison matrix. Tables 6–9 shows the mean judgement of two sub-groups (experts and farmers) calculated based on geometric mean (AIJ). Then, the final weights were calculated based on AIP to aggregate two sub groups.

Table 6. AHP pairwise comparison matrix for the maturity of maize sub-criteria.

Subgroup	Experts				Farmers				Mean
Criteria	FAO210	FAO240	FAO290	Weights	FAO210	FAO240	FAO290	Weights	Weights
FAO210	1.000	0.284	1.600	0.190	1.000	0.367	4.441	0.283	0.232
FAO240	3.519	1.000	6.382	0.696	2.728	1.000	6.921	0.640	0.668
FAO290	0.625	0.157	1.000	0.114	0.225	0.144	1.000	0.077	0.093
	Consistency Ratio (CR) = 0.002				Consistency Ratio (CR) = 0.034				

Table 7. AHP pairwise comparison matrix for the heat stress sub-criteria.

Subgroup	Experts			Farmers			Mean
Criteria	Tmax > 35 °C	Tmax > 40 °C	Weights	Tmax > 35 °C	Tmax > 40 °C	Weights	Weights
Tmax > 35 °C	1.000	3.956	0.798	1.000	5.471	0.845	0.822
Tmax > 40 °C	0.253	1.000	0.202	0.183	1.000	0.155	0.177
	Consistency Ratio (CR) = 0.000			Consistency Ratio (CR) = 0.000			

Table 8. AHP pairwise comparison matrix for the water stress sub-criteria.

Subgroup	Experts					Farmers					Mean
Criteria	Very Light	Light	Medium	Heavy	Weights	Very Light	Light	Medium	Heavy	Weights	Weights
Very light	1.000	0.186	0.126	0.156	0.044	1.000	0.259	0.135	0.123	0.042	0.043
Light	5.373	1.000	0.223	0.410	0.151	3.866	1.000	0.163	0.145	0.089	0.116
Medium	7.950	4.478	1.000	1.165	0.458	7.384	6.128	1.000	1.516	0.468	0.463
Heavy	6.423	2.440	0.859	1.000	0.346	8.106	6.903	0.660	1.000	0.401	0.372
	Consistency Ratio (CR) = 0.046					Consistency Ratio (CR) = 0.079					

Table 9. AHP pairwise comparison matrix for the main climate suitability criteria.

Subgroup	Experts					Farmers					Mean
Criteria	Maturity	Frost Stress	Heat Stress	Water Stress	Weights	Maturity	Frost Stress	Heat Stress	Water Stress	Weights	Weights
Maturity	1.000	2.460	4.004	0.678	0.3111	1.000	1.450	1.467	0.644	0.2092	0.2551
Frost stress	0.407	1.000	2.569	0.286	0.1431	0.690	1.000	1.217	0.138	0.1128	0.1271
Heat stress	0.250	0.389	1.000	0.136	0.0657	0.681	0.822	1.000	0.143	0.1029	0.0822
Water stress	1.476	3.497	7.331	1.000	0.4801	1.552	7.237	6.971	1.000	0.5751	0.5254
	Consistency Ratio (CR) = 0.007					Consistency Ratio (CR) = 0.064					

Tables 6–9 show consistency ratio (CR) values for main, maturity and water stress criteria, which were below the threshold value of 0.1, suggesting a reasonable level of consistency in the experts' and farmers' judgment. The judgement for heat stress was acceptable as there were only two sub-criteria compared (RI is 0). The analysed four climate factors that we indicated were as follows: probability of maturity, frost stress risk, heat stress risk and water stress risk. Water stress risk (53%) and probability of maturity (26%) were the main factors contributing to suitable areas' evaluation for maize cultivation in Poland according to respondents' opinion (Figure 9). The other two analysed climate factors, frost risk (13%) and heat stress risk (8%), were found to be less important but are

still considered by experts. They most often decided to use the medium-early varieties (67%) rather than early varieties (23%) and medium-late varieties (10%). The heat stress indicators, with an event when the maximum temperature exceeds 35 °C for 5 days or more, were the most important (83%) in relation to an event when the temperature is more than 40 °C for at least one day (17%). In order to avoid water stress, the respondents were considering cultivating maize on medium (47%) and heavy soils (37%). However they were also considering cultivating maize on light (12%) and very light soils (4%), although they see climate risks. It is important to note that the calculated weights of the sub-criteria and main criteria differ between the two sub-groups (experts and farmers). Therefore, if we do not aggregate them and produce suitability maps for different groups, the results may vary.

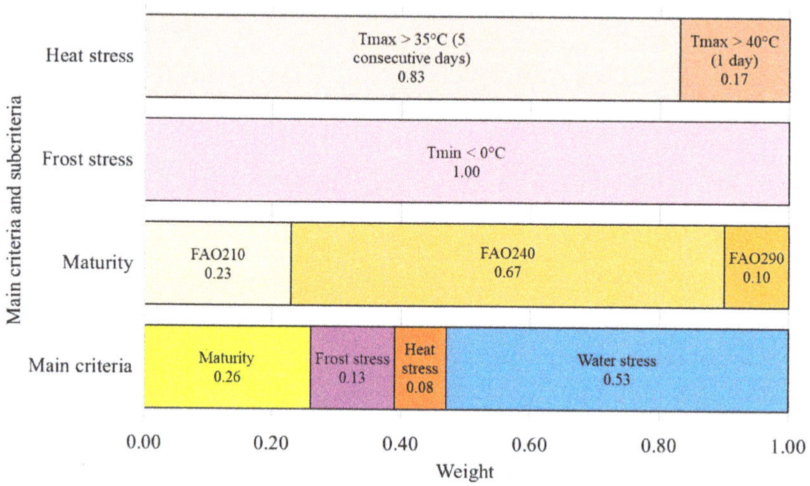

Figure 9. The relative importance (weights) for sub-criteria and main criteria according to pairwise comparisons.

3.4. *Climate Suitability for Maize in the Baseline and Future Scenarios*

The presented suitability maps of the main climatic criteria (maturity, frost stress, heat stress and water stress) as well as the final climatic suitability maps were classified into four suitability classes: high (S1), moderate (S2), marginal (S3) and not suitable (N). Table 10 and Figure 10 show the percentages of the suitability classes of Poland's area.

Table 10. Percentages of different climate suitability levels for maize cultivation in Poland according to maturity, frost stress, heat stress, water stress for the baseline (1981–2010) and projected climate according to low emissions (LE, RCP4.5) and high emissions (HE, RCP8.5) scenarios for the 2050s (2041–2070) and 2080s (2071–2100).

RCP					LE								HE							
Period	BL				2050s				2080s				2050s				2080s			
Rating Scale [1]	S1	S2	S3	N	S1	S2	S3	N	S1	S2	S3	N	S1	S2	S3	N	S1	S2	S3	N
Maturity	43	27	19	11	94	3	1	2	96	1	2	1	100	0	0	0	100	0	0	0
Frost stress	24	65	11	0	46	53	1	0	70	30	0	0	51	47	2	0	51	46	3	0
Heat stress	100	0	0	0	100	0	0	0	100	0	0	0	100	0	0	0	26	71	3	0
Water stress	81	18	1	0	67	26	7	0	64	24	12	0	69	22	8	0	44	24	21	11

[1] S1—highly suitable, S2—moderately suitable, S3—marginally suitable, N—not suitable.

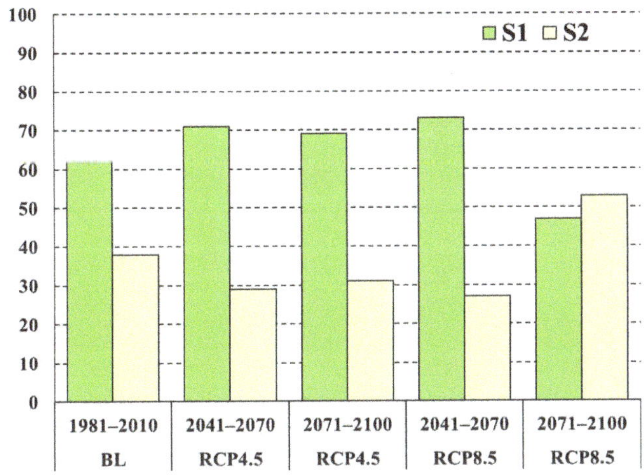

Figure 10. Percentages of different climate suitability levels (S1—highly suitable, S2—moderately suitable) for maize cultivation in Poland according to overall analyses for the baseline (1981–2010) and projected climate according to low emissions (LE, RCP4.5) and high emissions (HE, RCP8.5) scenarios for the 2050s (2041–2070) and 2080s (2071–2100).

From the reclassification results (Figure 11), taking into consideration maturity, it is shown that around 43% of the study area was highly suitable, whilst 27% was moderately suitable and 19% was marginally suitable for maize cultivation. However, about 11% of Poland was classified as unsuitable. Furthermore, as a result of the future increase in the accumulated temperature sum in both scenarios, the highly suitable areas will increase and it is projected that they will cover more than 90% of Poland, allowing cultivation of later maturing varieties than in the current climate. The obtained results show that frost stress limits climate suitability for maize production in Poland; for the baseline 24% of country area is highly suitable, 65% moderately suitable, while 11% is marginally suitable for maize cultivation. In the future, frost stress is predicted to decrease, and more than 50% of Poland is going to be highly suitable with the largest share of 70% of the area in the 2080s for RCP4.5. Considering heat stress, the entire analysed area was highly suitable under the current climatic conditions, and it will remain so in the future except for the HE scenario, in which heat stress will occur more frequently in the 2080s than in the past, thus reducing the highly suitable areas to moderately and marginally suitable in 71 and 3% of Poland, respectively. Due to the occurrence of water deficiency in the current climate, only central Poland is limited by water stress and has a moderate or marginal climate suitability (19%) for maize production. In the future, depending on the period and scenario analysed, these conditions are projected to extend to 30–45% of Poland. The increase in water stress will continue into the far future, and by the end of the century, according to the HE scenario, 21% of the area of Poland will be marginally suitable for maize production, and 11% will be unsuitable.

Finally, the climate suitability analysis (Figures 10 and 11) using the AHP weights shows that 62% of Poland was highly suitable and 38% was moderately suitable under current climate conditions. Furthermore, in the future climate conditions the highly suitable area will increase to 69–73%, except in the 2080s according to the HE scenario where this area will decrease to 47%. The rest of Poland is projected to be moderately suitable according to the selected climate factors.

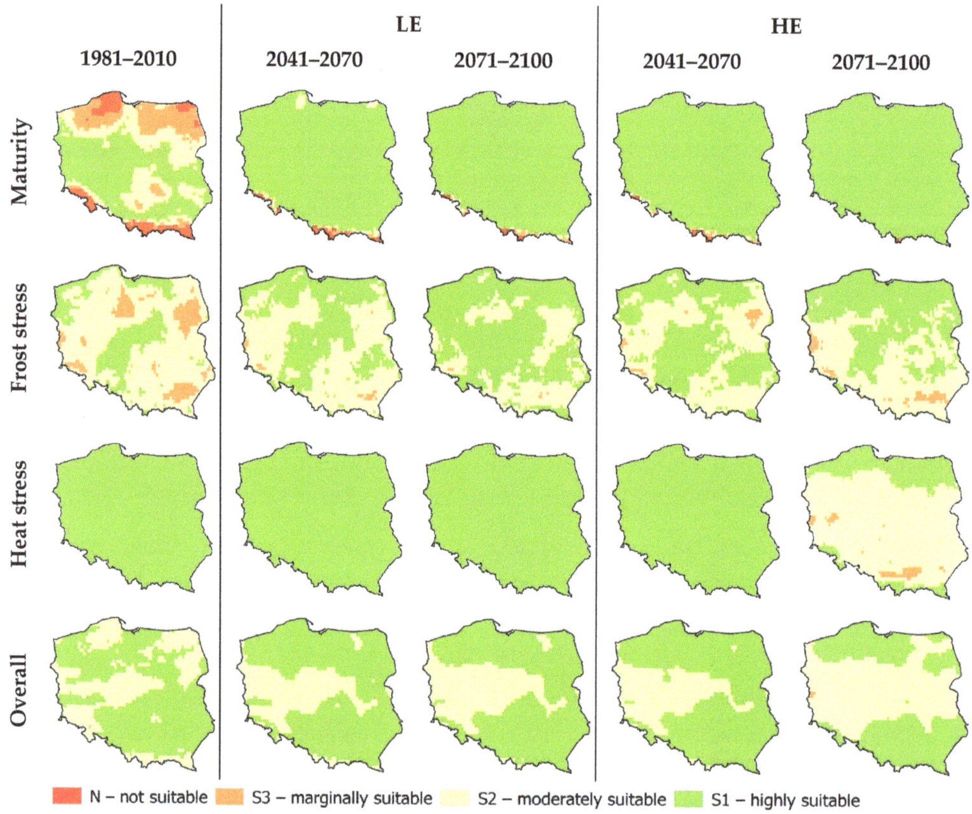

Figure 11. Average climate suitability for maize cultivation for baseline (1981–2010) and projected climate according to low-emissions (LE, RCP4.5) and high-emissions (HE, RCP8.5) scenario for 2050s (2041–2070) and 2080s (2071–2100).

4. Discussion

The climate in Poland and Europe is changing, with both average air temperatures and the frequency of extreme weather events increasing [68]. Observed climate changes may have both positive and negative effects on plant development. Projected warming may lead to a decrease in wheat yield, while it may increase in maize yields [69].

With future climate change and projected temperature increase, an acceleration of the beginning and delay of the end dates for the maize growing season are observed in the future reaching 15–25 days by the 2080s for the RCP8.5 scenario. These alterations will allow for the prolongation of the growing season, which will, in turn, allow the introduction of longer-duration cultivars [70]. A longer growing season can be associated with an increase in productivity and therefore an increase in nutrient requirements [70]. Furthermore, in both current and future climate conditions, the geographical diversity in the specific dates of the beginning and end of the growing season should be noted. The delayed beginning date and an accelerated end of the growing season in the northern part of the country was observed compared to the other parts of the country. This is also reflected in the study conducted by Nieróbca et al. [9], where the south-eastern part of Poland has more suitable climatic conditions for maize cultivation than northern regions [9]. Another benefit of climate change is the increase in thermal resources, allowing the cultivation of even medium-late varieties in a large part of Poland. However, even under future climate conditions, in a small area of the country in regions with a lower probability of maturity (less than 80%), the risk of a cold year should still be taken into account when choosing a variety [45].

Due to climate change and temperature increase, the number of frost days is projected to decrease in Poland [1,30]. Żarski et al. [71] assessed changes in climate risk indicators, such as the occurrence of late spring frost for grain maize in the Bydgoszcz region. They showed that late spring ground frost in May and June occurred 83% of the time in this region during the period analysed (between 1985 and 2014). However ground temperature below −2 °C occurred 30% of the time [71]. However, when the near-future was analysed (2026–2050), an increase in frost days compared to the historical period was predicted for central Poland [30]. Regarding frost risk that increased in the recent years, the phenological strategies (sowing later with respect to thermal requirements) that helped trees tolerate past frost frequencies will thus be increasingly mismatched with future conditions [55].

Despite the fact that maize is a thermophilic plant, the increased risk of heat that has caused plants injuries is reported by Ioannis Charalampopoulos [72]. For BL, the maximum temperature did not exceed 35 °C for 5 consecutive days and exceeded 40 °C for one day in just one year for a small area in Poland. However, the number of years with these extremes will increase in the future, and according to RCP8.5, heat stress will limit the suitability for maize cultivation in the 2080s. Previous studies have also assessed that the number of extremely hot days with Tmax above 35 °C will increase in the future in Poland [30]. In cooler zones of Europe (including Poland) lower heat stress is observed for maize than for wheat and barley whereas it is higher in warmer zones compared to current conditions [70]. Olesen et al. [70] noted that the occurrence of heat stress in the summer could be reduced by adopting earlier sowing dates.

Our study indicates that the most important element in assessing the climate suitability of areas for maize cultivation in Poland is water stress. A useful indicator for measuring this stress is climatic water balance, i.e., the difference between total precipitation and total evaporation. At present, evapotranspiration exceeds summer precipitation in almost all parts of Poland, resulting in an increase in water deficit [27]. A negative trend in availability of soil water during summer months and possible important effects on yield in central and eastern Europe is indicated by Pinke et al. [73], with warning about grain production stability and needs for intervention including implementation of water conservation agricultural practices. Olesen et al. [70] observed that water-conserving tillage practices can be a crucial adaptation measure in both warm and dry zones. In the preceding study [26], the impact of irrigation on water stress in maize cultivation was evaluated, and it was found that this practice is an appropriate method for adapting the increase in water stress related to climate change. According to work by Gobin et al. [74], the observed risk of drought for grain maize shifts earlier in the season. This is due to the fact that spring drought affects earlier vegetative stages, while summer drought poses a risk to anthesis. The development of smart irrigation systems and water management strategies is required, however relatively little research effort in this direction was noted by Lopes [75].

Our overall assessment of the consequences of climate change shows an improvement in growing conditions in northern Poland and a worsening in the central part of the country. However, the projected worsening of some climate conditions does not imply a decline in maize yields, at least according to the 2050 perspective, as pointed out by Parent et al. [76]. This trend may be positive or unchanged compared to the baseline if farms adopt upgraded practices regarding choosing varieties with optimal sowing dates and growing cycle lengths, and using the best of genetic variability suitable to the local environment. Ramirez-Cabral et al. [4] found crucial the importance of evaluating the regional climatic suitability for maize production variation under climate change. The ecoclimatic index (EI) was evaluated using a CLIMEX distribution model at the global level, assuming an A2 emissions scenario. In the current climate, the majority of Poland was identified as having medium climate suitability, with a small area in the centre indicated as having marginal suitability. In future climate conditions, the CSIRO model predicts that by 2050, the majority of Poland will be suitable for maize cultivation, with marginal climate suitability in central Poland that will increase in 2100. In contrast, the MIROC model indicates that the climate suitability for maize cultivation will be enhanced by climate change with the majority of Poland becoming optimal for this purpose by 2100 [4].

It should be noted that in our assessment we did not consider the spread of pests, weeds, or diseases among crops, which are also related to new climatic conditions, so it is relatively unknown how these major pathways of crop failure may amplify each other in the future [77]. Furthermore, our analysis did not consider non-climatic factors (e.g., topography, soil conditions and land use) or the potential genetic advancement of the species. Suitability may be significantly influenced by soil characteristics, particularly organic matter [78]. It would be beneficial for future studies to include such an analysis [4]. In our study we adapted the sowing date of maize to the predicted climatic conditions, which was not a fixed day of the year, therefore the predicted conditions for maize cultivation in relation to frost stress will improve in the future. In future studies it would be advisable to see how the sowing date affects these conditions. Karapetsas et al. [79] evaluated the effectiveness of SICSs (Soil-Improving Cropping Systems) as a long-term mitigation measure against the impacts of climate change on land suitability for maize, in addition to assessing current suitability for maize. There are some methods that can be used in combination in MCDA to provide more comprehensive assessments in agricultural land evaluation. Agrawal et al. [80] proposed a machine learning (ML)-based approach as a valuable method for assessing the suitability of agricultural land. Rangzan et al. [81] used an integration of Fuzzy logic, the Analytical Hierarchy Process (AHP) and satellite images in identifying optimal areas for wheat cultivation. AHP is a multi-criteria method based on pairwise comparison. It relies on expert judgement to determine priority scales, based on the principle that people's experience and knowledge are at least as valuable in decision making as the data they use [82]. More complexity and uncertainty can be brought to the process with large group of decision makers [83]. As the values of the criteria weights have an impact on the final suitability assessment, the choice of group and number of decision experts is challenging and varies between studies. Han et al. [23] noted that consultation with several experts is required to assign weights to selected criteria using an AHP method. Rodcha et al. [84] invited seven experts on crop plantations to determine the weights. Morales and de Vries [85] established the criteria weights with 20 experts. Some authors evaluated the weights based on previous studies, the literature and expert judgment [25,44]. Therefore, the group of experts could be expanded in future studies. A sensitivity analysis of the influence of the weights given by different respondents or different groups could be carried out in the future for a detailed analysis of the results [86].

5. Conclusions

The analysis of climate suitability for maize is important for increasing the capacity to cope with the impacts of climate change on maize production. In this study, climate suitability for maize was assessed for the baseline period and two future periods (2050s and 2080s). The climate risk indicators were assessed with possible sowing and harvesting dates set based on air temperature in order to adapt the crop to climate conditions and provide more realistic results.

A deeper understanding of the pros and cons of climate change impacts on maize development was achieved through the use of multi-criteria evaluation. The criteria chosen reflected the positive effects of climate change, i.e., the increase in thermal resources, and the negative effects related to the frequency of extreme weather events (frost, high temperatures, water stress). The use of a mean ensemble from six GCM-RCM climate simulations reduced the uncertainty in projections of further climate. Additionally, the participation of crop experts and farmers was important for incorporating their knowledge in the evaluation of criteria weights, having in mind that this evaluation is also subjective. Our main findings indicated that areas of high climate suitability for maize production in Poland are expected to increase due to increase in maturity achievement and a reduction in frost stress, with the exception of the 2080s forecast under scenario RCP8.5. In this scenario, a reduction in climate suitability is projected due to an increase in heat and water stress.

The results and methodology used can be useful in planning adaptation measures for climate change conditions. These changes in climatic conditions will require the implementation of adaptation agricultural techniques. Heat and water stress may limit the suitability of maize in the future, but these negative effects of climate change can be attenuated with new varieties that are resistant to increasingly frequent heat and drought stress. On the other hand, greater thermal resources as a result of climate change provide opportunities to grow maize varieties with higher heat requirements with less risk of not reaching maturity. Projected higher air temperatures will result in the acceleration of sowing dates and the delay of harvesting dates, affecting the potential scheduling of field operations, the possibility of introducing longer-duration cultivars, as well as increased productivity.

Author Contributions: Conceptualization, A.K.-B., J.K. and S.R.; methodology, A.K.-B., J.K. and S.R.; software, A.K.-B.; validation, A.K.-B.; formal analysis, A.K.-B.; investigation, A.K.-B. and J.K.; writing—original draft preparation, A.K.-B.; writing—review and editing, A.K.-B., J.K. and S.R.; visualization, A.K.-B.; funding acquisition, J.K. All authors have read and agreed to the published version of the manuscript.

Funding: This research was funded by Interreg Central Europe and the European Union in the framework of the project Clim4Cast (Central European Alliance for Increasing Climate Change Resilience to Combined Consequences of Drought, Heatwave, and Fire Weather through Regionally-Tuned Forecasting) grant number CE0100059.

Institutional Review Board Statement: Not applicable.

Informed Consent Statement: Not applicable.

Data Availability Statement: The raw data supporting the conclusions of this article will be made available by the authors on request.

Acknowledgments: The authors would like to thank Małgorzata Wydra for the linguistic revision of the article.

Conflicts of Interest: The authors declare no conflicts of interest.

References

1. Marcinkowski, P.; Piniewski, M.; Jefimow, M. Assessment of Projected Climate Change Impact on Agro-climatic Indicators in Poland. *Int. J. Climatol.* **2023**, *43*, 6003–6019. [CrossRef]
2. Holzkämper, A.; Calanca, P.; Fuhrer, J. Evaluating Climate Suitability for Agriculture in Switzerland. 2010. Available online: https://scholarsarchive.byu.edu/iemssconference/2010/all/610/ (accessed on 23 April 2024).
3. Gao, Y.; Zhang, A.; Yue, Y.; Wang, J.; Su, P. Predicting Shifts in Land Suitability for Maize Cultivation Worldwide Due to Climate Change: A Modeling Approach. *Land* **2021**, *10*, 295. [CrossRef]
4. Ramirez-Cabral, N.Y.; Kumar, L.; Shabani, F. Global Alterations in Areas of Suitability for Maize Production from Climate Change and Using a Mechanistic Species Distribution Model (CLIMEX). *Sci. Rep.* **2017**, *7*, 5910. [CrossRef] [PubMed]
5. Wang, X.; Li, X.; Lou, Y.; You, S.; Zhao, H. Refined Evaluation of Climate Suitability of Maize at Various Growth Stages in Major Maize-Producing Areas in the North of China. *Agronomy* **2024**, *14*, 344. [CrossRef]
6. Lopez-Blanco, J.; Pérez-Damián, J.L.; Conde-Álvarez, A.C.; Gómez-Díaz, J.D.; Monterroso-Rivas, A.I. Land Suitability Levels for Rainfed Maize under Current Conditions and Climate Change Projections in Mexico. *Outlook Agric.* **2018**, *47*, 181–191. [CrossRef]
7. Focker, M.; van Eupen, M.; Verweij, P.; Liu, C.; van Haren, C.; van der Fels-Klerx, H.J. Effects of Climate Change on Areas Suitable for Maize Cultivation and Aflatoxin Contamination in Europe. *Toxins* **2023**, *15*, 599. [CrossRef] [PubMed]
8. Jia, C.; Luo, X.; Zhou, W.; Chen, Y.; Sun, G. Evaluation of Suitability Areas for Maize in China Based on GIS and Its Variation Trend on the Future Climate Condition. In *Ecosystem Assessment and Fuzzy Systems Management. Advances in Intelligent Systems and Computing*; Cao, B.-Y., Ma, S.-Q., Cao, H.-H., Eds.; Springer: Cham, Switzerland, 2014; Volume 254, pp. 285–299.
9. Nieróbca, A.; Kozyra, J.; Pudełko, R. Evaluation of Meteorological Conditions for the Cultivation of Maize Grain in Poland. *Acta Agrophys* **2010**, *6*, 45–55.
10. The Food and Agriculture Organization (FAO) Crops and Livestock Products. Available online: https://www.fao.org/faostat/en/#data/QCL (accessed on 10 May 2024).
11. Statistics Poland Production of Agricultural and Horticultural Crops in 2023. Available online: https://stat.gov.pl/obszary-tematyczne/rolnictwo-lesnictwo/uprawy-rolne-i-ogrodnicze/produkcja-upraw-rolnych-i-ogrodniczych-w-2023-roku,9,22.html (accessed on 3 June 2024).

12. Trnka, M.; Olesen, J.E.; Kersebaum, K.C.; Skjelvåg, A.O.; Eitzinger, J.; Seguin, B.; Peltonen-Sainio, P.; Rötter, R.; Iglesias, A.; Orlandini, S.; et al. Agroclimatic Conditions in Europe under Climate Change. *Glob. Change Biol.* **2011**, *17*, 2298–2318. [CrossRef]
13. Nendel, C.; Reckling, M.; Debaeke, P.; Schulz, S.; Berg-Mohnicke, M.; Constantin, J.; Fronzek, S.; Hoffmann, M.; Jakšić, S.; Kersebaum, K.-C.; et al. Future Area Expansion Outweighs Increasing Drought Risk for Soybean in Europe. *Glob. Change Biol.* **2023**, *29*, 1340–1358. [CrossRef] [PubMed]
14. Kozyra, J.; Borek, R.; Matyka, M.; Wawer, R.; Pudełko, R.; Kozak, M.; Nieróbca, A.; Król, A.; Żyłowski, T.; Jadczyszyn, T.; et al. Praktyki Rolnicze Dla Rolnictwa Niskoemisyjnego Dostosowanego Do Obserwowanych i Prognozowanych Zmian Klimatu w Polsce Oceniane w Projekcie LCAgri. In *Wyznaczenie Uzupełniających i Nowych Obszarów Badawczych w Zakresie Ochrony Środowiska i Zmian Klimatu w Sektorze Rolnictwa*; Walczak, J., Krawczyk, W., Eds.; Instytut Zootechniki PIB: Kraków, Poland, 2017; pp. 31–53. ISBN 978-83-7607-255-5.
15. Grigorieva, E.; Livenets, A.; Stelmakh, E. Adaptation of Agriculture to Climate Change: A Scoping Review. *Climate* **2023**, *11*, 202. [CrossRef]
16. Akıncı, H.; Özalp, A.Y.; Turgut, B. Agricultural Land Use Suitability Analysis Using GIS and AHP Technique. *Comput. Electron. Agric.* **2013**, *97*, 71–82. [CrossRef]
17. Pilevar, A.R.; Matinfar, H.R.; Sohrabi, A.; Sarmadian, F. Integrated Fuzzy, AHP and GIS Techniques for Land Suitability Assessment in Semi-Arid Regions for Wheat and Maize Farming. *Ecol. Indic.* **2020**, *110*, 105887. [CrossRef]
18. Dedeoğlu, M.; Dengiz, O. Generating of Land Suitability Index for Wheat with Hybrid System Aproach Using AHP and GIS. *Comput. Electron. Agric.* **2019**, *167*, 105062. [CrossRef]
19. Malczewski, J. GIS-Based Land-Use Suitability Analysis: A Critical Overview. *Prog. Plan.* **2004**, *62*, 3–65. [CrossRef]
20. Ramamurthy, V.; Reddy, G.O.; Kumar, N. Assessment of Land Suitability for Maize (*Zea mays* L.) in Semi-Arid Ecosystem of Southern India Using Integrated AHP and GIS Approach. *Comput. Electron. Agric.* **2020**, *179*, 105806. [CrossRef]
21. Habibie, M.I.; Noguchi, R.; Shusuke, M.; Ahamed, T. Land Suitability Analysis for Maize Production in Indonesia Using Satellite Remote Sensing and GIS-Based Multicriteria Decision Support System. *GeoJournal* **2021**, *86*, 777–807. [CrossRef]
22. Ceballos-Silva, A.; Lopez-Blanco, J. Delineation of Suitable Areas for Crops Using a Multi-Criteria Evaluation Approach and Land Use/Cover Mapping: A Case Study in Central Mexico. *Agric. Syst.* **2003**, *77*, 117–136. [CrossRef]
23. Han, C.; Chen, S.; Yu, Y.; Xu, Z.; Zhu, B.; Xu, X.; Wang, Z. Evaluation of Agricultural Land Suitability Based on RS, AHP, and MEA: A Case Study in Jilin Province, China. *Agriculture* **2021**, *11*, 370. [CrossRef]
24. Rahaman, S.A.; Aruchamy, S. Land Suitability Evaluation of Tea (*Camellia Sinensis* L.) Plantation in Kallar Watershed of Nilgiri Bioreserve, India. *Geographies* **2022**, *2*, 701–723. [CrossRef]
25. Layomi Jayasinghe, S.; Kumar, L.; Sandamali, J. Assessment of Potential Land Suitability for Tea (*Camellia sinensis* (L.) O. Kuntze) in Sri Lanka Using a GIS-Based Multi-Criteria Approach. *Agriculture* **2019**, *9*, 148. [CrossRef]
26. Król-Badziak, A.; Kozyra, J.; Rozakis, S. Assessment of Suitability Area for Maize Production in Poland Related to the Climate Change and Water Stress. *Sustainability* **2024**, *16*, 852. [CrossRef]
27. Kundzewicz, Z.W.; Matczak, P. Climate Change Regional Review: Poland. *Wiley Interdiscip. Rev. Clim. Change* **2012**, *3*, 297–311. [CrossRef]
28. Ustrnul, Z.; Marosz, M.; Biernacik, D.; Walus, K.; Chodubska, A.; Wasielewska, K.; Kusek, K. *Climate of Poland in 2022*; IMGW-PIB: Warszawa, Poland, 2023.
29. Tomczyk, A.M.; Bednorz, E. *Atlas Klimatu Polski (1991–2020)*; Bogucki Wydawnictwo Naukowe: Poznań, Poland, 2022; ISBN 83-7986-415-8.
30. Ghazi, B.; Przybylak, R.; Pospieszyńska, A. Projection of Climate Change Impacts on Extreme Temperature and Precipitation in Central Poland. *Sci. Rep.* **2023**, *13*, 18772. [CrossRef] [PubMed]
31. Kundzewicz, Z.W.; Piniewski, M.; Mezghani, A.; Okruszko, T.; Pińskwar, I.; Kardel, I.; Hov, Ø.; Szcześniak, M.; Szwed, M.; Benestad, R.E. Assessment of Climate Change and Associated Impact on Selected Sectors in Poland. *Acta Geophys.* **2018**, *66*, 1509–1523. [CrossRef]
32. Mesta, B.; Kentel, E. Superensembles of Raw and Bias-adjusted Regional Climate Models for Mediterranean Region, Turkey. *Int. J. Climatol.* **2022**, *42*, 2566–2585. [CrossRef]
33. Benestad, R.; Buonomo, E.; Gutiérrez, J.M.; Haensler, A.; Hennemuth, B.; Illy, T.; Jacob, D.; Keup-Thiel, E.; Katragkou, E.; Kotlarski, S. Guidance for EURO-CORDEX Climate Projections Data Use. Version1.0—2017.08. 2017. Available online: https://www.euro-cordex.net/imperia/md/content/csc/cordex/euro-cordex-guidelines-version1.0-2017.08.pdf (accessed on 13 January 2020).
34. Yang, W.; Andréasson, J.; Phil Graham, L.; Olsson, J.; Rosberg, J.; Wetterhall, F. Distribution-Based Scaling to Improve Usability of Regional Climate Model Projections for Hydrological Climate Change Impacts Studies. *Hydrol. Res.* **2010**, *41*, 211–229. [CrossRef]
35. Landelius, T.; Dahlgren, P.; Gollvik, S.; Jansson, A.; Olsson, E. A High-resolution Regional Reanalysis for Europe. Part 2: 2D Analysis of Surface Temperature, Precipitation and Wind. *Q. J. R. Meteorol. Soc.* **2016**, *142*, 2132–2142. [CrossRef]
36. Schulzweida, U. CDO User's Guide. Climate Data Operators. Version 1.7.2; MPI for Meteorology: Hamburg, Germany, 2016.
37. Moss, R.H.; Edmonds, J.A.; Hibbard, K.A.; Manning, M.R.; Rose, S.K.; Van Vuuren, D.P.; Carter, T.R.; Emori, S.; Kainuma, M.; Kram, T. The next Generation of Scenarios for Climate Change Research and Assessment. *Nature* **2010**, *463*, 747–756. [CrossRef]
38. Thomson, A.M.; Calvin, K.V.; Smith, S.J.; Kyle, G.P.; Volke, A.; Patel, P.; Delgado-Arias, S.; Bond-Lamberty, B.; Wise, M.A.; Clarke, L.E.; et al. RCP4.5: A Pathway for Stabilization of Radiative Forcing by 2100. *Clim. Change* **2011**, *109*, 77–94. [CrossRef]

39. Kotlarski, S.; Keuler, K.; Christensen, O.B.; Colette, A.; Déqué, M.; Gobiet, A.; Goergen, K.; Jacob, D.; Lüthi, D.; Van Meijgaard, E. Regional Climate Modeling on European Scales: A Joint Standard Evaluation of the EURO-CORDEX RCM Ensemble. *Geosci. Model Dev.* **2014**, *7*, 1297–1333. [CrossRef]
40. Holzkämper, A.; Fossati, D.; Hiltbrunner, J.; Fuhrer, J. Spatial and Temporal Trends in Agro-Climatic Limitations to Production Potentials for Grain Maize and Winter Wheat in Switzerland. *Reg. Environ. Change* **2015**, *15*, 109–122. [CrossRef]
41. Zaliwski, A.; Górski, T. Prawdopodobieństwo Dojrzewania Kukurydzy-Aplikacja Internetowa. *Inż. Rol.* **2005**, *8*, 401–408.
42. FAO (Food and Agriculture Organization). *A Framework for Land Evaluation*; Soils Bulletin; FAO: Rome, Italy, 1976; Volume 32.
43. Moisa, M.B.; Merga, B.B.; Gemeda, D.O. Land Suitability Evaluation for Surface Irrigation Using Geographic Information System: A Case Study in Didessa River Sub-Basin, Western Ethiopia. *Sustain. Water Resour. Manag.* **2022**, *8*, 82. [CrossRef]
44. Muluneh, A.; Tadesse, T.; Girma, R. Assessing Potential Land Suitable for Surface Irrigation Using GIS and AHP Techniques in the Rift Valley Lakes Basin, Ethiopia. *Sustain. Water Resour. Manag.* **2022**, *8*, 46. [CrossRef]
45. Księżak, J.; Brzóska, F.; Kozyra, J.; Kęsik, K.; Madej, A.; Nieróbca, A.; Hołubowicz-Kliza, G.; Piecuch, K.; Janiak, W. *Uprawa Kukurydzy Na Kiszonkę z Całych Roślin*; Wydawnictwo IUNG-PIB: Puławy, Poland, 2015; ISBN 83-7562-183-8.
46. Zhang, Z.; Zhou, M.; Ou, G.; Tan, S.; Song, Y.; Zhang, L.; Nie, X. Land Suitability Evaluation and an Interval Stochastic Fuzzy Programming-Based Optimization Model for Land-Use Planning and Environmental Policy Analysis. *Int. J. Environ. Res. Public. Health* **2019**, *16*, 4124. [CrossRef] [PubMed]
47. Łabędzki, L. Potrzeby i Stan Nawodnień w Województwie Kujawsko-Pomorskim [Irrigation Needs and Condition in the Kuyavian-Pomeranian Voivodeship]. In Proceedings of the Warsztaty dla interesariuszy projektu OPERA [Workshops for the Stakeholders of the OPERA Project], Minikowo, Poland, 27 November 2017.
48. Marcinkowski, P.; Piniewski, M. Effect of Climate Change on Sowing and Harvest Dates of Spring Barley and Maize in Poland. *Int. Agrophysics* **2018**, *32*, 265–271. [CrossRef]
49. Dížková, P.; Bartošová, L.; Bláhová, M.; Balek, J.; Hájková, L.; Semerádová, D.; Bohuslav, J.; Pohanková, E.; Žalud, Z.; Trnka, M. Modeling Phenological Phases of Winter Wheat Based on Temperature and the Start of the Growing Season. *Atmosphere* **2022**, *13*, 1854. [CrossRef]
50. Sánchez, B.; Rasmussen, A.; Porter, J.R. Temperatures and the Growth and Development of Maize and Rice: A Review. *Glob. Change Biol.* **2014**, *20*, 408–417. [CrossRef] [PubMed]
51. Kaniuczak, Z.; Pruszyński, S. *Integrowana Produkcja Kukurydzy*; Instytut Ochrony Roślin: Poznań, Poland, 2007.
52. Sys, C.; Van Ranst, E.; Debaveye, J.; Beernaert, F. *Land Evaluation. Part III: Crop Requirements. Agricultural Publications N° 7*; GADC: Brussels, Belgium, 1993; 191p.
53. Doorenbos, J.; Kassam, A.H. Yield Response to Water. In *FAO Irrigation and Drainage Paper No. 33*; FAO: Rome, Italy, 1979; Volume 33.
54. Goswami, J.; Sharma, V.; Chaudhury, B.U.; Raju, P.L.N. Rapid Identification of Abiotic Stress (Frost) in in-Filed Maize Crop Using UAV Remote Sensing. *Int. Arch. Photogramm. Remote Sens. Spat. Inf. Sci.* **2019**, *42*, 467–471. [CrossRef]
55. Zohner, C.M.; Mo, L.; Renner, S.S.; Svenning, J.-C.; Vitasse, Y.; Benito, B.M.; Ordonez, A.; Baumgarten, F.; Bastin, J.-F.; Sebald, V.; et al. Late-Spring Frost Risk between 1959 and 2017 Decreased in North America but Increased in Europe and Asia. *Proc. Natl. Acad. Sci. USA* **2020**, *117*, 12192–12200. [CrossRef]
56. Birch, C.J.; Robertson, M.J.; Humphreys, E.; Hutchins, N. Agronomy of Maize in Australia: In Review and Prospect. In Proceedings of the Versatile Maize-Golden Opportunities: 5th Australian Maize Conference, Toowoomba, QLD, Australia, 18–20 February 2003; pp. 45–57.
57. Holzkämper, A.; Calanca, P.; Fuhrer, J. Identifying Climatic Limitations to Grain Maize Yield Potentials Using a Suitability Evaluation Approach. *Agric. For. Meteorol.* **2013**, *168*, 149–159. [CrossRef]
58. Łabędzki, L.; Bąk, B. Meteorological and Agricultural Drought Indices Used in Drought Monitoring in Poland: A Review. *Meteorol. Hydrol. Water Manag.* **2014**, *2*, 3–13. [CrossRef]
59. Doroszewski, A.; Górski, T. Prosty Wskaznik Ewapotranspiracji Potencjalnej. *Rocz. Akad. Rol. W Poznaniu Melior. Inż. Śr.* **1995**, *16*, 3–8.
60. Żyłowska, K. *Raport Końcowy z Tematu Badawczego Nr .4.03. Pt Warunki Agroklimatyczne w Polsce Według Scenariuszy Klimatycznych Na Lata 2030, 2050, 2080 Zrealizowanego w Ramach Działalności Statutowej w Podprogramie Badawczym Nr. 4.03*; IUNG-PIB: Puławy, Poland, 2015; pp. 1–107.
61. Mishra, A.K.; Deep, S.; Choudhary, A. Identification of Suitable Sites for Organic Farming Using AHP & GIS. *Egypt. J. Remote Sens. Space Sci.* **2015**, *18*, 181–193.
62. Saaty, T.L. How to Make a Decision: The Analytic Hierarchy Process. *Interfaces* **1994**, *24*, 19–43. [CrossRef]
63. Saaty, T.L.; Sodenkamp, M. The Analytic Hierarchy and Analytic Network Measurement Processes: The Measurement of Intangibles: Decision Making under Benefits, Opportunities, Costs and Risks. In *Handbook of Multicriteria Analysis*; Zopounidis, C., Pardalos, P.M., Eds.; Springer: Berlin, Germany, 2010; pp. 91–166.
64. Pauer, F.; Schmidt, K.; Babac, A.; Damm, K.; Frank, M.; von der Schulenburg, J.-M.G. Comparison of Different Approaches Applied in Analytic Hierarchy Process–An Example of Information Needs of Patients with Rare Diseases. *BMC Med. Inform. Decis. Mak.* **2016**, *16*, 117. [CrossRef]
65. Forman, E.; Peniwati, K. Aggregating Individual Judgments and Priorities with the Analytic Hierarchy Process. *Eur. J. Oper. Res.* **1998**, *108*, 165–169. [CrossRef]

66. Parra-Lopez, C.; Calatrava-Requena, J.; de-Haro-Gimenez, T. A Multi-Criteria Evaluation of the Environmental Performances of Conventional, Organic and Integrated Olive-Growing Systems in the South of Spain Based on Experts' Knowledge. *Renew. Agric. Food Syst.* **2007**, *22*, 189–203. [CrossRef]
67. Srdjevic, B. Linking Analytic Hierarchy Process and Social Choice Methods to Support Group Decision-Making in Water Management. *Decis. Support Syst.* **2007**, *42*, 2261–2273. [CrossRef]
68. Schipper, E.L.F. Maladaptation: When Adaptation to Climate Change Goes Very Wrong. *One Earth* **2020**, *3*, 409–414. [CrossRef]
69. Szyga-Pluta, K. Assessment of Changing Agroclimatic Conditions in Poland Based on Selected Indicators. *Atmosphere* **2022**, *13*, 1232. [CrossRef]
70. Olesen, J.E.; Trnka, M.; Kersebaum, K.C.; Skjelvåg, A.O.; Seguin, B.; Peltonen-Sainio, P.; Rossi, F.; Kozyra, J.; Micale, F. Impacts and Adaptation of European Crop Production Systems to Climate Change. *Eur. J. Agron.* **2011**, *34*, 96–112. [CrossRef]
71. Żarski, J.; Kuśmierek-Tomaszewska, R.; Dudek, S. Trends of Changes in Climate Risk of Grain Maize Cultivation in the Bydgoszcz Region. *Infrastruktura Ekol. Teren. Wiej.* **2016**, 725–735. [CrossRef]
72. Charalampopoulos, I. Agrometeorological Conditions and Agroclimatic Trends for the Maize and Wheat Crops in the Balkan Region. *Atmosphere* **2021**, *12*, 671. [CrossRef]
73. Pinke, Z.; Decsi, B.; Kardos, M.K.; Kern, Z.; Kozma, Z.; Pásztor, L.; Ács, T. Changing Patterns of Soil Water Content and Relationship with National Wheat and Maize Production in Europe. *Eur. J. Agron.* **2022**, *140*, 126579. [CrossRef]
74. Gobin, A.; Addimando, N.; Ramshorn, C.; Gutbrod, K. Climate Risk Services for Cereal Farming. *Adv. Sci. Res.* **2021**, *18*, 21–25. [CrossRef]
75. Lopes, M.S. Will Temperature and Rainfall Changes Prevent Yield Progress in Europe? *Food Energy Secur.* **2022**, *11*, e372. [CrossRef]
76. Parent, B.; Leclere, M.; Lacube, S.; Semenov, M.A.; Welcker, C.; Martre, P.; Tardieu, F. Maize Yields over Europe May Increase in Spite of Climate Change, with an Appropriate Use of the Genetic Variability of Flowering Time. *Proc. Natl. Acad. Sci. USA* **2018**, *115*, 10642–10647. [CrossRef]
77. Abramoff, R.Z.; Ciais, P.; Zhu, P.; Hasegawa, T.; Wakatsuki, H.; Makowski, D. Adaptation Strategies Strongly Reduce the Future Impacts of Climate Change on Simulated Crop Yields. *Earths Future* **2023**, *11*, e2022EF003190. [CrossRef]
78. Bazkiaee, P.A.; Kamkar, B.; Amiri, E.; Kazemi, H.; Rezaei, M.; Araji, H.A. Multi-Criteria GIS-Based Land Suitability Analysis for Rice Cultivation: A Case Study in Guilan Province, Iran. *Environ. Monit. Assess.* **2024**, *196*, 680. [CrossRef]
79. Karapetsas, N.; Gobin, A.; Bilas, G.; Koutsos, T.M.; Pavlidis, V.; Katragkou, E.; Alexandridis, T.K. Analysis of Land Suitability for Maize Production under Climate Change and Its Mitigation Potential through Crop Residue Management. *Land* **2024**, *13*, 63. [CrossRef]
80. Agrawal, N.; Govil, H.; Kumar, T. Agricultural Land Suitability Classification and Crop Suggestion Using Machine Learning and Spatial Multicriteria Decision Analysis in Semi-Arid Ecosystem. *Environ. Dev. Sustain.* **2024**, 1–38. [CrossRef]
81. Rangzan, K.; Kabolizadeh, M.; Zaheri Abdehvand, Z.; Karimi, D.; Jafarnejadi, A.; Mokarram, M. Optimized Land Suitability Mapping for Wheat Cultivation by Integrating Fuzzy Hierarchical Analysis and Satellite Images. *J. Indian Soc. Remote Sens.* **2024**, *52*, 1135–1151. [CrossRef]
82. Vargas, L.G. An Overview of the Analytic Hierarchy Process and Its Applications. *Eur. J. Oper. Res.* **1990**, *48*, 2–8. [CrossRef]
83. Chao, X.; Kou, G.; Peng, Y.; Viedma, E.H. Large-Scale Group Decision-Making with Non-Cooperative Behaviors and Heterogeneous Preferences: An Application in Financial Inclusion. *Eur. J. Oper. Res.* **2021**, *288*, 271–293. [CrossRef]
84. Rodcha, R.; Tripathi, N.K.; Prasad Shrestha, R. Comparison of Cash Crop Suitability Assessment Using Parametric, AHP, and FAHP Methods. *Land* **2019**, *8*, 79. [CrossRef]
85. Morales Jr, F.; de Vries, W.T. Establishment of Land Use Suitability Mapping Criteria Using Analytic Hierarchy Process (AHP) with Practitioners and Beneficiaries. *Land* **2021**, *10*, 235. [CrossRef]
86. Guchhait, R.; Sarkar, B. A Decision-Making Problem for Product Outsourcing with Flexible Production under a Global Supply Chain Management. *Int. J. Prod. Econ.* **2024**, *272*, 109230. [CrossRef]

Disclaimer/Publisher's Note: The statements, opinions and data contained in all publications are solely those of the individual author(s) and contributor(s) and not of MDPI and/or the editor(s). MDPI and/or the editor(s) disclaim responsibility for any injury to people or property resulting from any ideas, methods, instructions or products referred to in the content.

sustainability

MDPI

Article

Evaluating the Yields of the Rainfed Potato Crop under Climate Change Scenarios Using the AquaCrop Model in the Peruvian Altiplano

Jesus Puma-Cahua [1], Germán Belizario [1,2], Wilber Laqui [1,2,*], Roberto Alfaro [1,2], Edilberto Huaquisto [1,2] and Elmer Calizaya [2,3]

1 Escuela Profesional de Ingeniería Agrícola, Universidad Nacional del Altiplano, Puno 21001, Peru; jsyosef@gmail.com (J.P.-C.); gbelizario@unap.edu.pe (G.B.); ralfaro@unap.edu.pe (R.A.); ehuaquisto@unap.edu.pe (E.H.)

2 Escuela de Posgrado, Universidad Nacional del Altiplano, Puno 21001, Peru; ecalizaya@unap.edu.pe

3 Escuela Profesional de Ingeniería Topográfica y Agrimensura, Universidad Nacional del Altiplano, Puno 21001, Peru

* Correspondence: wlaqui@unap.edu.pe; Tel.: +51-990-088-187

Abstract: Ensuring global food security and adapting to the challenges posed by climate change, particularly in rainfed agriculture, are paramount concerns. This research investigates the impacts of climate change on the yield of the potato crop variety Imilla Negra (*Solanum tuberosum* spp.) under the extreme climatic conditions of the Peruvian Altiplano. From the experimentation in six crop plots under a rainfed agricultural system, periodic crop growth parameter measurements were obtained from 2017 to 2018. The results showed a good performance of the AquaCrop model in the calibration and validation, successfully simulating crop growth and yield parameters. Climate projections showed precipitation decreases and temperature and evapotranspiration increases for the representative concentration pathway (RCP), RCP 4.5, and RCP 8.5 scenarios in 2023–2050. A comparison of crop yields between the base period (2006–2021) and the period 2023–2037 showed no significant changes, whereas a more considerable decrease was observed for the period 2038–2050. It is concluded that climate change generates moderate impacts on potato crop yields under the rainfed agricultural system in the Peruvian Altiplano due to the average reduction in precipitation.

Keywords: climate change; crop yields; sustainable; AquaCrop; food security; rainfed agriculture; Altiplano

1. Introduction

Global warming is the increase in the Earth's temperature and has a significant impact on water supply; over the last 33 years, water demand has increased considerably, while water supply has been precarious [1]. This distribution of water resources will be further unbalanced at the river basin level, where agricultural production will be influenced by these complex movements of the water deficit [2], by significant projected climate changes, including unprecedented global mean surface temperature increases since the mid-20th century (1.5 to 5.0 °C) [3], and reduction in precipitation rate in summer (−25%) [4], leading to a higher frequency of these extreme weather events [5] attributed to climate change, which is defined as the change in climatic elements in a given period [6].

Climatic variables, mainly precipitation and temperatures, are changing [7], affecting crop growth and productivity [8]. These effects mostly turn out to be adverse, becoming a major global threat during the dry season due to increased temperatures and decreased precipitation [9]. In developing countries, this situation is aggravated by the fact that agricultural activities are more sensitive and vulnerable to climatic conditions and climate change, as well as by different resource, technological, and institutional constraints [10].

In addition, climate change and precipitation shortages limit agricultural production and significantly impact rainfed crop production [11]. The increase in temperature threshold and water scarcity affect the development of the different phenological phases [12,13] in potato crops, mainly in the formation of tubers [12].

Climate change is one of the world's most crucial concerns. It is a significant threat to the global food supply, mainly in family or subsistence farming [13], because it influences the primary conditions of agriculture [14], which are carried out with scarce resources, directly affecting the quality of life of farmers [15]. Agriculture is the mainstay of the economy and contributes to the food security and employment of rural households [16]; however, under the conditions of climate change and variability in the availability of water resources, adopting efficient irrigation systems, cultivating crops better suited to withstand drought conditions, implementing sustainable agriculture practices, adopting the strategy of climate-resilient sustainable agriculture (CRSA), and reducing our reliance on rainfall will be necessary to address this challenge to ensure crop production and guarantee food security [1,17,18]. Also, sustainable agriculture provides a potential solution to enable agricultural systems to feed a growing population while successfully operating within the changing environmental conditions [19]. Rainfed agriculture is currently one of the main economic activities most affected by climate change due to its significant social and economic consequences for human well-being [20].

Many approaches have been used to investigate the impacts of climate change on agriculture [21] using the AquaCrop model in a variety of seasons and locales; in all of them, good model performances were obtained [22,23]. However, the Peruvian Altiplano has particular characteristics, where rainfed agriculture is the principal agricultural production system and most vulnerable to climate change [24], so it is essential to study the impacts of climate change and to develop and improve practical assessment tools, which are crucial to reducing uncertainty in agriculture [25]. Therefore, in this research, we propose and evaluate the yields of the rainfed potato crop under climate change scenarios using the AquaCrop model in the conditions of the Peruvian Altiplano.

2. Materials and Methods

2.1. Study Area

The experiment was carried out in the rural community of San Martín, in the district and province of Azangaro, in the department of Puno in the Peruvian Altiplano. The field is located at 14° 04′ 05″ S and 70° 25′ 25.6″ W, with an altitude of 4315 m above sea level (Figure 1). It has a semi-arid climate typical of the northern Altiplano, characterized by a short rainy season (December–March) and a long dry season (April–November) [26,27]. According to Azangaro station data, the annual total precipitation is 597.8 mm, 67% of which occurs in the rainy season and 33% in the dry season. The yearly average temperature is 8.8 °C, with the minimum temperature recorded in July at 5.6 °C and the maximum in November at 10.8 °C (Figure 2).

2.2. Experimental Details

Six experimental plots of potato cultivation, variety Imilla Negra *(Solanum tuberosum* spp.), with the characteristics detailed in Table 1, were determined. The sowing date was 4 November 2017, and the harvest date was 2 April 2018, with a vegetative period of 151 days. Given that potato cultivation in the Peruvian Altiplano is mainly developed under the rainfed agriculture system, a single treatment for potato cultivation called rainfed potato crop (RPC) was determined, as was also carried out in previous research [22,28,29]. Sixteen canopy cover assessments were made during the growing season, with six replicates, obtaining 96 canopy cover data during the crop growing season. At the end of the growing season, potato crop yields in fresh weight (t ha^{-1}) were determined. The soil sample was collected from the study area, and the soil properties laboratory analysis is presented in Table 1.

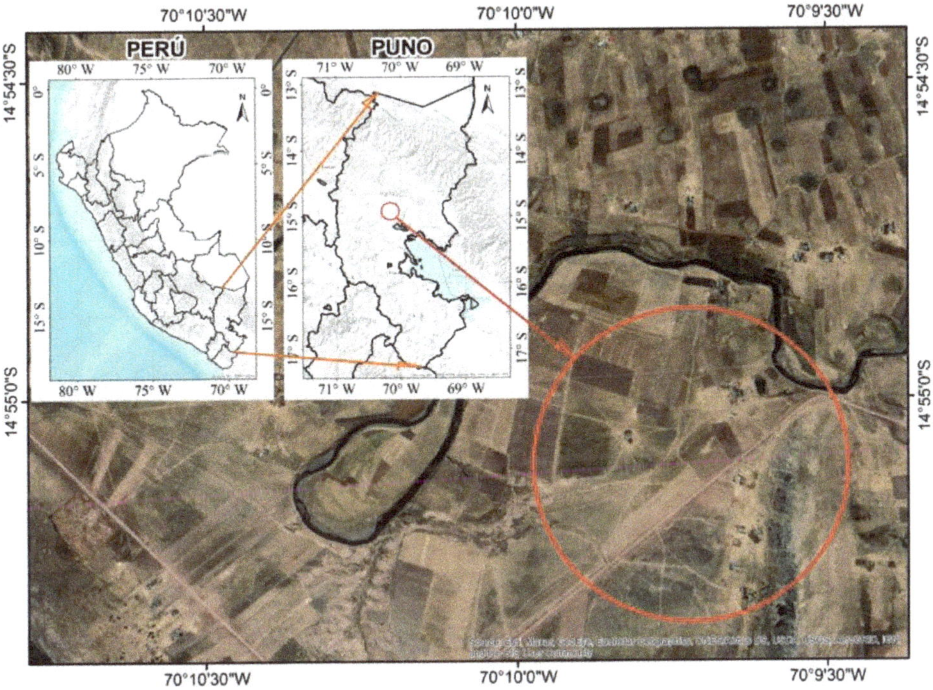

Figure 1. The location of the study area in Altiplano, Peru.

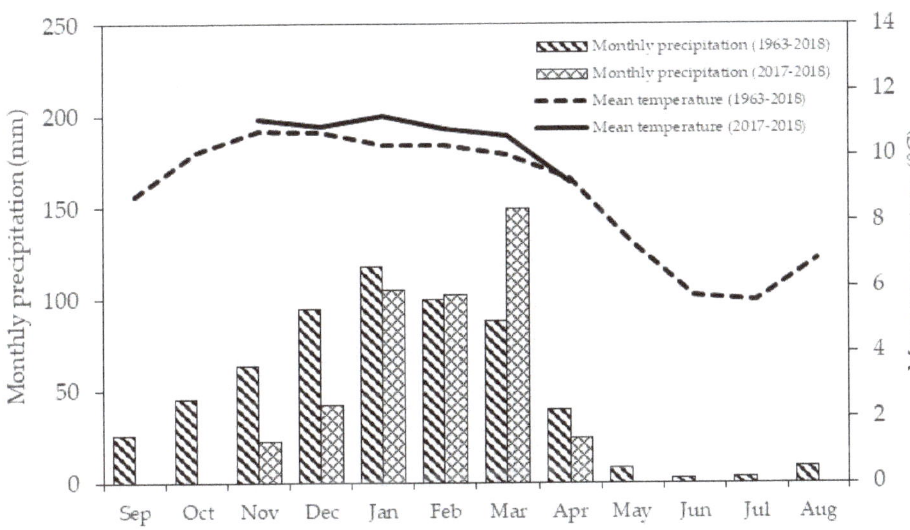

Figure 2. Historic monthly precipitation and mean temperature at the Azangaro station and data during crop growth period 2017–2018.

Table 1. Characteristics of experimental potato crop plots and soil properties.

Characteristic	Values
Total area	150.0 m^2
Number of plots	6
Plot area	25.0 m^2
Number of rows per plot	6
Furrow length	10.0 m
Row spacing	0.42 m
Plant spacing	0.40 m
Number of plants per row	25
Soil texture	sandy loam
Field capacity	27.6%
Permanent wilting point	14.2%
Bulk density	1.6 g cm^{-3}
Saturation percentage	42–49%
Hydraulic conductivity	400 mm day^{-1}
Soil depth	0–60 cm
pH	5.8
Organic matter	3.15%
Total nitrogen, phosphorus, and potassium	0.15%, 8.6 ppm, 119 ppm

2.3. Climate Data

Daily climatic data on precipitation and maximum and minimum temperature were obtained from the Azangaro station of the National Service for Meteorology and Hydrology of Peru (SENAMHI), located at 14° 54′ 52.2″ S and 70° 11′ 27.1″ W. Due to the limitation of climatic data, the Hargreaves method was used to estimate the reference evapotranspiration [30–33]. Figure 2 shows the monthly behavior of temperature and precipitation during the vegetative period of the potato crop (2017–2018).

2.4. Field Data Collection and AquaCrop Model Input Data

The setup of the AquaCrop model needs input data containing climatic parameters, crop, soil, field, and irrigation management data [23]. The daily precipitation and temperature values of Azangaro station were entered into the model, and evapotranspiration (ETo) values were also calculated using AquaCrop (ETo calculator) daily meteorological data. The canopy cover values were obtained through aerial images of plants from each repetition. The images were analyzed and processed in AutoCAD 24.2 to obtain the canopy cover. Final biomass and crop yield were obtained following maturity using samples from the experimental plot. The final aboveground biomass samples were collected by cutting the crop at the ground level and finally weighed digitally. The final yield samples were also harvested and weighed on the digital scale. Photographs of the process are shown in Supplementary Materials.

The physical and chemical properties of the soil characterized in samples taken from the study area (Table 1) were entered into the AquaCrop model. No irrigation management data was entered because it was considered a rainfed potato crop. The field management components include the soil fertility levels and weed infestation, which are the main agricultural activities in the study area [23]. Likewise, the model requires a set of soil and crop input parameters that were subsequently adjusted in the model's calibration.

2.5. AquaCrop Model Calibration and Validation

The AquaCrop model was calibrated with the observed field experiment values obtained from 2017–2018 for the rainfed potato crop (three experimental plots), according to the procedure described [34]. The field data were input variables to simulate rainfed potato crop growth and production. The fit of three growth parameters, such as canopy curve (CC), biomass (B), and crop yield (CY), obtained from the AquaCrop model, was evaluated with the field measurements made for the period 2017–2018, as also performed by [35], as well as statistics of the average annual yield of the potato crop obtained by the Regional Agrarian Directorate Puno for the period 2005–2017, using statistical evaluation criteria such as the Nash–Sutcliffe efficiency coefficient (NE), Pearson's correlation coefficient (r), normalized root mean squared error (NRMSE), and Willmott's aggregation index (d) [36].

The model was validated with the measurements of the field experiment for the 2017–2018 period (the other three experimental plots). The previously calibrated parameter values were used for the validation stage, and their evaluation was carried out using the NEC, r, NRMSE, and d performance criteria. The optimal value of NE is 1, representing a perfect fit; values of r close to 1 indicate a good association between the variables, while values below 30% are considered acceptable, and values of d close to 1 indicate a good fit [35,37,38].

2.6. Development of Climate Change Scenarios

Climate scenarios project gas emissions and temperatures for the future; the IPCC developed four climate scenarios called Representative Concentration Pathways (RCPs) (RCP2.6, RCP4.5, RCP6.0, and RCP8.5) [39]. For this research, two climate change scenarios were defined, RCP 4.5 and RCP 8.5, used in IPCC AR5. The RCP4.5 scenario corresponds to a medium stabilization scenario, and the RCP8.5 scenario corresponds to a very-high-emissions reference scenario. The precipitation and temperature projections used in this research correspond to data generated by CPTEC/INPE that are available on the climate change project platform for South America regionalized by the ETA Model (PROJETA), available at "https://projeta.cptec.inpe.br/ (accessed on 6 November 2022)" [40–42]. The projection database considers the CANESM2-ES, HadGEM2, and MIROC5 models regionalized at a 20 × 20 km spatial scale. The base period was defined as 1975–2005 and the projection period as 2023–2050.

Bias correction for daily precipitation maximum and minimum temperature projections was performed according to the procedure suggested by [43,44] for bias correction of regional climate models. From the corrected temperature projections, reference evapotranspiration was estimated using the Hargreaves method for the historical period (2006–2021) and the projected period (2023–2050). The reference evapotranspiration and precipitation were used as inputs to the AquaCrop model previously calibrated for the estimation of historical and projected potato crop production, as detailed in Figure 3.

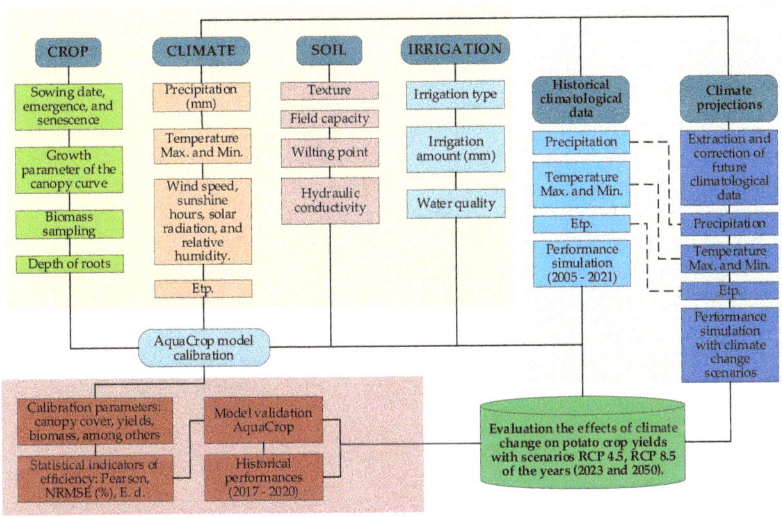

Figure 3. Methodological sequence for calibration, validation, and simulation of potato crop yield in the AquaCrop model.

3. Results

3.1. Calibration and Validation of the AquaCrop Model

The simulated canopy cover was compared with the potato crop canopy cover measurements recorded in the 2017–2018 season (Figure 4). It was observed that the values simulated with the AquaCrop model present consistency and a good fit with the field measurements of the canopy cover, obtaining an R^2 value of 0.95. Table 2 shows the performance statistics of the AquaCrop model for the calibration and validation phase. It was observed that the r, NRMSE, NEC, and d values had magnitudes of 0.98, 17.8, 0.86, and 0.96, respectively.

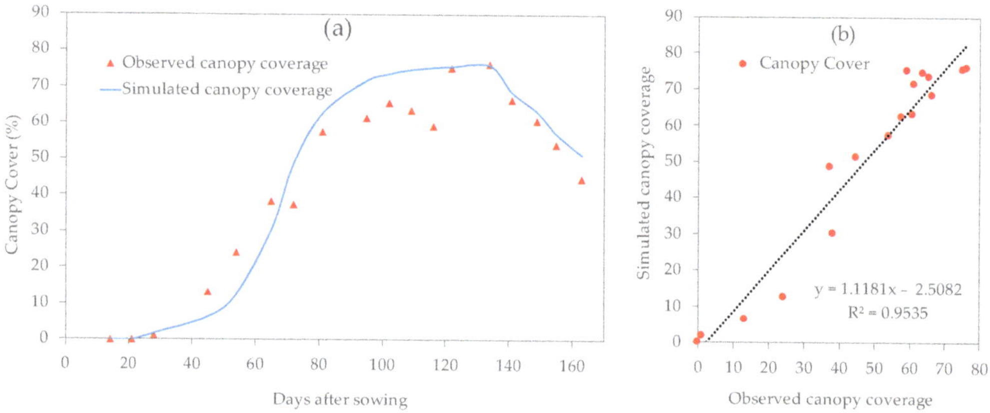

Figure 4. Calibration model: (**a**) comparison of the canopy cover simulated in the AquaCrop model and observed in the 2017–2018 period of the potato crop; (**b**) simulated vs. observed values of canopy cover.

Table 2. AquaCrop model performance statistics for canopy cover in calibration and validation phases.

Stage	r	NRMSE	NEC	d
Calibration	0.98	17.8	0.86	0.96
Validation	1.00	26.9	0.72	0.95

In the validation phase, good approximation was observed between simulated and observed canopy cover values for the 2017–2018 period recorded on 19 dates after planting. R^2 fit values of 0.99 were obtained, while the efficiency statistics of r, NRMSE, NEC, and d presented values of 1.00, 26.9, 0.72, and 0.95, respectively.

Concerning comparing observed and simulated dry aerial biomass with the AquaCrop model (Table 3), in the calibration stage, the performance statistics r, NRMSE, NE, and d presented values of 0.98, 6.4, 0.97, and 0.98, respectively, while in the validation stage, r, NRMSE, NE, and d were obtained with magnitudes of 0.96, 9.6, 0.96, and 0.98, respectively. Table 4 compares potato crop yield in dry weight observed and simulated by the AquaCrop model. In the calibration stage, a percentage difference of −0.44% was determined between the observed and simulated yield, while in the validation stage, the percentage difference was 0.87%.

Table 3. AquaCrop model performance statistics for dry aboveground biomass in the calibration and validation phases.

Stage	r	NRMSE	NEC	d
Calibration	0.98	6.4	0.97	0.98
Validation	0.96	9.6	0.96	0.98

Table 4. Comparison of observed and simulated crop yields at the harvest stage.

Stage	Performance (t ha^{-1})		Difference	
	Observed	**Simulated**	**Absolute**	**Percentual**
Calibration	11.73	11.78	−0.06	−0.44
Validation	11.12	10.57	0.53	0.87

In Figure 4, it can be seen that for the period between the emergence and tuberization stages (27 to 60 days after planting), a marked underestimation of the simulated canopy cover was observed. In comparison, an underestimation of the simulated canopy cover was observed for the period between the lodging and fruiting stages (69 to 126 days after planting). Likewise, the model slightly overestimated the canopy cover values for the period from the onset of senescence to crop maturity (134 to 175 days after sowing). Therefore, the AquaCrop model in the calibration stage overestimated canopy cover compared to the observed values. As for the validation stage (Figure 5), it was observed that in the hilling stage (69 days after planting) and the period between the yield formation and fruiting stages (91 to 126 days after planting), the AquaCrop model generated a slight overestimation of the canopy cover values. These results are consistent with the error statistics (NRMSE) obtained for the calibration and validation stages, which in both cases were positive, which is an indicator of overestimation of the simulated canopy cover values. For biomass (Table 3), the performance statistics (NRMSE) showed that the AquaCrop model slightly overestimated the simulated biomass values. At the same time, for yield, it was observed that in the calibration stage, the model slightly underestimated the simulated yield values, and the opposite for the validation stage, as can be seen in the difference values in Table 4.

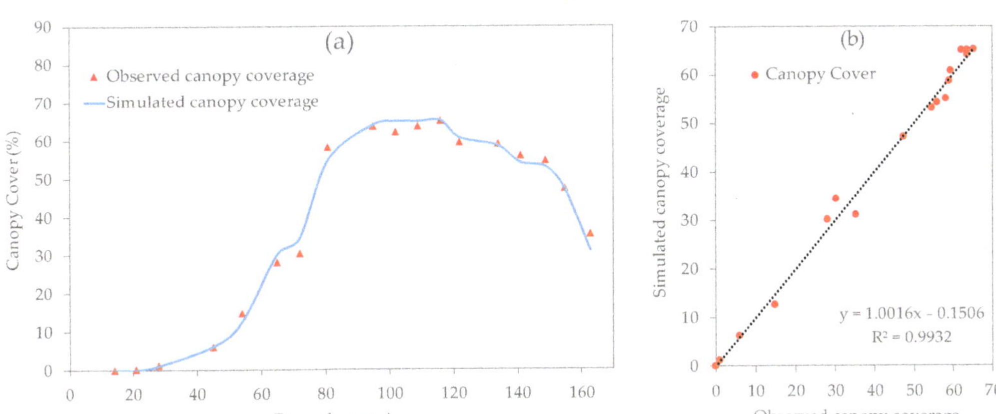

Figure 5. Validation model: (**a**) comparison of the canopy cover simulated in the AquaCrop model and observed in the 2017–2018 period of the potato crop; (**b**) simulated vs observed canopy cover values.

3.2. Precipitation Changes

The historical total annual precipitation (2006–2021) in the study area reached 610.5 mm, with yearly variations (Figure 6). For the projected period (2023–2050), using the CANESM2 model, it reached 687.07 mm and 743.66 mm per year, respectively, in the two scenarios RCP 4.5 (Figure 6a) and RCP 8.5 (Figure 6b). The average of the projected values shows a decreasing trend from December to April concerning the observed historical average values. It is deduced that the CANESM2 model behaves as an optimistic model in the two projected routes (RCP 4.5 and RCP 8.5) because its values favor the months of the agricultural season.

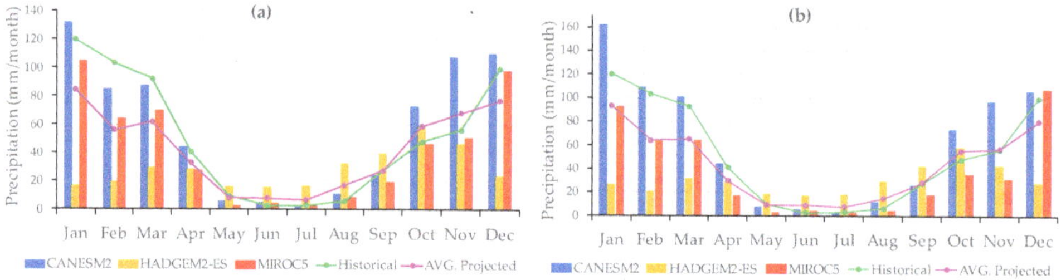

Figure 6. Projected average monthly precipitation (2023–2050) in scenarios (**a**) RCP 4.5 and (**b**) RCP 8.5. The curved lines show historical monthly precipitation summaries from January 2006 to December 2021 at the Azangaro station and the average of projected precipitation.

The values of the projected annual cumulative precipitation estimates correspond to the RCP 4.5 and RCP 8.5 scenarios (Figure 7); the former are represented by the solid lines and the latter by the dashed lines on the right margin, while the bar on the left side represents the observed historical precipitation. The vertical gray band marks the end of the observed historical precipitation graph and the beginning of the projections. The CANESM2 model in the two concentration paths (RCP 4.5 and RCP 8.5), which is represented as the most optimistic, shows relatively stable values above the historical by 120 mm per year throughout the projections; the highest peaks show the RCP 8.5 scenario. CANESM2 for the years 2023 and 2036 exceeds 900 mm per year; on the contrary, the RCP 4.5 CA-NESM2 scenario shows reductions for the year 2048, when it only reaches 350 mm per year (Figure 7). In general, the results of the average precipitation projections (2023–2050) made in the RCP 4.5 scenario estimate a decrease in precipitation from December to April, with reductions of 21.5%. For the RCP 8.5 scenario, the total rainfall for evaluating the period value reduces by 22.6% in relation to the base period.

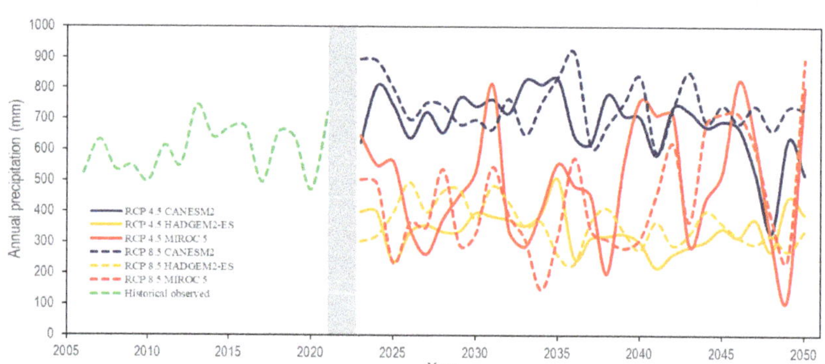

Figure 7. Historical annual precipitation (2006–2021) and projections with six climate change models for 2023–2050.

3.3. Analysis of the Change in Maximum and Minimum Temperature

Figure 8 shows the monthly behavior of the average monthly maximum temperature of the general circulation models under the representative concentration pathways RCP 4.5 (Figure 8a) and RCP 8.5 (Figure 8b) for the period 2023–2050 concerning the historical monthly maximum temperatures (2006–2021), noting that the overall projected maximum temperatures are in an average range of 17 °C to 19 °C in the case of the RCP 4.5 scenario and 17.5 °C to 19.3 °C for the representative concentration pathway RCP 8.5.

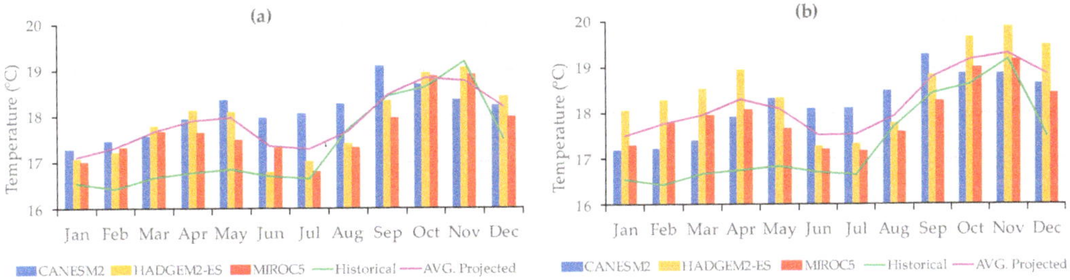

Figure 8. Average monthly maximum temperature for 2023–2050 for climate change scenarios (**a**) RCP 4.5 and (**b**) RCP 8.5, compared to historical values 2006–2021.

In general, they show more significant temperature variations because their scenarios project an increase in the average temperature of up to 13% compared to historical temperatures. The middle launched in the three models of the RCP 4.5 and RCP 8.5 routes show higher values than those recorded (Figure 8), except for November in the RCP 4.5.

For the minimum temperature, the RCP 4.5 and RCP 8.5 concentration pathways show greater increases between October and April; the CANESM2 model shows higher values in the RCP 8.5 pathway because its RCP 4.5 and RCP 8.5 scenarios project a higher average increase of up to 16% concerning the historical ones. This section indicates that all the scenarios under study are favorable since the minimum temperatures will have a relative increase, mainly in the months of the agricultural season (Figure 9). Regarding changes in mean temperature, it was observed that 2023–2050 temperature increases of 0.54 °C for the RCP 4.5 scenario and 0.88 °C for RCP 8.5 in the experimental area were observed.

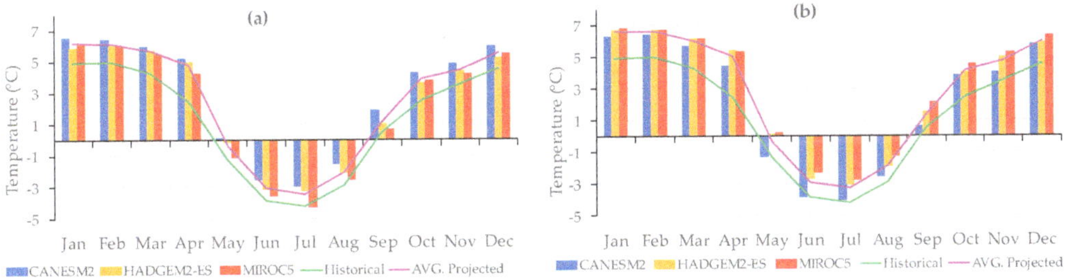

Figure 9. Average monthly minimum temperature in the period (2023–2050) in scenarios: (**a**) RCP 4.5 and (**b**) RCP 8.5 and historical (2006–2021) at the Azangaro station.

3.4. Crop Yields under Climate Change Scenarios

Figure 10 shows the simulation results using the AquaCrop model for the fresh weight yields in the concentration routes RCP 4.5 and RCP 8.5 for the period from 2023 to 2050; once these yields were adjusted in the process of calibration and validation of the AquaCrop model with the values observed during the 2017–2018 crop year, the fresh weight yields were simulated for the period 2006–2021. The new weight yields obtained remained relatively stable from 2023 to 2027 under two concentration pathways, RCP

4.5 and RCP 8.5, achieving an average annual result of 11.26 t ha^{-1}. For the second half (2038–2050), a decrease in potato crop yield is observed for the six future climate change scenarios, according to the AquaCrop model, reaching an annual average of 10.84 t ha^{-1}. In most years, it can be indicated that precipitation forecasts help to moderate the effects of temperature increases; however, there is a tendency for yields to fall in the most extreme years.

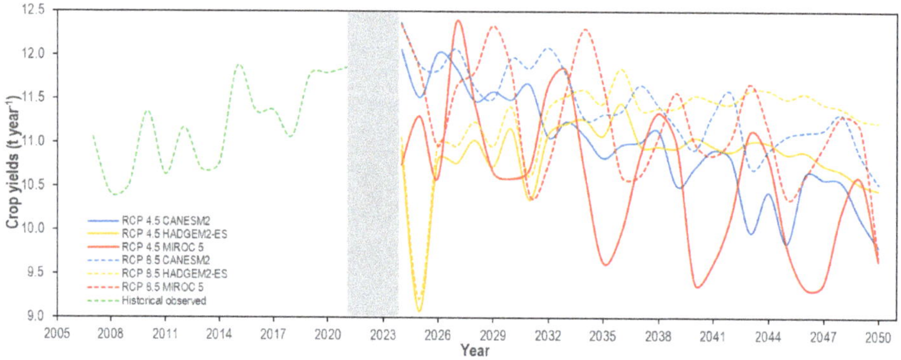

Figure 10. Annual crop yields for 2023–2050 in the climate change scenarios RCP 4.5 and RCP 8.5, compared to historical values 2006–2021.

The highest potato crop yield is in the RCP 4.5 MIROC 5 scenario (12.382 t ha^{-1}) and the lowest is for the RCP 4.5 HADGEM2-ES scenario (9.077 t ha^{-1}), where the trends of the results are favorable for the RCP 8.5 scenario and harmful for the RCP 4.5 scenario (Table 5). Under these future climate conditions, a slight negative impact on potato crop production in the Peruvian Altiplano was observed.

Table 5. Percentage change in simulated yields (2023–2050) compared to historical (2006–2021).

Routes	Scenarios		
	CANESM2	**HADGEM2-ES**	**MIROC 5**
RCP 4.5	−1.90%	−2.93%	−5.40%
RCP 8.5	2.30%	0.92%	0.49%

4. Discussion

The efficiency results of the simulation of rainfed potato crop development and production in the AquaCrop model in the calibration and validation stages showed satisfactory results. Despite slight over- and under-estimates of the model in canopy cover, biomass, and yield, which were also reported by other studies for the cultivation of potatoes [23,35], these results suggest that the AquaCrop model can satisfactorily simulate the development and production of rainfed potato crops under the extreme climatic conditions of the Peruvian Altiplano, since it simulates canopy cover, dry aerial biomass, and crop yield very closely to the observed values. The results obtained are consistent with previous research, which showed excellent performance of the AquaCrop model for simulating the work of different crops in different regions [34,35,45]. For example, in Ethiopia [23], coefficients of determination values (r^2) greater than 0.92 were found for the relation between observed and simulated values of canopy cover for potato crops under deficit irrigation, which are similar to those determined in this study (>0.95). In relation to the values of NRMSE, NE, and d, we found identical values to those obtained for Ethiopia, which suggests a high degree of adaptability of the AquaCrop model to different climatic conditions, soils, and crops.

Once the AquaCrop model was calibrated and validated, potato crop yields were simulated for the extended period (2005–2017). The simulated average annual yield results (Table 6), when compared with the yield statistics of the Dirección Regional Agraria Puno (DRA) for the province of Azangaro [46], show a good approximation. The mean difference analysis (t-Student) performed on the observed and simulated potato yields indicates no significant differences at a 5% significance level (t-statistic = 0.47 and t-tabular = 1.81). It suggests the AquaCrop model can adequately simulate potato crop yields for periods other than those used in the model calibration and validation. However, differences in potato crop yields can be attributed to various factors, including crop variety, local climatic conditions, irrigation rates, fertilization, and agricultural activities [34,47].

Table 6. Observed and simulated yields for the extended period 2005–2017.

Fuente	Average Annual Performance (t year^{-1})												
	2005	**2006**	**2007**	**2008**	**2009**	**2010**	**2011**	**2012**	**2013**	**2014**	**2015**	**2016**	**2017**
Simulated	10.21	10.39	10.49	10.57	10.66	10.77	10.91	11.00	11.12	11.23	11.33	12.91	12.49
DRA	9.58	9.69	9.08	7.13	8.56	9.38	10.11	9.53	10.34	9.66	11.57	10.60	11.07

Potato yield statistics for the province of Azangaro averaged 9.72 t yr^{-1} (Table 6), which is compared to regions with similar climatic and geographical conditions, such as the Mantaro Valley in the central Andes of Peru, where higher yield values of up to 18.9 t yr^{-1} were observed [48]. The vital difference in yields could be attributed to the annual rainfall amounts, which are higher in the Mantaro Valley (833 mm) compared to Azángaro (597.8 mm), as crop yield is directly related to biomass, and biomass is directly related to the amount of water available, which generates a more significant expansion and coverage of the foliage [49]. Therefore, dependent on annual rainfall amounts, potato cultivation under rainfed production systems causes crop water stress periods, significantly influencing the potato crop's phenology, growth, and productivity. Therefore, it is necessary to practice irrigated agriculture by implementing irrigation systems that allow the timely provision of water to crops and reduce periods of water stress to obtain better yield levels [23,28,50].

According to projections for the period 2023–2050, changes in precipitation in Azángaro were not as significant as in other regions of the Altiplano [51,52]; therefore, it is expected that the conditions of the Peruvian Altiplano, such as hailstorms, snowfalls, frosts, and droughts that are persistent [53], affect productive activities in a negative way [53]. However, contrary to the results obtained in this research [54], it found a positive trend for the projected period (2071–2100), with an increase of up to 2 mm day^{-1}, agreeing with the results of [55]. This behavior would be attributed to the scarcity of meteorological information in the region, the geographical conditions of the Altiplano, and the uncertainty of projections obtained from global climate models, among others [56], which could be improved with the use of regional climate models [57].

The changes in mean temperature projected for the experimental area are conservative in comparison to previous research that determined rates of increase of 0.8 to 2.7 °C century^{-1} [58]. Furthermore, in the projections made by [54], the increases would be between 2 °C and 4 °C and up to 6 °C in the north of Lake Titicaca (2071–2100). Likewise, Galera and González [59] indicate temperature increases of up to 4 °C, generating favorable conditions for certain crops, with production increases due to changes in the climate. The temperature has a significant effect on the yield of the potato crop. The optimum maximum and minimum temperatures are in the range of 13.52 °C and 3.75 °C, respectively. Values above and below this range would negatively affect the crop yield because these are extremely sensitive to environmental changes. Values above the optimum modify the growth mode according to the duration of the different phenological phases, which would significantly affect the yield achieved due to the decrease in the number of tubers [60,61].

The changes in potato crop yields found in this research for the RCP 4.5 and RCP 8.5 scenarios showed differentiated behaviors for the periods 2023–2037 and 2038–2050, with minimal reductions for the former and more significant decreases for the latter periods; these results are in line with those found in previous studies [60,62–64]. Changes in rainfall and temperature modify the different phenological stages of the crop, affecting yields by reducing the number and size of potato tubers in rainfed crops [7,60,64,65]. Since the potato crop requires 400 mm to 800 mm of rain per agricultural campaign for adequate production [29], values slightly lower than this range were obtained in this research: 382 mm for RCP 4.5 and 390 mm for RCP 8.5, which generated a minimal reduction in crop yield under a rainfed agriculture system. Likewise, droughts represent a severe threat to agricultural development and food security [45] due to the deficit in precipitation and the increase in temperature, generating a tendency for production to decrease in the coming years, which would require the implementation of irrigation systems, resulting in higher water demand, according to [66].

Under these current conditions in the Peruvian Altiplano, the introduction of heat-tolerant and drought-resistant potato genotypes [67,68], as well as the cultivation of bitter potato varieties [69], the development of irrigated agriculture reversing the loss of traditional knowledge [47], the adoption of sustainable agricultural practices [17], the adoption of strategies for climate-resilient sustainable agriculture (CRSA) [18], and techniques to reduce crop water loss [70], are required as an essential strategy for adapting to climate change and variability, as these are issues of current global concern due to their impact on agricultural production [71].

5. Conclusions

In general, the AquaCrop model performed well in simulating the development and production of the potato crop variety Imilla Negra (*Solanum tuberosum* spp.) under the extreme climatic conditions of the Peruvian Altiplano, adequately enacting the observed values of canopy cover, aerial dry biomass, and crop yield for the period 2017–2018.

Climate projections showed decreases in precipitation and increases in temperature and evapotranspiration for the RCP 4.5 and RCP 8.5 scenarios in 2023–2050. Projections for potato yields in the Peruvian Altiplano under the climate scenarios showed a minimal decrease in the first half of the projection period (2023–2037). In the second period (2038–2050), the reduction increased due to climate change.

In addition, further research and field experiments are needed to determine the influence of other factors, such as potato crop variety, local climatic conditions, irrigation doses, fertilization, and agricultural activities, on potato crop yields in the Peruvian Altiplano, from which to promote strategies to adapt to climate change and ensure the food security and well-being of the communities of the Peruvian Altiplano.

Supplementary Materials: The following supporting information can be downloaded at: https://www.mdpi.com/article/10.3390/su16010071/s1, Figure S1. We obtain aerial images of the potato crop in different phenological stages to calculate canopy coverage. Figure S2. Scaling and delimitation of canopy coverage of the potato crop in AutoCAD. Figure S3. Monitoring the development of the potato crop in different phenological phases. Figure S4. (A) quantification of aboveground biomass of potato crop, (B) quantification of root biomass (without tuber).

Author Contributions: Conceptualization, J.P.-C., G.B. and W.L.; methodology, J.P.-C., G.B., W.L. and R.A.; software, J.P.-C., E.H. and E.C.; validation, J.P.-C., G.B., W.L. and R.A.; formal analysis, J.P.-C., G.B., W.L. and E.H.; investigation, J.P.-C., G.B., W.L., R.A., E.H. and E.C.; resources, J.P.-C., G.B., W.L. and R.A.; data curation, J.P.-C., G.B., W.L. and E.C.; writing—original draft preparation, J.P.-C., G.B., W.L., R.A., E.H. and E.C.; writing—review and editing, J.P.-C., G.B., W.L., R.A., E.H. and E.C.; visualization, J.P.-C., W.L. and E.C.; supervision, J.P.-C., G.B. and W.L.; project administration, J.P.-C., G.B. and W.L.; funding acquisition, G.B. and R.A. All authors have read and agreed to the published version of the manuscript.

Funding: This research received no external funding.

Institutional Review Board Statement: Not applicable.

Informed Consent Statement: Not applicable.

Data Availability Statement: No new data were created or analyzed in this study. Data sharing is not applicable to this article.

Acknowledgments: The authors would like to thank Servicio Nacional de Meteorología e Hidrología–Perú (SENAMHI), and CPTEC/INPE by the climate information available. The authors have appreciated the Unidad de Posgrado de la Facultad de Ingeniería Agrícola of the Universidad Nacional del Altiplano for their support.

Conflicts of Interest: The authors declare no conflict of interest.

References

1. Peng, J.; Liu, T.; Chen, J.; Li, Z.; Ling, Y.; De Wulf, A.; De Maeyer, P. The Conflicts of Agricultural Water Supply and Demand under Climate Change in a Typical Arid Land Watershed of Central Asia. *J. Hydrol. Reg. Stud.* **2023**, *47*, 101384. [CrossRef]
2. Thi Nhung, T.; Le Vo, P.; Van Nghi, V.; Quoc Bang, H. Salt Intrusion Adaptation Measures for Sustainable Agricultural Development under Climate Change Effects: A Case of Ca Mau Peninsula, Vietnam. *Clim. Risk Manag.* **2019**, *23*, 88–100. [CrossRef]
3. Eum, H.-I.; Fajard, B.; Tang, T.; Gupta, A. Potential Changes in Climate Indices in Alberta under Projected Global Warming of 1.5–5 °C. *J. Hydrol. Reg. Stud.* **2023**, *47*, 101390. [CrossRef]
4. Drebenstedt, I.; Marhan, S.; Poll, C.; Kandeler, E.; Högy, P. Annual Cumulative Ambient Precipitation Determines the Effects of Climate Change on Biomass and Yield of Three Important Field Crops. *Field Crop. Res.* **2023**, *290*, 108766. [CrossRef]
5. Uleberg, E.; Hanssen-Bauer, I.; van Oort, B.; Dalmannsdottir, S. Impact of Climate Change on Agriculture in Northern Norway and Potential Strategies for Adaptation. *Clim. Chang.* **2014**, *122*, 27–39. [CrossRef]
6. Yu, L.; Liu, Y.; Li, X.; Yan, F.; Lyne, V.; Liu, T. Vegetation-Induced Asymmetric Diurnal Land Surface Temperatures Changes across Global Climate Zones. *Sci. Total Environ.* **2023**, *896*, 165255. [CrossRef]
7. Kaewthong, N.; Ditthakit, P. Effects of Climate Change on Agriculture Water Demand in Lower Pak Phanang River Basin, Southern Part of Thailand. *MATEC Web Conf.* **2018**, *192*, 03043. [CrossRef]
8. El-Bially, M.E.; Saudy, H.S.; El-Metwally, I.M.; Shahin, M.G. Sunflower Response to Application of L-Ascorbate Under Thermal Stress Associated with Different Sowing Dates. *Gesunde Pflanz.* **2022**, *74*, 87–96. [CrossRef]
9. Sharma, P.; Kattel, R.R.; Subedi, A.P. Climate Change and Agriculture Sector Frontier in Maize Cultivation among Chepang Communities of Nepal: Reviewed of Empirical Evidence. *J. Maize Res. Dev.* **2017**, *3*, 53–66. [CrossRef]
10. Galindo, L.M.; Samaniego, J.; Alatorre, J.E.; Ferrer, J.; Reyes, O. *Cambio Climático, Agricultura y Pobreza: Una Aproximación Empírica*; CEPAL: Santiago, Chile, 2014.
11. Kuzucu, M.; Dökmen, F.; Günes, A. Effects of Climate Change on Agriculture Production Under Rain-Fed Condition. *Int. J. Electron. Mech. Mechatron. Eng.* **2016**, *6*, 1057–1065. [CrossRef]
12. Wang, H.; Cheng, M.; Liao, Z.; Guo, J.; Zhang, F.; Fan, J.; Feng, H.; Yang, Q.; Wu, L.; Wang, X. Performance Evaluation of AquaCrop and DSSAT-SUBSTOR-Potato Models in Simulating Potato Growth, Yield and Water Productivity under Various Drip Fertigation Regimes. *Agric. Water Manag.* **2023**, *276*, 108076. [CrossRef]
13. Liaqat, W.; Barutçular, C.; Farooq, M.U.; Ahmad, H.; Jan, M.F.; Ahmad, Z.; Nawaz, H.; Li, M. Climate Change in Relation to Agriculture: A Review. *Span. J. Agric. Res.* **2022**, *20*, e03R01. [CrossRef]
14. Eulenstein, F.; Willms, M.; Schindler, U.; Müller, L.; Mirschel, W.; Dalchow, C.; Augustin, J.; Wiggering, H. What Are the Efects of Climate Change on Agriculture in North East Central Europe? *AgEcon Search* **2006**, *1*, 11.
15. Sayğı, H. Adverse Effects of Climate Change on Agriculture: An Evaluation of Fruit and Honey Bee Farming. *Asian J. Agric. Rural Dev.* **2020**, *10*, 504–514. [CrossRef]
16. Ochieng, J.; Kirimi, L.; Mathenge, M. Effects of Climate Variability and Change on Agricultural Production: The Case of Small Scale Farmers in Kenya. *NJAS Wagening. J. Life Sci.* **2016**, *77*, 71–78. [CrossRef]
17. Li, L.; Huang, Y. Sustainable Agriculture in the Face of Climate Change: Exploring Farmers' Risk Perception, Low-Carbon Technology Adoption, and Productivity in the Guanzhong Plain of China. *Water* **2023**, *15*, 2228. [CrossRef]
18. Simarmata, T.; Khais Proyoga, M.; Herdiyantoro, D.; Setiawati, M.R.; Adinata, K.; Stöber, S. Climate Resilient Sustainable Agriculture for Restoring the Soil Health and Increasing Rice Productivity as Adaptation Strategy to Climate Change in Indonesia. *IOP Conf. Ser. Earth Environ. Sci.* **2021**, *748*, 012039. [CrossRef]
19. Xie, L.; Zhao, H. Sustainable Agriculture and Climate Change. In *Innovations in Sustainable Agriculture*; Farooq, M., Pisante, M., Eds.; Springer International Publishing: Cham, Switzerland, 2019; pp. 441–468. ISBN 978-3-030-23168-2.
20. Semeraro, T.; Scarano, A.; Leggieri, A.; Calisi, A.; De Caroli, M. Impact of Climate Change on Agroecosystems and Potential Adaptation Strategies. *Land* **2023**, *12*, 1117. [CrossRef]
21. Gurgel, A.C.; Reilly, J.; Blanc, E. Challenges in Simulating Economic Effects of Climate Change on Global Agricultural Markets. *Clim. Chang.* **2021**, *166*, 29. [CrossRef]
22. Mibulo, T.; Kiggundu, N. Evaluation of FAO AquaCrop Model for Simulating Rainfed Maize Growth and Yields in Uganda. *Agronomy* **2018**, *8*, 238. [CrossRef]

23. Wale, A.; Dessie, M.; Kendie, H. Evaluating the Performance of AquaCrop Model for Potato Production Under Deficit Irrigation. *Air Soil Water Res.* **2022**, *15*, 117862212211082. [CrossRef]
24. Belizario, G. *Epistemología Ambiental Aplicada al Cambio Climático y su Impacto en la Agricultura*; ILAE, Primera, Ltd., Eds.; Instituto Latinoamericano de Altos Estudios: Bogotá, Colombia, 2021; ISBN 9789585367524.
25. Yao, F.; Qin, P.; Zhang, J.; Lin, E.; Boken, V. Uncertainties in Assessing the Effect of Climate Change on Agriculture Using Model Simulation and Uncertainty Processing Methods. *Chin. Sci. Bull.* **2011**, *56*, 729–737. [CrossRef]
26. Garreaud, R.; Vuille, M.; Clement, A.C. The Climate of the Altiplano: Observed Current Conditions and Mechanisms of Past Changes. *Palaeogeogr. Palaeoclimatol. Palaeoecol.* **2003**, *194*, 5–22. [CrossRef]
27. Pillco Zolá, R.; Bengtsson, L.; Berndtsson, R.; Martí-Cardona, B.; Satgé, F.; Timouk, F.; Bonnet, M.-P.; Mollericon, L.; Gamarra, C.; Pasapera, J. Modeling Lake Titicaca Daily and Monthly Evaporation. *Hydrol. Earth Syst. Sci.* **2018**, *23*, 657–668. [CrossRef]
28. Martínez-Maldonado, F.E.; Castaño-Marín, A.M.; Góez-Vinasco, G.A.; Marin, F.R. Gross Primary Production of Rainfed and Irrigated Potato (*Solanum tuberosum* L.) in the Colombian Andean Region Using Eddy Covariance Technique. *Water* **2021**, *13*, 3223. [CrossRef]
29. Adekanmbi, T.; Wang, X.; Basheer, S.; Nawaz, R.A.; Pang, T.; Hu, Y.; Liu, S. Assessing Future Climate Change Impacts on Potato Yields—A Case Study for Prince Edward Island, Canada. *Foods* **2023**, *12*, 1176. [CrossRef]
30. Allen, R.G. Using the FAO-56 Dual Crop Coefficient Method over an Irrigated Region as Part of an Evapotranspiration Intercomparison Study. *J. Hydrol.* **2000**, *229*, 27–41. [CrossRef]
31. Droogers, P.; Allen, R.G. Estimating Reference Evapotranspiration Under Inaccurate Data Conditions. *Irrig. Drain. Syst.* **2002**, *16*, 33–45. [CrossRef]
32. Laqui, W.; Zubieta, R.; Rau, P.; Mejía, A.; Lavado, W.; Ingol, E. Can Artificial Neural Networks Estimate Potential Evapotranspiration in Peruvian Highlands? *Model. Earth Syst. Environ.* **2019**, *5*, 1911–1924. [CrossRef]
33. Hargreaves, G.H.; Allen, R.G. History and Evaluation of Hargreaves Evapotranspiration Equation. *J. Irrig. Drain. Eng.* **2003**, *129*, 53–63. [CrossRef]
34. Umesh, B.; Reddy, K.S.; Polisgowdar, B.S.; Maruthi, V.; Satishkumar, U.; Ayyanagoudar, M.S.; Rao, S.; Veeresh, H. Assessment of Climate Change Impact on Maize (*Zea mays* L.) through Aquacrop Model in Semi-Arid Alfisol of Southern Telangana. *Agric. Water Manag.* **2022**, *274*, 107950. [CrossRef]
35. Montoya, F.; Camargo, D.; Ortega, J.F.; Córcoles, J.I.; Domínguez, A. Evaluation of Aquacrop Model for a Potato Crop under Different Irrigation Conditions. *Agric. Water Manag.* **2016**, *164*, 267–280. [CrossRef]
36. Willmott, C.J. Some Comments on the Evaluation of Model Performance. *Bull. Am. Meteorol. Soc.* **1982**, *63*, 1309–1313. [CrossRef]
37. Moriasi, D.N.; Arnold, J.G.; Van Liew, M.W.; Bingner, R.L.; Harmel, R.D.; Veith, T.L. Model Evaluation Guidelines for Systematic Quantification of Accuracy in Watershed Simulations. *Trans. ASABE* **2007**, *50*, 885–900. [CrossRef]
38. Jamieson, P.D.; Porter, J.R.; Wilson, D.R. A Test of the Computer Simulation Model ARCWHEAT1 on Wheat Crops Grown in New Zealand. *Field Crop. Res.* **1991**, *27*, 337–350. [CrossRef]
39. Van Vuuren, D.P.; Edmonds, J.; Kainuma, M.; Riahi, K.; Thomson, A.; Hibbard, K.; Hurtt, G.C.; Kram, T.; Krey, V.; Lamarque, J.-F.; et al. The Representative Concentration Pathways: An Overview. *Clim. Chang.* **2011**, *109*, 5–31. [CrossRef]
40. Chou, S.C.; Lyra, A.; Mourão, C.; Dereczynski, C.; Pilotto, I.; Gomes, J.; Bustamante, J.; Tavares, P.; Silva, A.; Rodrigues, D.; et al. Evaluation of the Eta Simulations Nested in Three Global Climate Models. *Am. J. Clim. Chang.* **2014**, *3*, 438–454. [CrossRef]
41. Chou, S.C.; Lyra, A.; Mourão, C.; Dereczynski, C.; Pilotto, I.; Gomes, J.; Bustamante, J.; Tavares, P.; Silva, A.; Rodrigues, D.; et al. Assessment of Climate Change over South America under RCP 4.5 and 8.5 Downscaling Scenarios. *Am. J. Clim. Chang.* **2014**, *3*, 512–527. [CrossRef]
42. Lyra, A.; Tavares, P.; Chou, S.C.; Sueiro, G.; Dereczynski, C.; Sondermann, M.; Silva, A.; Marengo, J.; Giarolla, A. Climate Change Projections over Three Metropolitan Regions in Southeast Brazil Using the Non-Hydrostatic Eta Regional Climate Model at 5-Km Resolution. *Theor. Appl. Climatol.* **2018**, *132*, 663–682. [CrossRef]
43. Teutschbein, C.; Seibert, J. Bias Correction of Regional Climate Model Simulations for Hydrological Climate-Change Impact Studies: Review and Evaluation of Different Methods. *J. Hydrol.* **2012**, *456–457*, 12–29. [CrossRef]
44. Willkofer, F.; Schmid, F.-J.; Komischke, H.; Korck, J.; Braun, M.; Ludwig, R. The Impact of Bias Correcting Regional Climate Model Results on Hydrological Indicators for Bavarian Catchments. *J. Hydrol. Reg. Stud.* **2018**, *19*, 25–41. [CrossRef]
45. Zhu, X.; Xu, K.; Liu, Y.; Guo, R.; Chen, L. Assessing the Vulnerability and Risk of Maize to Drought in China Based on the AquaCrop Model. *Agric. Syst.* **2021**, *189*, 103040. [CrossRef]
46. DRA. Serie Histórica de Producción Agrícola. 2018. Available online: https://www.agropuno.gob.pe/estadistica-agraria-informatica/agricola/ (accessed on 15 November 2022).
47. Sanabria, J.; Lhomme, J.P. Climate Change and Potato Cropping in the Peruvian Altiplano. *Theor. Appl. Climatol.* **2013**, *112*, 683–695. [CrossRef]
48. Grados, D.; García, S.; Schrevens, E. Assessing the Potato Yield Gap in the Peruvian Central Andes. *Agric. Syst.* **2020**, *181*, 102817. [CrossRef]
49. Raes, D.; Steduto, P.; Hsiao, T.C.; Fereres, E. AquaCrop—The FAO Crop Model to Simulate Yield Response to Water: II. Main Algorithms and Software Description. *Agron. J.* **2009**, *101*, 438–447. [CrossRef]
50. Rodríguez, P.L.; Sanjuanelo, C.D.; Ñústez, L.C.E.; Moreno-Fonseca, L.P. Growth and Phenology of Three Andean Potato Varieties (*Solanum tuberosum* L.) under Water Stress. *Agron. Colomb.* **2016**, *34*, 141–154. [CrossRef]

51. Ruiz, C. Ampliando la frontera agrícola de la papa (*Solanum tubrosum* L.) Para disminuir los efectos del cambio climático. *Univ. Soc. Rev. Científica Univ. Cienfuegos* **2018**, *10*, 46–51.
52. Tonconi, J. Producción Agrícola Alimentaria y Cambio Climático: Un Análisis Económico en el Departamento de Puno, Perú Food Production Agriculture and Climate Change: An Economic Analysis in the Department of Puno, Peru. *Idesia* **2015**, *33*, 119–136. [CrossRef]
53. Terrazas, O.A. Estrategias Ecológicas para el Manejo del Tizón Tardío de la Papa. *Phytophthora Infestans.* **2017**, *21*, 1–14.
54. Sanabria, J.; Marengo, J.; Valverde, M. Escenarios de Cambio Climático Con Modelos Regionales Sobre El Altiplano Peruano (Departamento de Puno). *Rev. Peru. Geo-Atmosférica RPGA* **2009**, *149*, 134–149.
55. Lujano, E.; Hidalgo, L.; Tapia, B.; Lujano, A. Cambios Proyectados de Los Recursos Hídricos Bajo Escenarios de Emisiones RCP4.5 y 8.5 de Modelos Climáticos Globales Del CMIP5 En El Altiplano Peruano. *Rev. Investig. Altoandinas J. High Andean Res.* **2016**, *18*, 195–204.
56. Wu, Y.; Miao, C.; Fan, X.; Gou, J.; Zhang, Q.; Zheng, H. Quantifying the Uncertainty Sources of Future Climate Projections and Narrowing Uncertainties with Bias Correction Techniques. *Earths Future* **2022**, *10*, e2022EF002963. [CrossRef]
57. Iles, C.E.; Vautard, R.; Strachan, J.; Joussaume, S.; Eggen, B.R.; Hewitt, C.D. The Benefits of Increasing Resolution in Global and Regional Climate Simulations for European Climate Extremes. *Geosci. Model Dev.* **2020**, *13*, 5583–5607. [CrossRef]
58. Magrin, G. El Cambio Climático En América Latina: Su Influencia en el Sector Agrícola y en la Producción del Cultivo de Papa. 2008. Available online: http://www.papaslatinas.org/home/wp-content/uploads/2017/08/Magrin.pdf (accessed on 20 November 2022).
59. Galera, K.; Gonzáñez, K. Efectos Del Cambio Climático Sobre La Producción de Papa En El Municipio de Villapinzón (Cundinamarca-Colombia) a Partir Del Enfoque Ricardiano. *Rev. Investig. Agrar. Ambient.* **2014**, *5*, 231–242.
60. Martin, R.; Jerez, E. Efecto De las Temperaturas en el Rendimiento de la Papa. *Cultiv. Trop.* **2017**, *38*, 75–80.
61. Martín, R.; Jeréz, E. Evaluación del Rendimiento en Papa de las Temperaturas. *Inca* **2015**, *36*, 93–97.
62. Alavi, G.; Diels, J.; Willems, P.; Magali, G. Simulacion de La Produccion de Quinua En el Altiplano Boliviano Con el Modelo de AQUACROP Con Escenarios Futuros Generados Por LARS-WG y QPM. *Investig. E Innov. Agropecu. Recur. Nat.* **2015**, *2*, 7–13.
63. Egerer, S.; Puente, A.F.; Peichl, M.; Rakovec, O.; Samaniego, L.; Schneider, U.A. Limited Potential of Irrigation to Prevent Potato Yield Losses in Germany under Climate Change. *Agric. Syst.* **2023**, *207*, 103633. [CrossRef]
64. Quiroz, R.; Ramírez, D.A.; Kroschel, J.; Andrade-Piedra, J.; Barreda, C.; Condori, B.; Mares, V.; Monneveux, P.; Perez, W. Impact of Climate Change on the Potato Crop and Biodiversity in Its Center of Origin. *Open Agric.* **2018**, *3*, 273–283. [CrossRef]
65. González-Jiménez, J.; Andersson, B.; Wiik, L.; Zhan, J. Modelling Potato Yield Losses Caused by Phytophthora Infestans: Aspects of Disease Growth Rate, Infection Time and Temperature under Climate Change. *Field Crop. Res.* **2023**, *299*, 108977. [CrossRef]
66. Rodrigues, D.; Fonseca, A.; Stolarski, O.; Freitas, T.R.; Guimarães, N.; Santos, J.A.; Fraga, H. Climate Change Impacts on the Côa Basin (Portugal) and Potential Impacts on Agricultural Irrigation. *Water* **2023**, *15*, 2739. [CrossRef]
67. Pradel, W.; Gatto, M.; Hareau, G.; Pandey, S.K.; Bhardway, V. Adoption of Potato Varieties and Their Role for Climate Change Adaptation in India. *Clim. Risk Manag.* **2019**, *23*, 114–123. [CrossRef] [PubMed]
68. Vandamme, E.; Manners, R.; Adewopo, J.; Thiele, G.; Friedmann, M.; Thornton, P. Strategizing Research and Development Investments in Climate Change Adaptation for Root, Tuber and Banana Crops in the African Great Lakes Region: A Spatial Prioritisation and Targeting Framework. *Agric. Syst.* **2022**, *202*, 103464. [CrossRef]
69. Meldrum, G.; Mijatović, D.; Rojas, W.; Flores, J.; Pinto, M.; Mamani, G.; Condori, E.; Hilaquita, D.; Gruberg, H.; Padulosi, S. Climate Change and Crop Diversity: Farmers' Perceptions and Adaptation on the Bolivian Altiplano. *Environ. Dev. Sustain.* **2018**, *20*, 703–730. [CrossRef]
70. El-Bially, M.E.; Saudy, H.S.; Hashem, F.A.; El–Gabry, Y.A.; Shahin, M.G. Salicylic Acid as a Tolerance Inducer of Drought Stress on Sunflower Grown in Sandy Soil. *Gesunde Pflanz.* **2022**, *74*, 603–613. [CrossRef]
71. Andati, P.; Majiwa, E.; Ngigi, M.; Mbeche, R.; Ateka, J. Effect of Climate Smart Agriculture Technologies on Crop Yields: Evidence from Potato Production in Kenya. *Clim. Risk Manag.* **2023**, *41*, 100539. [CrossRef]

Disclaimer/Publisher's Note: The statements, opinions and data contained in all publications are solely those of the individual author(s) and contributor(s) and not of MDPI and/or the editor(s). MDPI and/or the editor(s) disclaim responsibility for any injury to people or property resulting from any ideas, methods, instructions or products referred to in the content.

sustainability

MDPI

Article

Optimizing Qatar's Food Import Resilience: A Multi-Objective Framework Integrating Water Requirement Variability for Key Crops

Bashar Hassna, Farhat Mahmood, Sarah Namany, Adel Elomri and Tareq Al-Ansari *

College of Science and Engineering, Hamad Bin Khalifa University, Doha P.O. Box 34110, Qatar; baha33950@hbku.edu.qa (B.H.); fmahmood@hbku.edu.qa (F.M.); snamany@hbku.edu.qa (S.N.); aelomri@hbku.edu.qa (A.E.)

* Correspondence: talansari@hbku.edu.qa

Abstract: Global food supply chains face mounting vulnerabilities due to climate change and environmental variability, with particularly severe implications for import-dependent nations like Qatar, where over 90% of food supplies rely on international trade. This high import dependency creates unique challenges, including supply disruptions, price volatility, and food security risks, especially as climate variability increasingly affects major food-exporting regions. This study develops a multi-objective optimization framework to enhance the resilience of Qatar's food import system by integrating economic, environmental, and crop water requirement considerations, modeled as a stochastic variable. The framework addresses both average performance and worst-case scenarios using stochastic and robust optimization approaches, evaluating trade partners for three key crops—tomatoes, onions, and cucumbers. Results identify optimal suppliers that minimize costs, environmental emissions, and water usage variability, with Turkey contributing 42.10% of total imports, Iran 13.76%, and the Netherlands 9.52%. The findings demonstrate that a diversified import strategy significantly reduces vulnerability to climate-induced disruptions and improves supply chain stability. This research provides actionable insights for policymakers, including; (1) optimal supplier diversification targets to balance risk and efficiency, (2) specific trade partner recommendations based on multiple sustainability criteria, and (3) quantitative frameworks for assessing import portfolio resilience.

Keywords: food supply chain; stochastic optimization; climate variability; robust optimization

1. Introduction

Global food systems are highly interconnected with their surrounding environment. As a result, they are sensitive to various types of environmental changes. These changes can range from large-scale disruptions to gradual shifts in average conditions. For example, shifts in climate, nutrient cycles, and water availability can all have significant impacts on the production, storage, processing, distribution, retail, and consumption of food. In addition to gradual changes, food systems are also affected by sudden disruptions such as floods, droughts, and extreme heat, which are manifestations of climate variability. Furthermore, other environmental challenges like pest outbreaks, diseases, algal blooms, coral bleaching, and the presence of aflatoxins also pose significant threats to food security [1,2]. While natural variations have occurred within the historical range, today's environmental variability is increasingly characterized by extreme events. These shifts are largely driven by climate change and other human-induced factors, making food systems more vulnerable and highlighting the importance of developing resilience against these challenges. Anticipated shifts in climate and environmental conditions over the coming decades are expected to further exacerbate these vulnerabilities, impacting food security, nutrition, and both human and global health [3,4]. However, even in the present, extreme weather events and environmental variability are already leading to significant annual harvest losses and disruptions in food distribution. These immediate, short-term effects—those occurring within a single growing or production season—can be more severe than the long-term impacts of environmental and climate changes, highlighting the urgency for adaptive measures [5].

The impacts of environmental variability extend beyond individual harvests to the broader food supply chain (FSC), affecting both local and global scales. In regions with limited access to global markets, where food production and consumption are closely interlinked, often through short supply chains typical of subsistence-based food systems, environmental variability can quickly lead to local shortages of specific foods. These shortages can create food insecurity depending on the availability of substitutes, impacts on livelihoods, and how households adjust their diets and nutrient intake [6,7]. Moreover, the consequences of environmental variability are not confined to local producers but are increasingly propagated through longer supply chains. For example, in 2008, a drought in major grain-producing regions, combined with several other factors—including rising biofuel demand, high oil prices, declining grain stocks, and the depreciation of the US dollar—led to a spike in global grain prices. This price spike triggered a series of rice export bans, which exacerbated shortages and ultimately drove over 130 million people into poverty, with an additional 75 million experiencing malnourishment [8]. This example highlights how disruptions in one part of the supply chain can have cascading effects, illustrating the interconnectedness and fragility of global food systems.

Although the 2008 event was not entirely an environmental shock, it underscores the potential for disruptions to cascade through global food trade networks, affecting populations in geographically distant regions. As global food systems become increasingly interconnected, it is crucial to understand the emerging risks associated with environmental variability. Ongoing globalization and the expansion of multinational food enterprises have introduced greater complexity and efficiency in the relationship between producers and consumers. However, despite this growing interconnectedness, there is still a limited understanding of how these dynamics influence the impacts of environmental variability on both local and global FSCs. Moreover, the effects of major events—such as economic recessions, trade blockades, or pandemics—can amplify these vulnerabilities [9]. This amplification particularly affects the quality of food products, especially perishable items, which deteriorate over time and often lose all value, especially for fresh produce [10]. Perishability is defined by at least one of the following criteria: (1) a noticeable decline in product quality, (2) significant consequences from the loss of functionality, or (3) a gradual reduction in value [10,11]. These characteristics make fresh food products particularly vulnerable during various stages of the supply chain, including manufacturing, storage, and distribution [12]. Therefore, ensuring the timely delivery of fresh products to consumers is essential for maintaining the viability and efficiency of FSCs [13].

The design of perishable FSCs is, therefore, critical for mitigating the challenges presented by environmental variability. This process, known as Perishable Food Supply Chain Network Design (PFSCND), involves decisions related to facility location, manufacturing, storage, and transportation of food products that are subject to quality deterioration. PFSCND also requires consideration of the unique characteristics of perishable FSCs, which intersect with economic, social, and environmental factors. These unique traits stem from the nature of perishable items and the processes involved in their production, preparation, and distribution. As a result, decisions made within PFSCND not only affect product quality but also have broader implications for the sustainability of the entire supply chain, ultimately influencing consumer satisfaction and business success [14].

Global food systems are under increasing pressure from environmental and climate variability, posing significant risks to food security. This is particularly critical for nations like Qatar, where over 90% of food is imported due to limited agricultural capacity. Qatar's food supply chain faces unique challenges, including the following:

- High dependency on imports for staple crops like tomatoes, onions, and cucumbers.
- Vulnerability to climate-induced disruptions in exporting nations.
- The perishability of imported crops leads to quality degradation during transit.

Addressing these challenges requires a strategic framework that balances economic, environmental, and resource sustainability considerations. This study proposes a multi-objective optimization framework to strengthen Qatar's food import system against climate-induced risks. The framework integrates economic costs, environmental emissions, and variability in crop water requirements (CWR) to identify optimal trade partners. The key research objectives are framed as the following questions:

1. How can Qatar's food import system be optimized to reduce climate-related risks?
2. What are the trade-offs between economic, environmental, and resource sustainability considerations in food import decisions?
3. Which trade partners offer the most resilience against variability in CWR?

To support these objectives, this study utilizes historical climate and trade data, including statistics on crop production costs, water requirements, and transportation emissions. Turkey, a major trade partner, contributes 42.10% of Qatar's imports, demonstrating its critical role in supply chain stability. By combining stochastic and robust optimization approaches, this study provides actionable insights for policymakers to design resilient and sustainable food systems. This study also aims to propose a methodology that strengthens Qatar's food import system against climate-related risks. The proposed approach focuses on selecting optimal trade partners for key crops through a multi-objective optimization framework that balances economic and environmental considerations to ensure a resilient food import strategy. Furthermore, by integrating climate factors that affect water requirements for crop growth, the framework offers a holistic perspective on the challenges and opportunities within the global food system.

This study is structured over five sections. Section 2 provides a comprehensive literature review, examining previous research on food security, climate variability, and supply chain optimization. Section 3 outlines the methodology deployed in this study, including the multi-objective optimization framework and the factors considered in the model. Section 4 presents the results and discusses the findings, highlighting key insights related to Qatar's food import system and its resilience to climate risks. Finally, Section 5 concludes the paper with recommendations for future research and policy implications.

2. Literature Review

Food supply chains differ significantly from other product supply networks, primarily due to the continuous degradation in product quality along the entire chain, from production to final consumption. This distinct nature of FSCs, driven by the perishability of food items, price volatility, fluctuating demand, climate dependency, and increasing consumer concerns regarding food safety, makes them more complex than other supply chains [15]. A substantial portion of food is lost due to the perishability of agricultural products, emphasizing the importance of managing food sustainability, profitability, and quality throughout the FSC. Despite the best efforts, some post-harvest food losses are inevitable [15].

Given these unique challenges, the importance of perishable food supply chains (PFSCs) has become increasingly critical in the context of changing global demand patterns. As demand for fresh produce, such as milk, fruits, and vegetables, rises in response to the global shift towards healthier diets, managing PFSCs effectively is essential. Real-time management of perishable products and addressing factors like pricing, quality, and freshness are crucial for efficient PFSC operations. To maintain product integrity, continuous monitoring of these factors is necessary throughout the supply chain (SC). However, handling modern PFSCs remains challenging due to demand fluctuations, stringent environmental regulations, and unforeseen disruptions. Recent technological advancements have aided in overcoming some of these challenges, particularly in improving inventory control, transportation, and sustainability aspects [16].

Building on these advances in PFSCs, a broader perspective is needed to address the complexities of agri-food supply chains (AFSCs). In AFSCs, integrated strategies must be implemented, from raw material production to product distribution, aligning supply and demand in a competitive and dynamic environment. However, this process is fraught with challenges, primarily stemming from uncertainties inherent to the agri-food sector [17]. Effective management of these uncertainties is vital for improving supply chain performance. In PFSCs, value degradation is influenced by opportunity costs, shelf-life constraints, and product transportation dynamics. Singh et al. [18] proposed a location-allocation model considering value degradation and coordination using big data approximations. This model, framed as a mixed-integer linear programming (MILP) problem, was solved using a CPLEX solver. Additionally, Chen et al. [19] developed a method to classify fresh agricultural products' storage requirements to reduce consolidation costs and minimize value loss from shorter shelf lives [16].

A more detailed understanding of AFSCs is necessary to fully understand the process from farm production to consumer delivery. AFSCs include plant and animal based products, which are particularly vulnerable to uncertainties stemming from the nature of agricultural production. These uncertainties, if not managed effectively, can negatively impact the supply chain's performance and overall efficiency. Conversely, better management of uncertainties can enhance performance [20]. Various sources of uncertainty in crop-based AFSCs include shelf-life unpredictability, where the product's deterioration does not necessarily align with its marketable lifespan [21]. Product degradation, categorized as age-dependent or independent, further complicates supply chain management [11]. Additionally, the lack of product homogeneity, typical of the agri-food industry, creates additional challenges in maintaining consistent quality standards [22].

Supply lead times in AFSCs are generally long and impacted by seasonal and climatic factors, adding another layer of uncertainty [23]. Moreover, production uncertainty arises from the unpredictability of raw materials, necessitating flexibility in processing methods [24]. Demand uncertainty is influenced by consumer preferences, safety concerns, and promotional efforts, all of which fluctuate due to factors such as seasonality and weather [25]. In agriculture, weather patterns and land conditions heavily influence crop yields. This uncertainty in weather patterns poses significant challenges during the harvesting process [23]. Managing these uncertainties is particularly difficult when planning production in AFSCs. Many traditional planning models assume deterministic parameters, which is often unrealistic in agriculture due to the inherent variability [26].

In addition, recent studies emphasize the increasing role of climate change adaptation strategies in ensuring food security and supply chain resilience. Alhassan and Kuwornu [27] investigated the impacts of climate variability on food security in Sub-

Saharan Africa, highlighting the importance of integrating climate-responsive strategies in food supply chains. Similarly, Begashaw et al. [28] analyzed climate adaptation strategies and their effects on household vulnerability to food insecurity in Ethiopia, stressing the need for robust policy interventions in food trade networks. These findings reinforce the significance of incorporating climate resilience into food supply chain models, aligning with the objectives of this study. To address uncertainties in FSCs, robust optimization techniques are increasingly being applied. These methods can handle uncertain data, such as errors in data estimation or environmental factors, through strategies like the Robust Counterpart approach [29]. By incorporating robust optimization, supply chain managers can better manage unpredictable conditions in the agri-food industry, ensuring better outcomes despite uncertainties [30]. In addition to robust optimization, stochastic optimization approaches are also being explored to handle uncertainty in AFSCs. Unlike robust optimization, which aims to find solutions that perform well under worst-case scenarios, stochastic optimization incorporates probabilistic elements to model uncertainty, thereby enabling decision-making that accounts for a variety of possible future outcomes.

AFSCs face significant challenges due to the uncertainties inherent in production, demand, and environmental conditions. The growing use of advanced optimization methods, including robust and stochastic approaches, offers promising pathways to address these complexities and enhance the resilience of FSCs. As the demand for sustainable and efficient food systems continues to rise, these optimization strategies will play a crucial role in shaping the future of global agri-food networks.

3. Methodology

This study suggests a methodology designed to strengthen Qatar's food import system's resilience to climate risks affecting exporting nations. The framework aims to select the optimal trade partners to fulfill Qatar's yearly requirements for three primary crops, namely, tomatoes, cucumbers, and onions, which are the most widely consumed in the country. The methodology is centered on a multi-objective optimization approach that accounts for climate variability. This approach seeks to minimize both economic and environmental costs while also reducing the CWR, which is influenced by temperature fluctuations that affect crop growth cycles and trade partners' yields throughout the year. This study is an extension of the recent research conducted by the authors, Hassna et al. [31], wherein the risk factors impacting trade partners' selection are replaced by the CWR, which is a holistic representation of the climatic conditions of each country. The CWR, which is defined as the amount of water required by the crop to grow optimally, is estimated using the Penman-Monteith evapotranspiration equation that accounts for temperature, solar irradiance, humidity, and wind speed [32]. Detailed mathematical formulation is presented in Sections 3.1 and 3.2.

3.1. The Mathematical Formulation of the Optimization Model

The objective of the optimization framework suggested in this study is to determine the combination of trade partners (b) that can fulfill the demand (D) of Qatar for a combination of perishable crops (a) through diverse transportation means (c). The demand assumed in this study represents a partial demand from imported products and does not consider the total demand of the country, including the local production system. The model suggests three objective functions, wherein the first equation aims to minimize the total economic cost of supplying the products from the country (b) to Qatar. The second objective minimizes the environmental emissions associated with the trade decision, while the third objective has the goal of minimizing the water required to grow the crops in their country of origin. The detailed mathematical formulation of the multi-objective optimization is summarized in Tables 1–3.

Table 1. List of objective functions.

Objective Function	Description
$Min \sum_{a=1}^{3} \sum_{b=1}^{7} \sum_{c=1}^{3} X_{abc} \cdot D_a \cdot (P_{ab} + T_{bc})$	The purpose of this function is to minimize the overall cost of importing the crop from the set of exporting countries, counting production and transportation costs (USD).
$Min \sum_{a=1}^{3} \sum_{b=1}^{7} \sum_{c=1}^{3} X_{abc} \cdot D_a \cdot \left(E_{ab}^{g} + E_{bc}^{t}\right)$	The purpose of this function is to minimize the carbon emissions associated with the trade activity (kg of CO_{2eq}).
$Min \sum_{a=1}^{3} \sum_{b=1}^{7} \sum_{c=1}^{3} X_{abc} \cdot D_a \cdot w_{ab}$	The purpose of this function is to minimize the total CWR associated with importing crops from the set of selected countries to Qatar.

Table 2. List of constraints.

Constraints	Description
$\sum_{a=1}^{3} \sum_{b=1}^{7} \sum_{c=1}^{3} X_{abc} \cdot D_a \cdot F_{abc} = D_a$	This constraint ensures that the local demand from each crop is fulfilled.
$0.1 \leq X_{abc} \leq 0.5$	This constraint ensures diversification in selecting trade partners.
If $F_{abc} \geq 0.2,\ X_{abc} \leq 0.2$	This constraint ensures the quality of the crops imported given their degradation factor.
$X_{abc} \cdot D_a \leq Cap_{ab}$	This constraint is linked to the supplying exporting capacity that is predetermined in this context.

Table 3. List of variables used in the mathematical optimization model.

Variable	Definition
X_{abc}	The percentage contribution of each trade partner b to the satisfaction of the demand from crop a using shipping method c (%).
D_a	The total demand from crop a (kg).
P_{ab}	The unit cost of growing crop a in exporting country b. This cost is inclusive of the cost of seeds, water, energy, fertilizers, and operational costs (USD/kg).
T_{bc}	The unit cost of transportation from country b using transportation means c and it is computed as a function of the distance between Qatar and country c (USD/kg).
E_{ab}^{g}	The unit environmental emissions associated with cultivating crop a in country b (kg CO_{2eq}/kg).
E_{bc}^{t}	The unit environmental emissions associated with transporting the crops from country b to Qatar using the shipping method c (kg CO_{2eq}/kg).
w_{ab}	The unit CWR for the cultivation of crop a in country b (m^3/kg).
F_{abc}	The degradation factor represented as a percentage weight loss due to transporting crop a from country b using shipping method c (%).
Cap_{ab}	The allowed exportable quantity by supplier b from crop a (kg).

Multiple scenarios were conducted to dynamically quantify the impact of the variability of CWR on the growth patterns of crops in exporting countries and hence on the results of the optimization model. Two different scenarios were conducted in this regard, and they are detailed in Sections 3.1.1 and 3.1.2. Section 3.2 presents the case study data utilized in this optimization problem.

The methodology integrates stochastic and robust optimization through a systematic process. First, historical climate data are analyzed to determine the mean (μ) and variance

(σ^2) of CWR for each crop-country combination, from which multiple climate scenarios are generated following a normal distribution. The stochastic optimization model is then solved to optimize average performance across all scenarios, providing an initial solution that balances expected costs, emissions, and water requirements. Subsequently, the robust optimization model addresses worst-case scenarios by identifying and minimizing the maximum impact of extreme conditions. This dual approach enables decision-makers to understand both the average system performance and its resilience under extreme scenarios, with the final recommendations incorporating insights from both methods to suggest a balanced import strategy that considers both routine operations and extreme events.

3.1.1. Optimization Under Uncertainty

The CWR is a critical factor in agricultural planning and FSC optimization. The CWR is affected by various climate factors such as temperature, humidity, solar radiation, and wind speed, all of which fluctuate due to natural variability and climate change. These factors directly impact how much water is needed for optimal crop growth. Given the uncertainty in future climate conditions, the CWR is represented as a stochastic variable following a normal distribution:

$$\mathrm{CWR}_{ab}(\omega) \sim \mathcal{N}\left(\mu_{ab}, \sigma^2_{ab}\right),$$

where $\mathrm{CWR}_{ab}(\omega)$ represents the CWR for crop a in country b under scenario ω; μ_{ab} is the mean CWR for crop a in country b, based on historical data; σ^2_{ab} is the variance in CWR for crop a in country b; and ω represents the individual climate scenario.

The present study adopts two approaches to handle uncertainty in CWRs: stochastic optimization and min–max robust optimization. Stochastic optimization addresses the uncertainty in climate factors by modeling CWRs as a probabilistic variable. Incorporating multiple climate scenarios allows for decision-making that is, on average, optimal across a range of possible future outcomes. While min–max robust optimization focuses on ensuring the best possible outcome under the worst-case scenario. Rather than averaging over all potential outcomes, it plans for the most extreme conditions within a defined uncertainty set.

3.1.2. Stochastic Optimisation

In stochastic optimization, the goal is to minimize the objective function considering the expected value of the CWR, while the total cost and emissions remain deterministic. The CWR, denoted by $w_{ab}(\omega)$, depends on the scenario ω where ω represents one of the possible climate scenarios. By simulating these scenarios, the expected CWR can be estimated. The objective minimization is expressed as follows:

$$\text{minimize } f_1(x) + f_2(x) + E[f_3(x, \omega)],$$

where $f_1(x)$ represents the deterministic total cost, $f_2(x)$ represents the deterministic total emissions, $E[f_3(x, \omega)]$ represents the expected total water requirements. The stochastic nature of the CWR is captured through the inclusion of ω, which represents different climate scenarios. By incorporating these scenarios, the model accounts for the uncertainty and variability in future climate conditions. A high number of climate scenarios are generated based on historical climate data and current trends. Each scenario represents a possible realization of future climate conditions, including variations in temperature, precipitation patterns, solar irradiance, and other relevant climatic variables that influence CWR. The expected value of the water requirement function can be expressed as follows:

$$E[f_3(x,\omega)] = \frac{1}{|\Omega|}\sum_{\omega\in\Omega}\sum_{a=1}^{A}\sum_{b=1}^{B}\sum_{c=1}^{C}[X_{abc}\cdot D_{ab}\cdot w_{ab}(\omega)],$$

where ω is the total number of scenarios; a, b, and c are the total numbers of crops, countries, and transportation modes, respectively; X_{abc} is the decision variable representing the proportion of crop a, sourced from country b via transport mode c; D_{ab} is the demand for crop a from country b; and $w_{ab}(\omega)$ is the CWR for crop a in country b under scenario ω. By minimizing the expected value, the model seeks a solution that performs across the entire range of potential future conditions, while ensuring optimal performance with respect to both cost and emissions under deterministic conditions.

3.1.3. Robust Optimisation

In robust optimization, the goal is to minimize the worst possible outcome of the total cost, emissions, and CWR. This method focuses on minimizing the maximum value of these objectives across all potential climate scenarios. By doing so, it ensures that the solution is resilient even under the most extreme climate conditions. In robust optimization, the CWR is modeled as the worst-case scenario, which corresponds to the scenario ω that results in the highest water requirement represented as following:

$$f_3(x,\omega) = \max_{\omega\in\Omega}[X_{abc}\cdot D_{ab}\cdot w_{ab}(\omega)],$$

where $\max\limits_{\omega\in\Omega}$ denotes the maximum value across all possible scenarios. This ensures that the solution accounts for the worst possible climate conditions, preventing the FSC from being disrupted by extreme water shortages. Thus, the overall objective function in robust optimization is to minimize the maximum total outcome, including costs, emissions, and water requirements:

$$\text{minimize} \max_{\omega\in\Omega}[f_1(x) + f_2(x) + f_3(x,\omega)].$$

This min–max approach ensures that the optimization model accounts for the worst-case scenario, making the system resilient against extreme variations in climate conditions. While this approach may lead to more conservative solutions, it is crucial for ensuring the long-term stability and sustainability of Qatar's FSC, especially under conditions of climate uncertainty.

3.2. *Available Data*

The optimization model in this study aims to enhance the resilience of Qatar's food import system by focusing on minimizing economic costs, reducing environmental emissions, and optimizing the variability of CWRs for imported crops under uncertainty. CWR, a critical metric in agricultural planning, represents the water needed for optimal crop growth and is influenced by climatic factors such as temperature, solar radiation, humidity, and wind speed. In this model, CWR values are modeled as stochastic variables following a normal distribution characterized by a mean (μ) and standard deviation (σ), capturing the inherent variability in climate conditions for each exporting country. The specific values of the mean and standard deviation will be determined based on historical climate data and included in the final analysis. Data used in the model are derived from the datasets developed in a previous study conducted by the authors (Namany et al. [33]), where the crop water of each country was estimated using the evapotranspiration equation of

Penman–Monteith. The choice of the countries is based on the historical trade activity with Qatar. Only countries with frequent trade trends were selected. Table 4 below provides placeholders for the mean (μ) and standard deviation (σ) of CWR for each crop (tomatoes, onions, and cucumbers) across the selected exporting countries.

Table 4. Mean (μ) and standard deviation (σ) of CWR by country.

Crop	Statistic	USA	Iran	Lebanon	India	Turkey	Morocco	Netherlands
Tomatoes	μ	816.84	798.67	1008.92	1264.28	560.52	811.62	900.45
	σ	138.46	512.94	256.55	174.40	417.26	183.51	300.87
Onions	μ	547.2828	535.1089	675.9764	847.0676	375.5484	543.7854	603.3015
	σ	92.7682	343.6698	171.8885	116.848	279.5642	122.9517	201.5829
Cucumbers	μ	653.472	638.936	807.136	1011.424	448.416	649.296	720.36
	σ	110.768	410.352	205.24	139.52	333.808	146.808	240.696

The second dataset focuses on economic and transportation parameters. Production costs include expenses related to seeds, water, energy, fertilizers, and operational activities. Transportation costs are calculated based on the mode of shipment (ground, sea, or air freight) and the distance from each exporting country to Qatar. These factors are vital in determining the overall logistics costs. Table 5 presents the estimated distances and shipment times for each mode of transport.

Table 5. Estimated shipment distances and times (from Hassna et al. [31]).

Country	Distance (km)	Ground (Days)	Sea (Days)	Air (Days)
USA	12,229	NA	16,000	6–8
Iran	800	1900	500	1
Lebanon	1686	2200	3900	2
India	2856	NA	1600	2–3
Turkey	2128	2800	4800	2–3
Morocco	5638	NA	5800	2–3
Netherlands	4984	NA	7500	1–2

NA: Not applicable.

Environmental impact data includes CO_2-equivalent emissions from both cultivation and transportation stages. Emissions factors are determined based on the type of transportation used and the shipping distances. This data enables the model to prioritize suppliers with lower environmental footprints, aligning with sustainable import strategies. The demand for tomatoes, onions, and cucumbers was obtained from the Planning and Statistics Authority of Qatar. The effective demand used in this study is set at 158,008 tons for tomatoes, 5396 tons for onions, and 87,392 tons for cucumbers. These figures account for potential losses during transportation, ensuring that the actual demand is met reliably. Given the perishable nature of these crops, the model includes data on quality degradation during transportation. Weight loss rates are influenced by factors such as shipment duration and handling, affecting the effective quantity available upon arrival in Qatar. Table 6 details the weight loss rates for each crop across the different exporting countries.

Table 6. Weight loss rates by country (% per Day) (modified from Hassna et al. [31]).

Country	Tomatoes	Onions	Cucumbers
USA	0.27%	0.03%	0.50%
Iran	0.53%	0.06%	1.00%
Lebanon	0.80%	0.09%	1.50%
India	1.06%	0.12%	2.00%
Turkey	1.33%	0.15%	2.50%
Morocco	1.59%	0.18%	3.00%
Netherlands	1.86%	0.21%	3.50%

4. Results and Discussion

4.1. Stochastic Optimization

The stochastic optimization provides the optimal import quantities for tomatoes, onions, and cucumbers from seven countries: the USA, Iran, Lebanon, India, Turkey, Morocco, and the Netherlands. The optimization aims to minimize total cost, emissions, and expected CWR under uncertainty.

4.1.1. Optimal Import Networks: Stochastic Scenario

The total quantity imported for tomatoes was 167,494.16 tons, with Turkey being the largest supplier, contributing 49,344.14 tons (29.5% of the total tomato imports), as illustrated in Figure 1. This significant contribution can be attributed to Turkey's competitive position in the market. Iran was the second-largest supplier, with 29,231.72 tons (17.5%). The Netherlands and Morocco supplied similar amounts, with 19,324.48 tons (11.50%) and 19,265.42 tons (11.50%), respectively. The USA contributed 17,724.73 tons (10.60%), followed by Lebanon with 16,802.87 tons (10.00%), and India with 15,800.80 tons (9.40%). Their inclusion results from acceptable waste and loss factors (all $\leq$ 20%) and compliance with constraints requiring each country to supply between 10% and 50% of the demand for items with acceptable waste and loss factors.

For onions, the total import quantity was 5430.09 tons. Turkey was again the largest supplier, providing 1491.23 tons (27.40% of total onion imports). Iran contributed 953.25 tons (17.60%), while the Netherlands supplied 645.03 tons (11.90%). Morocco provided 632.98 tons (11.70%), followed by the USA with 598.71 tons (11.00%), Lebanon with 569.29 tons (10.50%), and India with 539.60 tons (9.90%).

In the case of cucumbers, the total imported quantity was 95,257.11 tons. Turkey maintained its position as the largest supplier, contributing 30,263.27 tons (31.80% of total cucumber imports). Iran provided 20,495.98 tons (21.50%), followed by the Netherlands with 13,044.15 tons (13.70%) and the USA with 12,558.54 tons (13.20%). Lebanon supplied 10,135.94 tons (10.60%), and India contributed 8759.23 tons (9.20%). Morocco did not contribute to cucumber imports due to its high waste and loss factor for cucumbers (27%), exceeding the acceptable threshold of 20% set in the constraints.

4.1.2. Overall Country Contributions

Considering total imports across all three crops, Turkey was the largest contributor, supplying 81,099 tons (30.24% of the overall total), as demonstrated in Figure 2. This significant contribution is due to Turkey's substantial import contribution in all three crops, being the leading supplier for tomatoes (29.46%), onions (27.46%), and cucumbers (31.77%). Iran was the second largest supplier with 50,681 tons (18.90%), with notable contributions in cucumbers (21.52%), tomatoes (17.45%), and onions (17.55%). The Netherlands contributed

33,014 tons (12.31%), maintaining consistent contributions across tomatoes (11.54%), onions (11.88%), and cucumbers (13.69%). The USA supplied 30,882 tons (11.52%), with balanced contributions in tomatoes (10.58%), onions (11.03%), and cucumbers (13.18%). Lebanon, India, and the Netherlands each contribute approximately 25,100 to 33,014 tons, accounting for around 10% of the overall total each. Their consistent contributions across all three crops reflect the constraints ensuring diversification and limiting dependence on any single supplier.

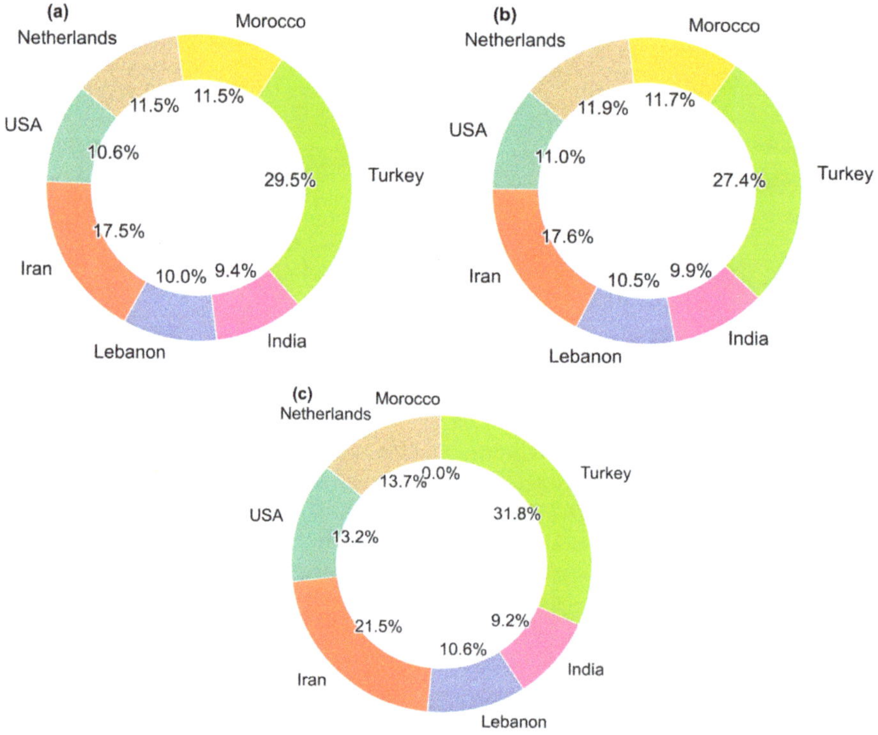

Figure 1. Optimal import networks derived using stochastic optimization for key crops ((**a**) tomatoes, (**b**) onions, and (**c**) cucumbers).

4.1.3. Effective and Initial Quantities: Stochastic Scenario

The optimization model accounted for waste and loss factors during transportation, impacting the initial quantities to be imported to meet effective demand. The total quantities before accounting for waste and loss were 167,494.16 tons for tomatoes, 5430.10 tons for onions, and 95,257.10 tons for cucumbers as demonstrated in Figure 3, to meet the respective demands. The initial quantities before degradation are higher due to anticipated losses. For instance, Turkey's initial quantities were 49,344.14 tons of tomatoes, 1491.23 tons of onions, and 30,263.27 tons of cucumbers. After applying their specific waste and loss factors (9.28% for tomatoes, 1.05% for onions, and 17.50% for cucumbers), the effective quantities would be approximately 44,763.01 tons for tomatoes, 1475.57 tons for onions, and 24,967.20 tons for cucumbers. This adjustment process was applied across all suppliers to ensure actual demand fulfillment despite losses during transportation and handling.

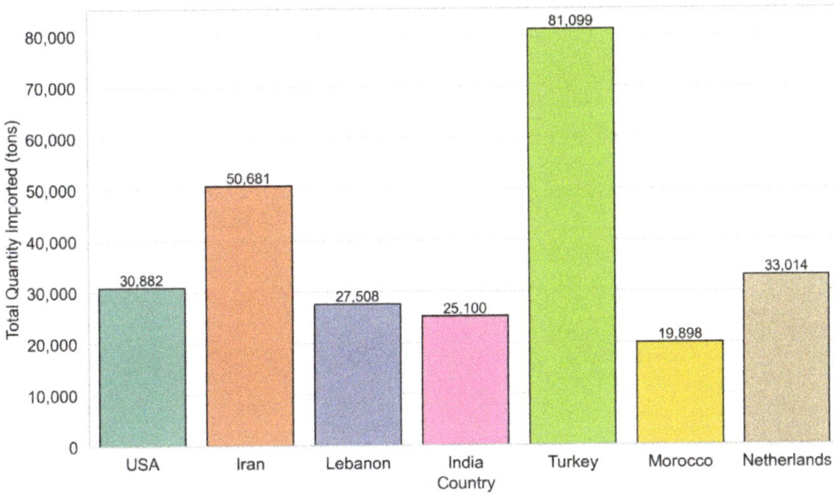

Figure 2. Overall country contributions were derived using stochastic optimization for the supply of key crops (tomatoes, onions, and cucumbers).

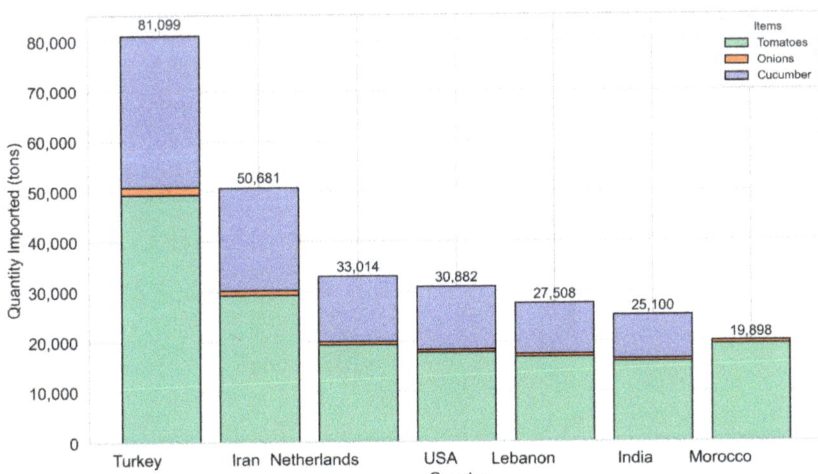

Figure 3. Quantity of imported items across countries derived using stochastic optimization for key crops: tomatoes (green), onions (orange), and cucumbers (blue).

4.2. Robust Optimization

4.2.1. Optimal Import Networks: Robust Scenario

A min-max approach was adopted for the robust optimization to determine the optimal import quantities of tomatoes, onions, and cucumbers from seven countries—USA, Iran, Lebanon, India, Turkey, Morocco, and the Netherlands—while minimizing the worst-case scenarios of CWR as illustrated in Figure 4.

The total effective quantity imported for tomatoes was 158,008 tons after accounting for waste and loss. Turkey was the largest supplier, contributing an effective quantity of 67,349.33 tons, which accounted for 42.70% of the total tomato imports. This significant share is due to Turkey's competitive costs, low emissions, acceptable waste and loss factors, and favorable CWR. The USA, Iran, Lebanon, India, and the Netherlands each contributed an effective quantity ranging from 15,047.10 to 15,717.06 tons, representing approximately 9.52% to 9.95% of the total effective tomato imports each. Morocco contributed 13,539.71 tons, accounting for 8.60% of the total tomato imports. These contributions are close to the lower limit of the supply constraint, which requires each acceptable supplier to contribute between 10% and 50% of the total demand for each item. However, some contributions are slightly below 10%. This discrepancy can be attributed to computational tolerances and rounding errors in the optimization process. In practice, optimization solvers often accept small deviations from constraints due to numerical precision limits and the need to balance multiple competing objectives.

Figure 4. Optimal import networks derived using robust optimization for key crops ((**a**) tomatoes, (**b**) onions, and (**c**) cucumbers).

For onions, the total effective quantity imported is 5396 tons, exactly meeting the demand after waste and loss adjustments. Turkey again is the largest supplier, providing an effective quantity of 2174.43 tons, accounting for 40.30% of the total effective onion imports. This reinforces Turkey's favorable position in the import network due to its competitive costs, low emissions, acceptable waste and loss factors, and manageable CWR. The USA, Iran, Lebanon, India, Morocco, and the Netherlands each supplied an effective quantity of approximately 530.86 to 539.28 tons, representing around 9.90% to 10.00% of the total effective onion imports each. These contributions are close to the lower limit of the supply constraint but are slightly below 10% in some cases.

In the case of cucumbers, the total effective quantity imported is 87,392 tons, fulfilling the demand precisely after accounting for waste and loss. Turkey is the dominant supplier, contributing an effective quantity of 36,049.20 tons, which is 41.30% of the total effective cucumber imports. The significant share from Turkey can be attributed to its acceptance of waste and loss factor (17.50% for cucumbers) and competitive costs. Iran provided an effective quantity of 18,264.93 tons (20.90%), while the USA contributed 8608.11 tons (9.80%), India supplied 8389.63 tons (9.60%), the Netherlands provided 8127.46 tons (9.30%), and Lebanon contributes 7952.67 tons (9.10%) of the total effective cucumber imports. Morocco does not supply cucumbers due to its high waste and loss factor (27% for cucumbers), exceeding the acceptable threshold of 20%.

4.2.2. Overall Country Contributions: Robust Scenario

Considering the total effective quantities across all three crops, Turkey is the largest contributor, supplying a total of 105,573 tons, which accounts for 42.10% of the overall quantity, as demonstrated in Figure 5. This dominance is primarily due to its substantial contribution to tomatoes, onions, and cucumbers. Iran contributes 34,521 tons (13.76%), followed by the USA with 24,822 tons (9.90%), India with 24,394 tons (9.73%), the Netherlands with 23,879 tons (9.52%), Lebanon with 23,536 tons (9.38%), and Morocco with 14,071 tons (5.61%).

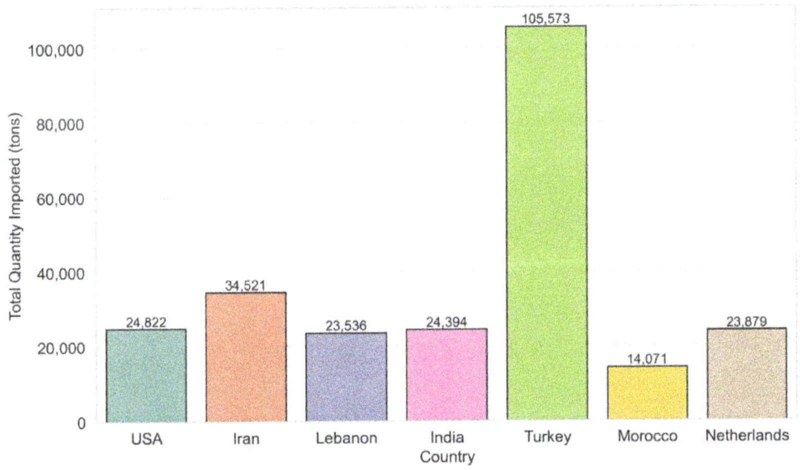

Figure 5. Overall country contributions derived using robust optimization for supply of key crops (tomatoes, onions, and cucumbers).

While Turkey emerges as the dominant supplier (42.10% of total imports), its long-term reliability under shifting climatic conditions warrants careful consideration. Analysis of Turkey's historical CWR data shows significant variability ($\sigma = 417.26$ m^3/kg for tomatoes), and climate projections for the Mediterranean region indicate increasing water stress and temperature extremes. This could affect Turkey's agricultural productivity in two ways: first, through direct impacts on crop yields due to changing temperature and precipitation patterns, and second, through increased competition for water resources between agricultural and urban uses. However, Turkey's robust agricultural infrastructure, diverse growing regions, and significant investments in climate-adaptive farming technologies partially mitigate these risks. The country's geographic proximity to Qatar also provides logistical advantages that could help maintain supply chain stability even under stressed conditions. Nevertheless, the high reliance on Turkey as a primary supplier highlights the significance of sustaining the diversified supplier network identified in our optimization model, as it serves as a buffer against possible climate-induced disruptions in any single source country.

4.2.3. Effective and Initial Quantities

The initial quantities (before accounting for waste and loss) are higher to compensate for anticipated losses during transportation and handling as depicted in Figure 6. For instance, Turkey's initial quantity of tomatoes is approximately 74,253.15 tons, which, after a waste and loss adjustment of 9.28%, results in an effective quantity of 67,349.33 tons. Similarly, the USA's initial quantity of cucumbers is approximately 8738.40 tons, leading to an effective quantity of 8608.11 tons after a 1.50% waste and loss adjustment. This ensures that the actual demand is satisfied despite losses, maintaining supply chain reliability.

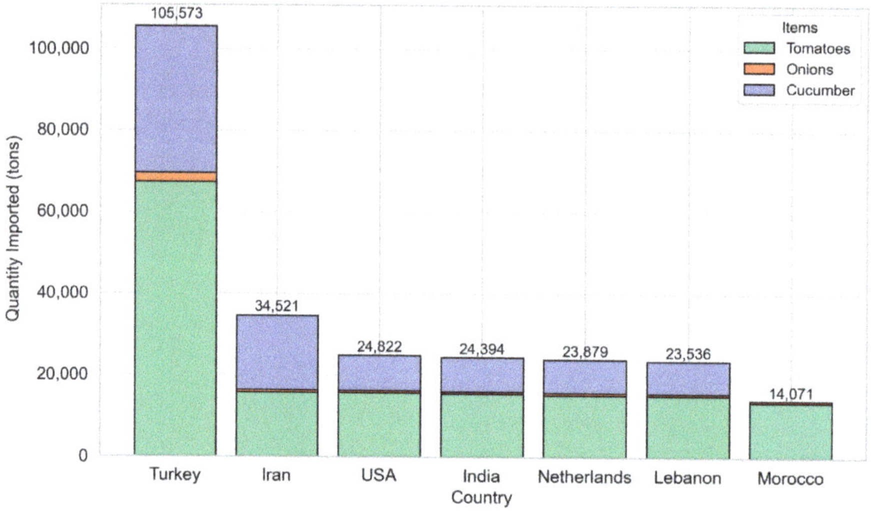

Figure 6. Quantity of imported items across countries derived using robust optimization for key crops: tomatoes (green), onions (orange), and cucumbers (blue).

The robust optimization results exhibit a strategic allocation of import quantities to minimize the worst-case total cost, emissions, and water consumption. The results differ from stochastic optimization due to the nature of the min–max approach, which seeks to minimize the worst-case outcomes across all scenarios. This leads to decisions that are conservative and resilient against extreme conditions. The emphasis on worst-case CWR means that countries and crops with lower variability and lower maximum water usage under adverse scenarios are preferred. Turkey's significant contribution, particularly in tomatoes, onions, and cucumbers, suggests that it offers favorable conditions even under worst-case scenarios, with manageable water consumption, costs, and emissions. The relatively uniform contributions from Iran, Lebanon, India, USA, and the Netherlands across all three crops highlight the model's emphasis on diversification to mitigate risks associated with uncertainties in water consumption and other factors. Each of these countries contributes approximately 9% to 13% of the total effective imports for various crops. This diversification ensures that the supply chain is not overly dependent on a single source, enhancing resilience against potential disruptions.

The robust optimization's conservative allocation reflects a trade-off between optimizing for average expected outcomes (as in stochastic optimization) and safeguarding against the worst possible scenarios. This approach is particularly valuable when dealing with uncertainties in environmental factors like CWR, where worst-case conditions could have significant impacts on supply chain reliability. By prioritizing suppliers with lower worst-case CWR and consistent performance across scenarios, the model ensures that the supply chain remains robust even under adverse conditions.

The comparative visualization of stochastic and robust optimization results reveals distinct patterns in supplier allocation strategies for Qatar's food import system. Under the stochastic approach, imports are more evenly distributed among suppliers, especially for tomatoes and onions, while the robust method concentrates a larger share of imports from Turkey. This difference reflects the core objectives of each approach: stochastic optimization seeks to maximize average performance across multiple scenarios, whereas robust optimization prioritizes stability under worst-case conditions. Turkey's dominant role is evident in both methods, but its influence is markedly greater under robust optimization. For instance, Turkey supplies 29.46% of tomato imports, 27.46% of onions, and 31.77% of cucumbers under stochastic optimization; these figures increase to 42.62%, 40.30%, and 41.25%, respectively, under robust optimization. This increased reliance on Turkey in the robust scenario underscores its reliability as a supplier—due to its geographical proximity, well-established trade relationships, and strong agricultural infrastructure. The landscape for secondary suppliers also varies between the two methods. Iran consistently ranks as the second-largest supplier, though its allocations differ; for example, Iran receives 17.45% of tomatoes and 21.52% of cucumbers under the stochastic model, compared to approximately 9.95% and 20.90% under the robust model. Other suppliers, such as the Netherlands and the USA, exhibit relatively stable contributions, albeit with slightly reduced shares in the robust scenario. Notably, Morocco displays significant variation, particularly for tomatoes, where its allocation diverges considerably between the two methods. Overall, these results highlight the inherent trade-offs between supply chain efficiency and resilience. The stochastic approach, with its balanced allocations, offers potential cost efficiency and market flexibility, whereas the robust method's focus on concentrating imports from reliable suppliers enhances supply chain stability under adverse conditions, albeit potentially at higher operational costs. This analysis provides valuable insights for policymakers in Qatar, demonstrating how different optimization strategies can be used to balance cost efficiency, supply chain resilience, and climate risk mitigation in food import systems (Figure 7).

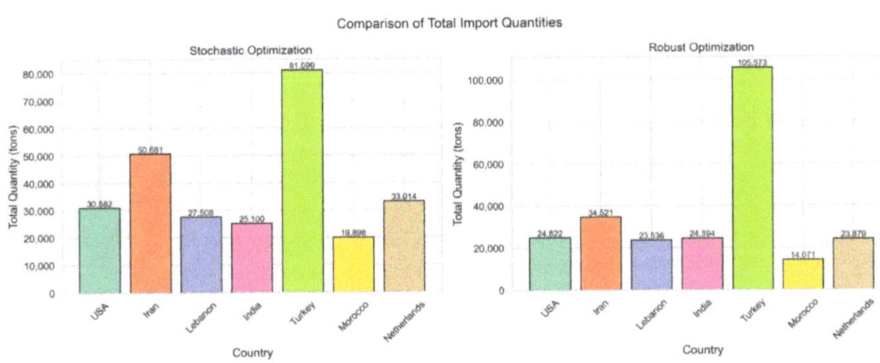

Figure 7. Comparison of total import quantities under stochastic and robust scenarios.

4.3. *Analysis of Trade-Offs in Supplier Selection*

The optimization results demonstrate significant trade-offs between economic costs, environmental impact, and water use variability when selecting suppliers. Turkey's dominant position (42.10% of total imports) demonstrates these trade-offs: while offering competitive economic costs (averaging 15–20% lower than other suppliers), its selection comes with moderate environmental emissions from transportation and varying levels of water requirement uncertainty. Iran's significant contribution (13.76%) represents a different balance, with slightly higher economic costs but lower transportation emissions due to geographical proximity. However, Iran shows higher water requirement variability (σ = 512.94 m^3/kg for tomatoes) compared to other suppliers.

The Netherlands, contributing 9.52% of imports, exemplifies the trade-off between high environmental impact due to long-distance transportation and stable water requirements (σ = 300.87 m^3/kg for tomatoes) owing to advanced agricultural technologies. This stability in water requirements comes at a premium, with costs approximately 25–30% higher than regional suppliers. These trade-offs become more pronounced under the robust optimization approach, where protecting against worst-case scenarios leads to an average cost increase of 18% compared to the stochastic optimization results but provides a 23% reduction in maximum water requirement variability.

The diversification strategy itself represents a trade-off between system efficiency and resilience. While concentrating imports among fewer suppliers could reduce operational costs by approximately 12–15%, the current diversified approach provides essential risk mitigation against both climate variability and supply chain disruptions. This is particularly evident in the case of Morocco, where despite competitive pricing, its high degradation factors limit its role in the supply chain, demonstrating how quality preservation considerations can outweigh pure economic advantages.

4.4. Implementation Considerations and Policy Recommendations

The implementation of a diversified import strategy in Qatar faces several practical challenges and barriers that policymakers must address. First, existing long-term contracts and established trade relationships may limit Qatar's immediate ability to adjust import proportions according to the optimization model's recommendations. Transitioning from current import patterns (where some suppliers may hold dominant positions) to the proposed diversified structure requires careful diplomatic and commercial negotiations. Second, infrastructure limitations, particularly in port capacity and specialized storage facilities for perishable goods, could constrain the ability to handle multiple simultaneous shipments from different suppliers effectively. A phased approach is necessary for successful implementation. In the short term (1–2 years), efforts should focus on developing additional cold storage infrastructure and upgrading port facilities to accommodate diverse shipment schedules. Medium-term actions (2–5 years) should include establishing new trade agreements with recommended suppliers while maintaining existing relationships and developing quality control protocols specific to each supplier's products and transportation routes. Long-term strategies (5+ years) should prioritize building resilience through investments in suppliers' agricultural infrastructure and establishing joint early warning systems for climate-related disruptions. Financial implications must also be considered. While diversification enhances resilience, it may initially increase operational costs by 15–20% due to the complexity of managing multiple suppliers and smaller shipment volumes. However, these costs can be partially offset by reduced risk premiums and lower emergency procurement expenses during supply disruptions. To facilitate implementation, a dedicated task force within Qatar's food security apparatus should be established to coordinate among multiple stakeholders, including port authorities, customs officials, quality control agencies, and private sector importers.

5. Limitations and Future Research Endeavor

While this study provides valuable insights for optimizing Qatar's food import system, several limitations must be acknowledged. First, the model assumes a normal distribution of CWR, which may not fully capture extreme climate events or non-linear climate impacts. Moreover, by focusing primarily on CWR as the key climate variable, the model overlooks other critical factors, such as soil conditions, pest prevalence, and extreme weather events, that can significantly influence crop production and supply chain resilience. Data-related uncertainties further complicate the analysis; variability in data accuracy, measurement methodologies, and reporting standards across exporting countries may affect the reliability of both climate variability and trade condition data. Additionally, other important risk factors, such as the political stability of supplier countries and short-term price fluctuations, have not been fully incorporated into the framework. Second, the study's scope is limited to three crops (tomatoes, onions, and cucumbers) and seven major supplier countries, which restricts the generalizability of the findings to other agricultural products and regions. The framework's focus on optimizing imports from established trade partners, without considering potential new sourcing regions or alternative crops, may overlook emerging markets that could enhance supply chain resilience, especially for perishable goods with different storage requirements and market dynamics. Third, the model relies on historical data for estimating degradation factors and transportation times, which may not accurately reflect future supply chain disruptions or technological advancements in preservation methods. Furthermore, economic parameters such as production and transportation costs are subject to market fluctuations and require regular updates, while the model's static nature does not capture long-term shifts in trade relationships and agricultural production patterns driven by climate change.

Future research should address these limitations by expanding the scope of the framework to include a broader range of agricultural products, such as grains, dairy, and protein sources, and by incorporating data on emerging supplier regions. Enhancements to the model should also consider additional risk factors like geopolitical stability and short-term price volatility and adopt dynamic modeling approaches that account for evolving trade relationships, technological advancements in logistics and food preservation, and long-term climate change impacts. Moreover, establishing robust data collection and verification protocols through collaboration with international organizations, such as the FAO, can help improve data reliability and mitigate reporting biases. These extensions will not only enhance the model's practical utility for strategic planning and policy development but also provide a more comprehensive understanding of global food supply chain resilience in the face of climate variability and market uncertainties.

6. Conclusions

This study presents a comprehensive optimization framework aimed at enhancing the resilience of Qatar's food import system against climate-induced uncertainties. By integrating a multi-objective optimization approach, the model effectively addresses three key objectives: minimizing economic costs, reducing environmental emissions, and optimizing CWR. The innovative use of CWR as a stochastic variable, modeled using a normal distribution, offers a nuanced representation of climate variability, providing deeper insights into the water needs of crops across different exporting countries. The findings demonstrate that strategic diversification of trade partners is crucial for mitigating risks associated with CWR variability. The results from both stochastic and robust optimization approaches indicate that Turkey, Iran, and the Netherlands consistently emerge as top suppliers across the three key crops (tomatoes, onions, and cucumbers), owing to their favorable cost structures, manageable emissions, and efficient water usage. Important trade-offs emerge between average performance and system resilience. While the stochastic optimization approach yields better average economic performance, the robust optimization provides greater protection against extreme climate events at the cost of higher operational expenses. This highlights the critical balance policymakers must strike between cost efficiency and supply chain resilience. The implementation of these findings faces several potential risks, including political instability in supplier regions, logistical bottlenecks, and rapid market price fluctuations, which could affect the practical application of the proposed diversification strategy. Additionally, the dynamic nature of international trade relationships and evolving climate patterns may require periodic reassessment of supplier allocations. Before full implementation, these findings should be validated through detailed simulations and pilot studies that test the framework's performance under various real-world scenarios. Such validation would help identify practical challenges and refine the model's parameters for better applicability. The robust optimization highlights the importance of planning for worst-case scenarios, which ensures stability in Qatar's FSC even under extreme climatic conditions. Overall, this study underscores the value of integrating climate variability considerations into food import strategies. The proposed framework provides a practical tool for policymakers and stakeholders, offering data-driven insights that support resilient decision-making in the face of increasing environmental uncertainties. Future research could build on this work by exploring dynamic adaptation strategies and incorporating additional climate variables, further enhancing the predictive capabilities of the model. Additionally, collaboration with exporting countries to improve data accuracy and share climate risk information could strengthen the resilience of global food trade networks, benefiting both importers and exporters alike.

Author Contributions: Conceptualization, B.H., F.M., S.N., A.E. and T.A.-A.; Methodology, B.H., S.N., F.M., A.E. and T.A.-A.; Investigation, B.H., F.M. and S.N.; Data curation, B.H., F.M. and S.N.; Writing—original draft, B.H.; Writing—review and editing, F.M., S.N., A.E. and T.A.-A.; Supervision, A.E. and T.A.-A.; Validation, A.E. All authors have read and agreed to the published version of the manuscript.

Funding: This research was made possible by an Award (GSRA7-1-0521-20080) and supported by proposal number MME01-0922-190049 from Qatar National Research Fund (a member of Qatar Foundation). The contents herein are solely the responsibility of the authors.

Data Availability Statement: Data are contained within the article.

Conflicts of Interest: The authors declare no conflicts of interest.

References

1. Lesk, C.; Rowhani, P.; Ramankutty, N. Influence of extreme weather disasters on global crop production. *Nature* **2016**, *529*, 84–87. [CrossRef] [PubMed]
2. Myers, S.S.; Smith, M.R.; Guth, S.; Golden, C.D.; Vaitla, B.; Mueller, N.D.; Dangour, A.D.; Huybers, P. Climate Change and Global Food Systems: Potential Impacts on Food Security and Undernutrition. *Annu. Rev. Public. Health* **2017**, *38*, 259–277. [CrossRef] [PubMed]
3. Schmidhuber, J.; Tubiello, F.N. Global food security under climate change. *Proc. Natl. Acad. Sci. USA* **2007**, *104*, 19703–19708. [CrossRef] [PubMed]
4. Springmann, M.; Mason-D'Croz, D.; Robinson, S.; Garnett, T.; Godfray, H.C.J.; Gollin, D.; Rayner, M.; Ballon, P.; Scarborough, P. Global and regional health effects of future food production under climate change: A modelling study. *Lancet* **2016**, *387*, 1937–1946. [CrossRef]
5. Fanzo, J.; Davis, C.; McLaren, R.; Choufani, J. The effect of climate change across food systems: Implications for nutrition outcomes. *Glob. Food Secur.* **2018**, *18*, 12–19. [CrossRef]
6. Badjeck, M.-C.; Allison, E.H.; Halls, A.S.; Dulvy, N.K. Impacts of climate variability and change on fishery-based livelihoods. *Mar. Policy* **2010**, *34*, 375–383. [CrossRef]
7. Vermeulen, S.J.; Aggarwal, P.K.; Ainslie, A.; Angelone, C.; Campbell, B.M.; Challinor, A.J.; Hansen, J.W.; Ingram, J.S.I.; Jarvis, A.; Kristjanson, P.; et al. Options for support to agriculture and food security under climate change. *Environ. Sci. Policy* **2012**, *15*, 136–144. [CrossRef]
8. Headey, D. Rethinking the global food crisis: The role of trade shocks. *Food Policy* **2011**, *36*, 136–146. [CrossRef]
9. Davis, K.F.; Downs, S.; Gephart, J.A. Towards food supply chain resilience to environmental shocks. *Nat. Food* **2021**, *2*, 54–65. [CrossRef]

10. Osvald, A.; Stirn, L.Z. A vehicle routing algorithm for the distribution of fresh vegetables and similar perishable food. *J. Food Eng.* **2008**, *85*, 285–295. [CrossRef]
11. Amorim, P.; Meyr, H.; Almeder, C.; Almada-Lobo, B. Managing perishability in production-distribution planning: A discussion and review. *Flex. Serv. Manuf. J.* **2013**, *25*, 389–413. [CrossRef]
12. Govindan, K.; Jafarian, A.; Khodaverdi, R.; Devika, K. Two-echelon multiple-vehicle location–routing problem with time windows for optimization of sustainable supply chain network of perishable food. *Int. J. Prod. Econ.* **2014**, *152*, 9–28. [CrossRef]
13. Musavi, M.; Bozorgi-Amiri, A. A multi-objective sustainable hub location-scheduling problem for perishable food supply chain. *Comput. Ind. Eng.* **2017**, *113*, 766–778. [CrossRef]
14. Yakavenka, V.; Mallidis, I.; Vlachos, D.; Iakovou, E.; Eleni, Z. Development of a multi-objective model for the design of sustainable supply chains: The case of perishable food products. *Ann. Oper. Res.* **2020**, *294*, 593–621. [CrossRef]
15. Lemma, Y.; Kitaw, D.; Gatew, G. Loss in Perishable Food Supply Chain: An Optimization Approach Literature Review. *Int. J. Sci. Eng. Res.* **2014**, *5*, 302–311.
16. Karanam, M.; Krishnanand, L.; Manupati, V.K.; Antosz, K.; Machado, J. Identification of the Critical Enablers for Perishable Food Supply Chain Using Deterministic Assessment Models. *Appl. Sci.* **2022**, *12*, 4503. [CrossRef]
17. Esteso, A.; Alemany, M.; Bas, A. Deterministic and Uncertain Methods and Models for Managing Agri-Food Supply Chains. *Dir. Organ.* **2017**, *62*, 41–46.
18. Singh, A.K.; Subramanian, N.; Pawar, K.S.; Bai, R. Cold chain configuration design: Location-allocation decision-making using coordination, value deterioration, and big data approximation. *Ann. Oper. Res.* **2018**, *270*, 433–457. [CrossRef]
19. Chen, M.-C.; Lu, C.-C.; Liu, Y.-C. Optimal consolidation of fresh agricultural products in a multi-temperature joint distribution system. *Int. J. Logist. Manag.* **2018**, *29*, 887–901. [CrossRef]
20. Esteso, A.; Alemany, M.M.E.; Ortiz, A. Conceptual Framework for Managing Uncertainty in a Collaborative Agri-Food Supply Chain Context. In *Collaboration in a Data-Rich World*; Camarinha-Matos, L.M., Afsarmanesh, H., Fornasiero, R., Eds.; Springer International Publishing: Cham, Switzerland, 2017; pp. 715–724.
21. Amorim, P.; Günther, H.O.; Almada-Lobo, B. Multi-objective integrated production and distribution planning of perishable products. *Int. J. Prod. Econ.* **2012**, *138*, 89–101. [CrossRef]
22. Grillo, H.; Alemany, M.M.E.; Ortiz, A. A review of mathematical models for supporting the order promising process under Lack of Homogeneity in Product and other sources of uncertainty. *Comput. Ind. Eng.* **2016**, *91*, 239–261. [CrossRef]
23. Kusumastuti, R.D.; Donk, D.P.V.; Teunter, R. Crop-related harvesting and processing planning: A review. *Int. J. Prod. Econ.* **2016**, *174*, 76–92. [CrossRef]
24. Van der Vorst, J.; Beulens, A. Identifying sources of uncertainty to generate supply chain redesign strategies. *Int. J. Phys. Distrib. Logist. Manag.* **2002**, *32*, 409–430. [CrossRef]
25. Baghalian, A.; Rezapour, S.; Farahani, R.Z. Robust supply chain network design with service level against disruptions and demand uncertainties: A real-life case. *Eur. J. Oper. Res.* **2013**, *227*, 199–215. [CrossRef]
26. Soto-Silva, W.E.; Nadal-Roig, E.; González-Araya, M.C.; Pla-Aragones, L.M. Operational research models applied to the fresh fruit supply chain. *Eur. J. Oper. Res.* **2016**, *251*, 345–355. [CrossRef]
27. Alhassan, H.; Kuwornu, J.K.M. The impact of climate variability and change on food security in Sub-Saharan Africa: Empirical evidence from panel data analysis. *Sustainability* **2022**, *14*, 759.
28. Begashaw, A.; Ketema, M.; Mehare, A.; Yami, M.; Feleke, S.; Abdoulaye, T. Climate change adaptation strategies and their impact on household vulnerability to food insecurity: Micro-level evidence from Southwest Ethiopia. *Sustainability* **2024**, *16*, 5766. [CrossRef]
29. Ben-Tal, A.; Nemirovski, A. Robust optimization—Methodology and applications. *Math. Program.* **2002**, *92*, 453–480. [CrossRef]
30. Rahbari, M.; Arshadi Khamseh, A.; Mohammadi, M. Robust optimization and strategic analysis for agri-food supply chain under pandemic crisis: Case study from an emerging economy. *Expert. Syst. Appl.* **2023**, *225*, 120081. [CrossRef]
31. Hassna, B.; Namany, S.; Alherbawi, M.; Elomri, A.; Al-Ansari, T. Multi-Objective Optimization for Food Availability under Economic and Environmental Risk Constraints. *Sustainability* **2024**, *16*, 4336. [CrossRef]
32. Monteith, J.L. Evaporation and environment. *Symp. Soc. Exp. Biol.* **1965**, *19*, 205–234. [PubMed]
33. Namany, S.; Govindan, R.; Alfagih, L.; McKay, G.; Al-Ansari, T. Sustainable food security decision-making: An agent-based modelling approach. *J. Clean. Prod.* **2020**, *255*, 120296.

Disclaimer/Publisher's Note: The statements, opinions and data contained in all publications are solely those of the individual author(s) and contributor(s) and not of MDPI and/or the editor(s). MDPI and/or the editor(s) disclaim responsibility for any injury to people or property resulting from any ideas, methods, instructions or products referred to in the content.

sustainability

MDPI

Article

Combining Zinc Biofortification and Native *Trichoderma* Inoculation Strategies for Subterranean Clover

Carlos García-Latorre [1,*], Rocío Velázquez [1], Alejandro Hernández [2], Paula Tejero [2] and Maria J. Poblaciones [1,*]

[1] Department of Agronomy and Forest Environment Engineering, University of Extremadura, Avenida Adolfo Suárez s/n, 06007 Badajoz, Spain; rvotero@unex.es

[2] Department of Animal Production and Food Science, School of Agricultural Engineering, Universidad de Extremadura, Avenida Adolfo Suárez s/n, 06007 Badajoz, Spain; ahernandez@unex.es (A.H.); paulatc@unex.es (P.T.)

* Correspondence: cgarcialn@unex.es (C.G.-L.); majops@unex.es (M.J.P.)

Abstract: Using beneficial microorganisms along with sustainable strategies such as agronomic biofortification offers eco-friendly alternatives to combat climate change in ecosystems like dehesas. This study analyzes the combined effects of four wild *Trichoderma* spp. isolated from Extremadura, Spain (*T. koningiopsis*, two *T. gamsii*, and *T. koningii*, with negative and positive controls) and four Zn biofortification treatments (no Zn application; soil application of 5 mg of $ZnSO_4{\cdot}7H_2O$ per kg of soil, labeled soil Zn; two foliar applications of 5 mL 0.5% $ZnSO_4{\cdot}7H_2O$, labeled foliar Zn; and soil + foliar combination, labeled SF) on *Trifolium subterraneum* performance. The combination of *T. koningiopsis* and *T. gamsii* with foliar Zn improved plant growth by up to 34.4%. Zinc accumulation was about 30% higher when *T. gamsii* and *T. koningii* were applied with SF, and their inoculation resulted in a 2.5-fold increase in ash. *Trichoderma* spp. affected nodulation differently; both *T. gamsii* inhibited nodulation by 24%, whereas neither *T. koningiopsis* nor *T. koningii* showed differences from the controls. These results highlight the potential of combining beneficial microorganisms with biofortification strategies to address nutrient deficiencies and improve agricultural sustainability. However, the complex interactions between both factors underscore the importance of strain selection and call for further research to optimize application strategies and elucidate underlying mechanisms.

Keywords: bioinoculation; pasture management; zinc sulfate; nutrient uptake

1. Introduction

Pastures are integral parts of the ecological and agricultural landscapes of the Southwestern Iberian Peninsula, covering approximately 2.07 million hectares in the Spanish region of Extremadura, which constitutes about 70% of the region's total land area [1]. This extensive land area comprises a variety of ecosystems, most notably dehesas, grasslands, and areas characterized by dense tree cover [2]. These ecosystems provide a variety of ecosystem services that are critical to both agricultural sustainability and environmental well-being. Well-managed pastures play a key role in fire prevention and control through proper grazing planning that includes the establishment of firebreaks [3]. Moreover, these pastures provide other valuable resources to nearby agricultural areas, including the commercialization of by-products such as crop residues and straw. Beyond their economic contributions, these pastures also serve as habitat and food sources for a wide range of faunal species, including storks, vultures, lynx, eagles, lizards, and cranes, among others, contributing to the rich biodiversity of the region [4]. Agroforestry systems, including dehesas, are vital for carbon sequestration and biodiversity preservation [5–9]. Appropriate agroforestry management practices, including sustainable grassland–livestock systems and legume incorporation, are essential for increasing productivity and diversity, especially under increased climate variability [10]. These practices could play an important role in climate change mitigation [9]. However, pastures in the southwest of the Iberian Peninsula have been frequently perceived as being of low quality due to poor soil conditions and irregular rainfall patterns [11,12]. As a result, these pastures have been used directly as livestock feed regardless of their growth stage. This traditional approach has resulted in minimal management and capital investment, mainly for livestock feeding purposes [13], leading to widespread pasture degradation, reduced soil fertility, and economic losses [14,15].

Among active management strategies to improve pasture conditions in the southwest of the Iberian Peninsula, the promotion of *Trifolium subterraneum*, commonly known as subterranean clover, plays a key role. Its adaptability to the Mediterranean climate and the high quality of its forage makes it essential for supporting livestock [16]. Furthermore, its ability to self-regenerate and adapt to varying climatic conditions, as demonstrated in studies on climatic adaptation and salt tolerance mechanisms, further underlines its significance in sustainable pasture management [17]. Another promising approach to improve pasture sustainability is represented by the adoption of biofortification strategies. Biofortification aims to increase the nutritional value of crops, particularly with micronutrients that are essential to the ecosystem, such as zinc (Zn), in soils that are naturally deficient or have low bioavailability [18]. This addresses Zn deficiency in animal and human diets by increasing its concentration and bioavailability in food crops [19,20]. Agronomic biofortification programs have demonstrated the potential to improve pasture sustainability by increasing the nutritional quality of forage crops [21–23]. This improvement not only benefits the livestock consuming the fortified feed but also contributes to enhancing overall soil health and ecosystem sustainability [24,25]. It represents a promising way to improve pasture sustainability by enhancing the nutritional quality of forage crops like *Trifolium subterraneum*, thereby promoting more resilient and nutrient-rich pasture ecosystems. Finally, the use of native *Trichoderma* isolates has been highlighted to enhance nutrient uptake and crop yield, promoting sustainable agricultural practices [26].

The combination of Zn biofortification and the use of native *Trichoderma* isolates represents a holistic approach to addressing nutrient deficiencies and improving crop productivity. The simultaneous biofortification of crops with micronutrients through microbial interventions has already demonstrated additive and synergistic effects on plant quality [27]. For example, the application of Zn biofortification, along with seed co-inoculations of plant-growth-promoting bacteria, has been found to enhance growth, yield, biofortification, and Zn use efficiencies in legumes such as common bean [28]. However, there is still limited information available on the efficacy of combining both techniques, especially concerning subterranean clover. Recent studies have demonstrated the positive effect of selenium biofortification in natural pastures containing *T. subterraneum*, among other species [29,30], as well as the impact of fungal endophytes on the performance of this legume [31]. However, to our knowledge, no studies have incorporated both strategies for *T. subterraneum*. Therefore, this research holds significant importance due to the scarcity of data in this area. Additionally, this integrated approach is consistent with the goal of sustainable agriculture as it promotes soil health, reduces the dependence on external inputs, and improves the nutritional quality of crops. Integrating agronomic biofortification to increase the nutrient content of *T. subterraneum* with the application of native *Trichoderma* isolates to stimulate plant growth and manage soil-borne pathogens could enhance pasture sustainability. Thus, the use of native *Trichoderma* isolates as natural biofortification agents underscores the potential for environmentally friendly and cost-effective agricultural strategies [32]. This approach would not only improve the nutritional quality of forage crops but also increase plant resilience and productivity through biocontrol mechanisms.

Therefore, based on the premise that the improvement of pasturelands can favor their sustainability and resilience to climate change, this study aimed to evaluate the effects of soil and foliar Zn biofortification, coupled with the simultaneous use of different *Trichoderma* isolates obtained from soils of Extremadura, on plant growth and nutrient concentration in *Trifolium subterraneum*, a legume that is well adapted to dehesa ecosystems.

2. Materials and Methods

2.1. Site Details

The experiment was conducted in a naturally lit greenhouse at the School of Agricultural Engineering, University of Extremadura, situated in Badajoz, Spain (38°89′ N, 6°97′ W; elevation: 186 m above sea level). Throughout the duration of the experiment (from 9 October 2023 to 26 January 2024), the greenhouse maintained average temperatures of 18 ± 6 °C during the day and 12 ± 4 °C during the night.

2.2. Soil Characterization

The soil used in this experiment was collected from the area of Elvas in the Alentejo region of Portugal. This soil was first air-dried under laboratory conditions and then sieved through a 2 mm mesh sieve. Following this, four aliquots underwent oven-drying at 50 °C until they reached a constant weight, and they were then analyzed to determine their physicochemical characteristics using standardized procedures. The soil was characterized by a sandy loam texture with constituent proportions of 14.9% clay, 57.1% sand, and 28.0% silt. The soil pH was 6.5 ± 0.1 (average ± standard error). Organic matter content was 1.2 ± 0.1 g/kg [33], carbonate content was <1%, available phosphorus was 19.5 ± 3.4 mg/kg, and potassium content was 0.8 ± 0.1 meq/kg [34]. The extractable Zn in the soil, 0.34 ± 0.04 mg/kg, was extracted with DTPA (diethylenetriamine penta acetic acid) according to the method of Lindsay and Norvell [35] and determined by inductively coupled plasma mass spectrometry (ICP-MS) (Thermo Fisher Scientific iCAPQ, Bremen, Germany). Internal references and blanks were included in each batch of samples, and all the results were reported on a dry weight basis.

2.3. Fungal and Plant Material

In this study, four different fungal isolates of the genus *Trichoderma* were selected to evaluate their potential to enhance subterranean clover growth and nutrient uptake. These isolates were specifically obtained from soils at different locations in the Valle del Jerte region of Extremadura, Spain, where cherry cultivation is prevalent. Selection criteria included factors such as isolation frequency and observed bioactivity in previous assays in relation to biocontrol and plant growth promotion in vitro. Table 1 shows the identification details of these isolates. In addition, a commercially available product (Bactiva®, Bactiva GmbH, Straelen, Germany), referred to as Tc, was used as a positive control. This product contained a combination of three different *Trichoderma* species (*T. harzianum*, *T. eesei* and *T. viride*) that have already been shown to be effective, allowing us to assess the efficacy of our wild isolates. Sterilized distilled water (designated as T0) was used as the negative control in the experiment.

Table 1. *Trichoderma* isolate identification and experimental controls in this study.

Code	Identification	GenBank Accession Number
T05	*Trichoderma koningiopsis*	MT520626
T09	*Trichoderma gamsii*	MT557537
T14	*Trichoderma koningii*	KT715712
T18	*Trichoderma gamsii*	MT557537
T0	Negative control: Sterilized distilled water	
Tc	Positive control: *Trichoderma* consortium (Bactiva®): *T. harzianum*, *T. eesei*, *T. viride*	

Prior to plant inoculation, fungal spores were obtained from *Trichoderma* isolates. Spore samples from previous plates of each fungal isolate were seeded on autoclaved potato dextrose agar (PDA; 39 g/L of distilled water) and incubated in a germination chamber at 23 °C for seven days in the dark, followed by exposure to direct sunlight and room temperature for another seven days to induce sporulation. The same procedure was used for the commercial product (Bactiva®), internally designated as Tc. Once sufficient growth and spore presence were confirmed on the plates, spore collection was performed by adding 3–5 mL of sterile distilled water to the plates. The spore concentration was determined by using a hemocytometer (Neubauer chamber). From these initial concentrations, each concentration was diluted to obtain the final application concentration of 10^6 spores per mL for the different *Trichoderma* spp. isolates used in the experiments.

Trifolium subterraneum var. Antas seeds were surface-sterilized by successive immersion in a solution of 70% ethanol and a solution of 2% sodium hypochlorite, and then washed three times with sterile distilled water, according to the method described by Zabalgogeazcoa et al. [36].

2.4. Experimental Design and Treatments

The experiment followed a completely randomized block design with four Zn treatments combined with six fungal treatments, and four replicates per combination, resulting in a total of 96 pots. Each pot was seeded with 3 seeds of *T. subterraneum* pretreated with one of the fungal treatments. The Zn biofortification treatments included the following: no Zn application (no Zn); soil application of 5 mg of $ZnSO_4{\cdot}7H_2O$ per kg of soil (soil Zn); foliar application of 5 mL 0.5% $ZnSO_4{\cdot}7H_2O$ per pot performed once, three weeks after sowing, and again four weeks later (foliar Zn); and the combination of both soil and foliar $ZnSO_4{\cdot}7H_2O$ applications (SF Zn). Foliar treatments were applied by continuous spraying to wet the plants evenly. Soil moisture content was maintained at an adequate level by daily irrigation with 50 mL of deionized water. No pests or disease incidents were observed during this study.

2.5. Microorganism Re-Isolation

In greenhouse trials, the efficacy of fungal treatments on *Trifolium subterreum* was assessed at two critical time points. The first assessment occurred just before sowing and focused on the viability of the fungal inoculum applied to the seed, establishing a baseline for the study and confirming the initial success of seed inoculation. At harvest, two assessment methods were used. First, fungi were re-isolated from soil samples to study their persistence and adaptation within the soil ecosystem. Second, plant tissues were examined to determine the presence and development of fungi within the plants and their potential behavior as endophytes. In all cases, seed inoculation with each isolate proved effective, resulting in fungal growth and development both in the soil and within the plants. These results are significant because they provide a solid basis for establishing a confident link between the effects observed in subsequent results and the action of the selected fungal strains.

2.6. Measurements

Plant Growth Parameters

Plants were harvested three weeks after the second foliar application of Zn, approximately 10 weeks after sowing, and carefully washed with deionized water. Prior to harvest, the number of plants per pot was recorded, and indirect chlorophyll concentration was measured using a SPAD meter (Minolta SPAD 502). During plant collection, many *Rhizobium* sp. nodules were observed in most of the roots. An evaluation based on the number and activity of nodules was conducted, with samples categorized on a scale of 1 to 4: 1 represents samples without nodules; 2 represents samples with few green or white nodules, indicating low N-fixing activity; 3 represents samples with numerous pink nodules, indicating good fixation activity; and 4 represents samples with roots practically covered with nodules, indicating high fixation capacity. After harvesting, the herbage and root parts were separated, and their fresh weights were recorded. Both parts were then air-dried for three days and then oven-dried at 50 °C until a constant weight was reached to determine their dry weights. Following complete drying of the samples, determinations of neutral detergent fiber (NDF), acid detergent fiber (ADF), acid detergent lignin (ADL), and total ash were conducted. NDF, ADF, and ADL analyses were performed using a fiber analyzer

(ANKOM 8-98, ANKOM Technology, Macedon, NY, USA), while total ash analysis was carried out by ignition in a muffle furnace at 600 °C. In all cases, the determinations were made by following the official procedures [37]. Dry matter digestibility was then calculated from the ADF percentage as follows [38]:

$$Dry\ matter\ digestibility,\ DMD\ (\%) = 88.9 - (0.779 * ADF)$$

2.7. Nutrient Concentration

The concentrations of iron (Fe), magnesium (Mg), phosphorus (P), selenium (Se), and Zn were measured in the herbage dry matter of the plants, as well as in samples obtained from the potted soils for each treatment. For this, 20 mg of dried soil or herbage dry matter from each sample was digested using a mixture of nitric acid (2 mL, 70%) and hydrogen peroxide (2 mL, 50%) in a closed-vessel microwave system (Anton Paar Gmbh, Graz, Austria). Each digestion run included two blanks and two certified reference materials (CRM: tomato leaf SRM 1573a NIST, Gaithersburg, MD, USA). The digested samples were analyzed using ICP-MS (Thermo Fisher Scientific iCAPQ, Thermo Fisher Scientific, Bremen, Germany).

2.8. Statistical Analysis

Two-way analysis of variance (ANOVA) was used to evaluate the effect of *Trichoderma* spp. isolates, biofortification treatment, and the interaction of both factors on the yield and nutritional parameters of *Trifolium subterraneum* plants. Fisher's protected least significant difference (LSD) test at a significance level of $p < 0.05$ was used for mean comparisons when significant differences were detected. Normality was assessed using the Shapiro–Wilk test, while homoscedasticity was assessed using the Levene test. The relationship between each parameter was evaluated by Pearson correlation at a significance level of $p < 0.05$. All statistical analyses were performed with the STATISTIX 8.1 statistical package (Analytical Software, version 8.1, Tallahassee, FL, USA, 2005). The correlogram (shown as Figure S1) was performed with the R 4.3.3 statistical package (R Core Team, Vienna, Austria, 2024) [39].

3. Results

3.1. Influence on Growth and Quality Parameters of *Trifolium subterraneum*

The results of the two-way ANOVA are summarized in Table 2. The inoculation with the different *Trichoderma* spp. isolates revealed significant effects on all studied parameters except for the herbage dry matter (HDM), neutral detergent fiber (NDF), acid detergent fiber (ADF), and dry matter digestibility (DMD). Regarding the effect of the biofortification treatment on the yield and quality parameters, Zn application significantly affected the root dry matter (RDM) and the rhizobia nodulation index, as well as the ADF, acid detergent lignin (ADL), and DMD, while its effects on the HDM, the SPAD value, as well as the NDF and ash percentages were not statistically significant.

When examining the effect of *Trichoderma* isolates combined with biofortification treatments on the HDM (Figure 1a), it can be noticed that the plants inoculated with T09 (*Trichoderma gamsii*), T14 (*Trichoderma koningii*), and T18 (*Trichoderma gamsii*) showed increased plant yield in the absence of Zn applications (no Zn) compared to the negative control without Zn (T0 and no Zn), although no significant differences were observed compared to the T0 group treated with SF Zn. In addition, the combinations of T05 (*T. koningiopsis*) with a double Zn application (SF) and T18 (*T. gamsii*) with a foliar Zn application significantly increased the HDM compared to the T0 control with no Zn (23.4% and 34.4%, respectively), reaching similar levels to those obtained with the sole Zn application by soil and foliar methods (T0 and SF).

Regarding the effect on the RDM, the improvements were more limited. In particular, the only *Trichoderma* spp. isolate that significantly improved this parameter without Zn application was T14 (*T. koningii*) compared to both controls (10.64 mg). In addition, the

This varying effect on nutrient uptake could have been affected by the combined effects of *Trichoderma* spp. with other beneficial microbes, such as the soil microbial consortium, as evidenced by studies such as that of Moradtalab et al. [58]. Considering the natural nodulation with Rhizobia recorded in the soil, the lack of a significant effect of the interaction between *Trichoderma* inoculation and Zn biofortification on this parameter could be due to the independent mechanisms through which *Trichoderma* spp. and Zn biofortification influence plant nodulation, as well as to the complex interactions among plants, microbes, and soil in biofortification processes [43]. As for the independent effect of both factors on nodulation, the inhibitory effect in the plants inoculated with T09 and T18 (both identified as *T. gamsii*) may be a result of the colonization dynamics in the rhizosphere, considering that nodulation success may be influenced by the timing and sequence in rhizosphere interactions between different microorganisms [59]. These results are consistent with a related study by Khan et al. [60], where some *Trichoderma* strains enhanced plant growth and nutrient uptake by promoting root development and nutrient solubilization, whereas others inhibited plant growth and nodulation. This underscores the potential influence of combining different beneficial microbes to enhance plant growth. On the other hand, the biofortification treatments did not show significant differences in nodulation compared to the plants of the negative control without added Zn (T0 and no Zn), except for the combination of soil and foliar Zn application, which notably increased root nodulation. This different impact from the biofortification treatments on nodulation could be due to the higher application rate in the SF treatment, as it has been identified in previous studies as one of the interacting factors between biofortification and nodulation [61].

In this sense, the combination of T05 and T14 (together with the positive control, Tc) with the soil Zn application, and the combination of T05 and T18 with the foliar application of Zn led to similar levels of Mg to those of the negative control without Zn (T0 and no Zn) being maintained. In the first case, these fungi did not affect the rhizobial nodulation index, and therefore, they could have modulated the soil nutrient dynamics together with microbial communities [62] and buffered the negative effect of the increased Zn concentration in the soil colloid on the Mg content. Conversely, in the cases of T05 and T18 (both *T. gamsii*), which significantly reduced rhizobial nodulation, the same buffering effect only occurred with foliar applications since the beneficial effect with the soil microbial consortium seemed to have been less important. This underscores the potential additive benefits of combining different beneficial microbes to enhance plant growth.

Additionally, the use of different wild strains of *Trichoderma*, which may have different abilities to promote plant growth and nutrient uptake, as well as the complexities of Zn uptake pathways in plants [63,64], may have contributed to the observed variability in the results. The same variable results, depending on the specific combination of fungal strain and biofortification treatment, were found by Marra et al. in their study on the biofortification of lentils [53]. At the same time, this result emphasizes the importance of the correct formulation and timing of the application of these biological products containing living microorganisms, as well as the importance of optimizing the application of inorganic Zn fertilizers, which can be costly and whose efficacy may vary depending on the plant being fertilized, rendering it insufficient as a standalone strategy for improving zinc deficiency and enhancing crop yield [65]. Thus, based on the preliminary results of our study, future experiments should consider the optimization of both the microbial inoculation and the Zn application conditions of the most promising combinations. In this regard, the results obtained by Anwar et al. [42] show that the application of *Trichoderma harzianum* together with copper as a biofortification strategy, using a lower dosage of the mineral, produced a higher accumulation of said nutrient in the treated wheat plants.

Notably, the combination of *T. koningiopsis* (T05) inoculation with Zn significantly increased the Fe levels, with similar results being observed for other isolates such as T14 (*T. koningii*). The most effective strategy for Fe uptake was the combination of soil Zn with T05 and T14, resulting in substantial increases compared to the control. In addition, the SF treatment, particularly in combination with the inoculation of T09 (*T. gamsii*) or T14

(*T. koningii*), significantly increased the Zn levels in the plants, surpassing those of the negative control that underwent the same Zn treatment. The combined treatment of Zn biofortification with *Trichoderma* spp. may have enhanced the nutrient uptake efficiency of subterranean clover. *Trichoderma* spp. can solubilize nutrients in the soil, making them more accessible to plants [64]. By inoculating subterranean clover with different *Trichoderma* spp., the plants may have experienced improved root colonization and nutrient mobilization, leading to increased Zn and Fe uptake. Furthermore, the interactions between *Trichoderma* spp. and plant roots can trigger systemic responses in plants, including the activation of defense mechanisms and stress tolerance pathways [66]. These responses can optimize nutrient uptake processes and mitigate the negative effects of Zn application on plant growth. The combined treatment of Zn biofortification and *Trichoderma* spp. inoculation may have induced a more robust plant defense and nutrient acquisition system, resulting in the observed increase in Zn and Fe uptake in subterranean clover.

Regarding Se uptake, the different effects of the biofortification treatments may be explained by prior studies that have shown that foliar Zn application can notably impact the Se content in plants [21,67], whereas the positive effect of isolate T18 (*T. gamsii*) may be related to a potential ability to mobilize this nutrient more effectively than the other isolates.

The observed reduction in the soil bioavailable Zn content when *Trichoderma* spp. was inoculated without a Zn application, as well as with a foliar application, could be attributed to the ability of *Trichoderma* to alter soil conditions [68]. It has been reported that *Trichoderma* spp. promotes the solubilization of complex compounds by stimulating the secretion of organic acids from plant roots, leading to a significant decrease in the soil pH [69]. This change in soil pH may have influenced the availability and mobility of the Zn that is naturally present in the soil, as none of these plants received additional Zn from the soil, potentially resulting in decreased Zn levels in the soil.

On the other hand, when Zn was applied to the soil (soil and SF) in combination with *Trichoderma* inoculation, a general positive effect on bioavailable Zn accumulation was observed, especially for T09 (*T. gamsii*) and T14 (*T. koningii*). This could be due to the beneficial effects of soil Zn applications and *Trichoderma* inoculation on plant nutrient uptake, which may have facilitated the uptake and translocation of Zn within the plant, leading to increased Zn accumulation despite the initial reduction in the soil Zn content. Overall, the contrasting effects of *Trichoderma* inoculation and Zn application on the soil Zn content highlight the complex interactions between microbial activity, nutrient availability, and plant responses. Further research into the specific mechanisms underlying these interactions, such as changes in the soil pH, nutrient solubilization, and plant–microbe interactions, could provide valuable insights into optimizing agricultural practices for improved nutrient uptake and biofortification.

5. Conclusions

In conclusion, this study highlights several key findings that contribute to our understanding of the beneficial effects of *Trichoderma* spp. inoculation and Zn biofortification on plant growth and nutrient uptake. Firstly, the combined application of *Trichoderma* spp. and Zn biofortification led to increased Fe and Zn accumulation in plants, demonstrating the potential of these strategies to enhance nutrient uptake efficiency. However, different *Trichoderma* strains exhibited variable effects on plant growth and nodulation, emphasizing the importance of strain selection in agricultural practices. While some strains (T05, *T. koningiopsis* and T14, *T. koningii*) promoted plant growth and nutrient uptake, others (T09 and T18, both identified as *T. gamsii*) inhibited nodulation.

Moreover, this study revealed the complex interactions between *Trichoderma* spp., Zn biofortification, and soil dynamics, leading to variable outcomes depending on factors such as soil conditions and application methods. These results suggest that optimization opportunities exist to maximize the benefits of microbial inoculants and biofortification in sustainable agriculture. Additionally, integrating locally adapted microorganisms like

Trichoderma spp. with Zn biofortification presents a promising approach to enhance crop productivity and sustainability, particularly in vulnerable ecosystems like dehesas. To this end, their long-term survival may be a key factor that needs to be taken into account in future studies.

In summary, these findings underscore the potential of combining beneficial microorganisms with biofortification strategies to address nutrient deficiencies and improve agricultural sustainability. However, further research is needed to elucidate the underlying mechanisms and optimize application strategies to maximize these benefits.

Supplementary Materials: The following supporting information can be downloaded at https://www.mdpi.com/article/10.3390/su16093730/s1, Table S1. Effect of the interaction between the application of the *Trichoderma* spp. Isolates and the biofortification treatment on fiber content, measured through Neutral Detergent Fiber (NDF) Acid Detergent Fiber, Acid Detergent Lignin (ADL) and the content of total ashes. Additionally, the percentage of digestibility of the treated plants is shown; Figure S1. Correlation Matrix showing the significant Pearson correlations ($p \leq 0.05$) between parameters investigated in the study. The robustness of the correlations is depicted using a gradation of colors, with darker colors indicating stronger correlations.

Author Contributions: Conceptualization, C.G.-L. and M.J.P.; Data Curation, C.G.-L., R.V. and A.H.; Formal Analysis, C.G.-L. and M.J.P.; Funding Acquisition, M.J.P., R.V. and A.H.; Investigation, C.G.-L. and P.T.; Methodology, C.G.-L. and M.J.P.; Project Administration, M.J.P., R.V. and A.H.; Resources, C.G.-L., M.J.P., R.V., P.T. and A.H.; Supervision, C.G.-L. and M.J.P.; Validation, R.V., A.H. and M.J.P.; Visualization, C.G.-L. and M.J.P.; Writing—Original Draft, C.G.-L.; Writing—Review and Editing, C.G.-L., M.J.P., R.V., P.T. and A.H. All authors have read and agreed to the published version of the manuscript.

Funding: Paula Tejero has been financed by a pre-doctoral grant from the Fundación Valhondo (Cáceres, Spain).

Institutional Review Board Statement: Not applicable.

Informed Consent Statement: Not applicable.

Data Availability Statement: All data are included in the present study.

Acknowledgments: We would like to thank Teodoro Garcia-White, Carlota Ortíz Benítez, and Carlos Moraga Lozano for their invaluable help with the laboratory work.

Conflicts of Interest: The authors declare no conflicts of interest.

References

1. MAPA. *Encuesta Sobre Superficies y Rendimientos de Cultivos de 2023*; MAPA: Madrid, Spain, 2023.
2. Ferrer, C.; Miguel, A.S.; Olea, L. Nomenclátor Básico De Pastos En España. *Pastos* **2001**, *31*, 7–44.
3. Ruiz-Mirazo, J.; Robles, A.B.; González-Rebollar, J.L. Two-Year Evaluation of Fuelbreaks Grazed by Livestock in the Wildfire Prevention Program in Andalusia (Spain). *Agric. Ecosyst. Environ.* **2011**, *141*, 13–22. [CrossRef]
4. González, F.; Maya, V. Los Pastos y Su Importancia En La Comunidad de Extremadura. Métodos de Mejora. In *Los Pastos: Nuevos Retos, Nuevas Oportunidades, Proceedings of the LII Reunión Científica de la Sociedad Española para el Estudio de los Pastos, Badajoz, Spain, 8–12 April 2013*; Olea, L., Poblaciones, M.J., Morales, S., Santamaria, O., Eds.; Sociedad Española Para el Estudio de los Pastos: Madrid, Spain, 2013; pp. 83–105.
5. López-Carrasco, C.; López-Sánchez, A.; San Miguel, A.; Roig, S. The Effect of Tree Cover on the Biomass and Diversity of the Herbaceous Layer in a Mediterranean Dehesa. *Grass Forage Sci.* **2015**, *70*, 639–650. [CrossRef]
6. Andreu, A.; Kustas, W.P.; Polo, M.J.; Carrara, A.; González-Dugo, M.P. Modeling Surface Energy Fluxes over a Dehesa (Oak Savanna) Ecosystem Using a Thermal Based Two-Source Energy Balance Model (TSEB) I. *Remote Sens.* **2018**, *10*, 567. [CrossRef]
7. Hidalgo-Galvez, M.D.; Matías, L.; Cambrollé, J.; Gutiérrez, E.; Pérez-Ramos, I.M. Impact of Climate Change on Pasture Quality in Mediterranean Dehesas Subjected to Different Grazing Histories. *Plant Soil* **2023**, *488*, 465–483. [CrossRef]
8. Hidalgo-Galvez, M.D.; Barkaoui, K.; Volaire, F.; Matías, L.; Cambrollé, J.; Fernández-Rebollo, P.; Carbonero, M.D.; Pérez-Ramos, I.M. Can Trees Buffer the Impact of Climate Change on Pasture Production and Digestibility of Mediterranean Dehesas? *Sci. Total Environ.* **2022**, *835*, 155535. [CrossRef]
9. Seidl, R.; Rammer, W.; Jäger, D.; Currie, W.S.; Lexer, M.J. Assessing Trade-Offs between Carbon Sequestration and Timber Production within a Framework of Multi-Purpose Forestry in Austria. *For. Ecol. Manag.* **2007**, *248*, 64–79. [CrossRef]

10. Lüscher, A.; Mueller-Harvey, I.; Soussana, J.F.; Rees, R.M.; Peyraud, J.L. Potential of Legume-Based Grassland–Livestock Systems in Europe: A Review. *Grass Forage Sci.* **2014**, *69*, 206–228. [CrossRef]
11. Serrano, J.M.; Peça, J.O.; da Silva, J.R.M.; Shahidian, S.; Carvalho, M. Phosphorus Dynamics in Permanent Pastures: Differential Fertilizing and the Animal Effect. *Nutr. Cycl. Agroecosystems* **2011**, *90*, 63–74. [CrossRef]
12. Hernández-Esteban, A.; López-Díaz, M.L.; Cáceres, Y.; Moreno, G. Are Sown Legume-Rich Pastures Effective Allies for the Profitability and Sustainability of Mediterranean Dehesas? *Agrofor. Syst.* **2019**, *93*, 2047–2065. [CrossRef]
13. López-Sánchez, A.; San Miguel, A.; Dirzo, R.; Roig, S. Scattered Trees and Livestock Grazing as Keystones Organisms for Sustainable Use and Conservation of Mediterranean Dehesas. *J. Nat. Conserv.* **2016**, *33*, 58–67. [CrossRef]
14. Pulido, M.; Schnabel, S.; Contador, J.F.L.; Lozano-Parra, J.; Gómez-Gutiérrez, Á. Selecting Indicators for Assessing Soil Quality and Degradation in Rangelands of Extremadura (SW Spain). *Ecol. Indic.* **2017**, *74*, 49–61. [CrossRef]
15. Díaz-Pereira, E.; Romero-Díaz, A.; de Vente, J. Sustainable Grazing Land Management to Protect Ecosystem Services. *Mitig. Adapt. Strateg. Glob. Chang.* **2020**, *25*, 1461–1479. [CrossRef]
16. Nichols, P.G.H.; Foster, K.J.; Piano, E.; Pecetti, L.; Kaur, P.; Ghamkhar, K.; Collins, W.J. Genetic Improvement of Subterranean Clover (*Trifolium Subterraneum* L.). 1. Germplasm, Traits and Future Prospects. *Crop Pasture Sci.* **2013**, *64*, 312–346. [CrossRef]
17. Llobat, L.; Marín-García, P.J. Application of Protein Nutrition in Natural Ecosystem Management for European Rabbit (*Oryctolagus Cuniculus*) Conservation. *Biodivers. Conserv.* **2022**, *31*, 1435–1444. [CrossRef]
18. Wei, Y.; Shohag, M.J.I.; Yang, X. Biofortification and Bioavailability of Rice Grain Zinc as Affected by Different Forms of Foliar Zinc Fertilization. *PLoS ONE* **2012**, *7*, e45428. [CrossRef]
19. Cakmak, I.; Kutman, U.B. Agronomic Biofortification of Cereals with Zinc: A Review. *Eur. J. Soil Sci.* **2018**, *69*, 172–180. [CrossRef]
20. Rehman, A.; Farooq, M.; Ullah, A.; Nadeem, F.; Im, S.Y.; Park, S.K.; Lee, D.J. Agronomic Biofortification of Zinc in Pakistan: Status, Benefits, and Constraints. *Front. Sustain. Food Syst.* **2020**, *4*, 591722. [CrossRef]
21. Cakmak, I.; Kalayci, M.; Kaya, Y.; Torun, A.A.; Aydin, N.; Wang, Y.; Arisoy, Z.; Erdem, H.; Yazici, A.; Gokmen, O.; et al. Biofortification and Localization of Zinc in Wheat Grain. *J. Agric. Food Chem.* **2010**, *58*, 9092–9102. [CrossRef]
22. Melash, A.A.; Mengistu, D.K.; Aberra, D.A. Linking Agriculture with Health through Genetic and Agronomic Biofortification. *Agric. Sci.* **2016**, *7*, 295–307. [CrossRef]
23. Rivera-Martin, A.; Reynolds-Marzal, D.; Martin, A.; Velzazquez, R.; Poblaciónes, M.J. Combined Foliar Zinc and Nitrogen Application in Broccoli (*Brassica Oleracea* Var. *Italica* L.): Effects on Growth, Nutrient Bioaccumulation, and Bioactive Compounds. *Agronomy* **2021**, *11*, 548.
24. Santos, S.A.; Takahashi, F.; Cardoso, E.L.; Flores, C.; de Oliveira, L.O.F.; Souza, G.D.S.; Gomes, E.G.; Ortega, E. An Emergy-Based Approach to Assess and Valuate Ecosystem Services of Tropical Wetland Pastures in Brazil. *Open J. Ecol.* **2020**, *10*, 303–319. [CrossRef]
25. Chugh, G.; Siddique, K.H.M.; Solaiman, Z.M. Iron Fortification of Food Crops through Nanofertilisation. *Crop Pasture Sci.* **2022**, *73*, 736–748. [CrossRef]
26. Upadhayay, V.K.; Singh, A.V.; Khan, A. Cross Talk Between Zinc-Solubilizing Bacteria and Plants: A Short Tale of Bacterial-Assisted Zinc Biofortification. *Front. Soil Sci.* **2021**, *1*, 788170. [CrossRef]
27. Bamdad, H.; Papari, S.; Lazarovits, G.; Berruti, F. Soil Amendments for Sustainable Agriculture: Microbial Organic Fertilizers. *Soil Use Manag.* **2022**, *38*, 94–120. [CrossRef]
28. Jalal, A.; Mortinho, E.S.; da Silva Oliveira, C.E.; Fernandes, G.C.; Junior, E.F.; de Lima, B.H.; Moreira, A.; Nogueira, T.A.R.; Galindo, F.S.; Filho, M.C.M.T. Nano-Zinc and Plant Growth-Promoting Bacteria Is a Sustainable Alternative for Improving Productivity and Agronomic Biofortification of Common Bean. *Chem. Biol. Technol. Agric.* **2023**, *10*, 77. [CrossRef]
29. Hall, J.A.; Bobe, G.; Filley, S.J.; Bohle, M.G.; Pirelli, G.J.; Wang, G.; Davis, T.Z.; Bañuelos, G.S. Impact of Selenium Biofortification on Production Characteristics of Forages Grown Following Standard Management Practices in Oregon. *Front. Plant Sci.* **2023**, *14*, 1121605. [CrossRef]
30. Hall, J.A.; Bobe, G.; Filley, S.J.; Pirelli, G.J.; Bohle, M.G.; Wang, G.; Davis, T.Z.; Bañuelos, G.S. Effects of Amount and Chemical Form of Selenium Amendments on Forage Selenium Concentrations and Species Profiles. *Biol. Trace Elem. Res.* **2023**, *201*, 4951–4960. [CrossRef]
31. Lledó, S.; Rodrigo, S.; Poblaciones, M.J.; Santamaria, O. Biomass Yield, Nutritive Value and Accumulation of Minerals in *Trifolium Subterraneum* L. as Affected by Fungal Endophytes. *Plant Soil* **2016**, *405*, 197–210. [CrossRef]
32. Porteous-Álvarez, A.J.; Fernández-Marcos, A.; Ramírez-Lozano, D.; Mayo-Prieto, S.; Cardoza, R.E.; Gutiérrez, S.; Casquero, P.A. Native *Trichoderma* Isolates from Soil and Rootstock to Fusarium Spp. Control and Growth Promotion of *Humulus Lupulus* L. Plantlets. *Agriculture* **2023**, *13*, 720. [CrossRef]
33. Walkley, A.J.; Black, I.A. Estimation of Soil Organic Carbon by the Chromic Acid Titration Method. *Soil Sci.* **1934**, *37*, 29–38. [CrossRef]
34. Colwell, J.D. An Automatic Procedure for the Determination of Phosphorus in Sodium Hydrogen Carbonate Extract of Soil. *Chem. Ind.* **1965**, *21*, 893–895.
35. Lindsay, W.L.; Norvell, W.A. Development of a DTPA Soil Test for Zinc, Iron, Manganese, and Copper. *Soil Sci. Soc. Am. J.* **1978**, *42*, 421–428. [CrossRef]
36. Zabalgogeazcoa, Í.; Ciudad, A.G.; Vázquez de Aldana, B.R.; Criado, B.G. Effects of the Infection by the Fungal Endophyte Epichloë Festucae in the Growth and Nutrient Content of *Festuca rubra*. *Eur. J. Agron.* **2006**, *24*, 374–384. [CrossRef]

37. AOAC. *Official Methods of Analysiss. Association of Official Analytical Chemists*; AOAC: Rockville, MD, USA, 2006.
38. Kuehn, C.S.; Jung, H.G.; Linn, J.G.; Martin, N.P. Characteristics of Alfalfa Hay Quality Grades Based on the Relative Feed Value Index. *J. Prod. Agric.* **1999**, *12*, 681–684. [CrossRef]
39. R Core Team. *R: A Language and Environment for Statistical Computing*; R Foundation for Statistical Computing: Vienna, Austria, 2024.
40. Parra-López, C.; Sayadi, S.; Garcia-Garcia, G.; Ben Abdallah, S.; Carmona-Torres, C. Prioritising Conservation Actions towards the Sustainability of the Dehesa by Integrating the Demands of Society. *Agric. Syst.* **2023**, *206*, 103613. [CrossRef]
41. Ali, I.; Khan, A.; Ali, A.; Ullah, Z.; Dai, D.Q.; Khan, N.; Khan, A.; Al-Tawaha, A.R.; Sher, H. Iron and Zinc Micronutrients and Soil Inoculation of *Trichoderma harzianum* Enhance Wheat Grain Quality and Yield. *Front. Plant Sci.* **2022**, *13*, 960948. [CrossRef]
42. Anwar, S.; Ali, A.; Ullah, Z.; Binjawhar, D.N.; Sher, H.; Ali, R.; Iqbal, R.; Ali, B.; Ali, I. The Impact of *Trichoderma harzianum* Together with Copper and Boron on Wheat Yield. *ACS Agric. Sci. Technol.* **2023**, *3*, 517–527. [CrossRef]
43. Ku, Y.S.; Rehman, H.M.; Lam, H.M. Possible Roles of Rhizospheric and Endophytic Microbes to Provide a Safe and Affordable Means of Crop Biofortification. *Agronomy* **2019**, *9*, 764. [CrossRef]
44. Aishwarya, S.; Viswanath, H.S.; Singh, A.; Singh, R. Biosolubilization of Different Nutrients by *Trichoderma* spp. and Their Mechanisms Involved: A Review. *Int. J. Adv. Agric. Sci. Technol.* **2020**, *7*, 34–39.
45. Cao, Q.; Liang, Y.; Tian, Y.; Lian, H.; Jiang, X.; Li, M. Survival Dynamics of *Trichoderma longibrachiatum* Tr58 in Conidia- and Chlamydospore-Amended Soils with Different Moisture Levels. *Agric.* **2023**, *13*, 238. [CrossRef]
46. Longa, C.M.O.; Savazzini, F.; Tosi, S.; Elad, Y.; Pertot, I. Evaluating the Survival and Environmental Fate of the Biocontrol Agent *Trichoderma atroviride* SC1 in Vineyards in Northern Italy. *J. Appl. Microbiol.* **2009**, *106*, 1549–1557. [CrossRef] [PubMed]
47. Sood, M.; Kapoor, D.; Kumar, V.; Sheteiwy, M.S.; Ramakrishnan, M.; Landi, M.; Araniti, F.; Sharma, A. *Trichoderma*: The "Secrets" of a Multitalented Biocontrol Agent. *Plants* **2020**, *9*, 762. [CrossRef] [PubMed]
48. Nandini, B.; Puttaswamy, H.; Saini, R.K.; Prakash, H.S.; Geetha, N. Trichovariability in Rhizosphere Soil Samples and Their Biocontrol Potential against Downy Mildew Pathogen in Pearl Millet. *Sci. Rep.* **2021**, *11*, 9517. [CrossRef] [PubMed]
49. Szerement, J.; Szatanik, A.; Jakub, K.; Monika, M.; Hersztek, M. *Agronomic Biofortification with Se, Zn, and Fe: An Effective Strategy to Enhance Crop Nutritional Quality and Stress Defense—A Review*; Springer: Berlin/Heidelberg, Germany, 2022; ISBN 0123456789.
50. Lata-Tenesaca, L.F.; de Mello Prado, R.; Ajila-Celi, G.E.; da Silva, D.L.; Junior, J.S.P.; Mattiuz, B.H. Iron Biofortification in Quinoa: Effect of Iron Application Methods on Nutritional Quality, Anti-Nutrient Composition, and Grain Productivity. *Food Chem.* **2023**, *404*, 134573. [CrossRef] [PubMed]
51. Tan, B.; Li, Y.; Deng, D.; Pan, H.; Zeng, Y.; Tan, X.; Zhuang, W.; Li, Z. Role of Zinc Nutrition for Increasing Zinc Availability, Uptake, Yield, and Quality of Maize (*Zea Mays* L.) Grains: An Overview. *Commun. Soil Sci. Plant Anal.* **2020**, *51*, 2001–2021. [CrossRef]
52. Umair Hassan, M.; Aamer, M.; Umer Chattha, M.; Haiying, T.; Shahzad, B.; Barbanti, L.; Nawaz, M.; Rasheed, A.; Afzal, A.; Liu, Y.; et al. The Critical Role of Zinc in Plants Facing the Drought Stress. *Agriculture* **2020**, *10*, 396. [CrossRef]
53. Marra, R.; Lombardi, N.; Piccolo, A.; Bazghaleh, N.; Prashar, P.; Vandenberg, A.; Woo, S. Mineral Biofortification and Growth Stimulation of Lentil Plants Inoculated with Trichoderma Strains and Metabolites. *Microorganisms* **2022**, *10*, 87. [CrossRef] [PubMed]
54. Taghavi Ghasemkheili, F.; Jenabiyan, M.; Ghadirnezhad Shiade, S.R.; Pirdashti, H.; Ghanbari, M.A.T.; Emadi, M.; Yaghoubian, Y. Screening of Some Endophytic Fungi Strains for Zinc Biofortification in Wheat (*Triticum aestivum* L.). *J. Soil Sci. Plant Nutr.* **2023**, *23*, 5196–5206. [CrossRef]
55. Baligah, H.U.; Mir, S.A.; Sofi, P.; Mir, A.H. Grain Zinc Phytic Acid and Nutrient Uptake in Common Bean Is Influenced by Sources and Concentrations of Zinc Fertilization. *Int. Res. J. Pure Appl. Chem.* **2020**, *21*, 9–17. [CrossRef]
56. Ledgard, S.F.; Steele, K.W. Biological Nitrogen Fixation in Mixed Legume/Grass Pastures. *Plant Soil* **1992**, *141*, 137–153. [CrossRef]
57. Liu, D.Y.; Liu, Y.M.; Zhang, W.; Chen, X.P.; Zou, C.Q. Zinc Uptake, Translocation, and Remobilization in Winter Wheat as Affected by Soil Application of Zn Fertilizer. *Front. Plant Sci.* **2019**, *10*, 443999. [CrossRef] [PubMed]
58. Moradtalab, N.; Ahmed, A.; Geistlinger, J.; Walker, F.; Höglinger, B.; Ludewig, U.; Neumann, G. Synergisms of Microbial Consortia, N Forms, and Micronutrients Alleviate Oxidative Damage and Stimulate Hormonal Cold Stress Adaptations in Maize. *Front. Plant Sci.* **2020**, *11*, 396. [CrossRef] [PubMed]
59. Burr, A.A.; Woods, K.D.; Cassidy, S.T.; Wood, C.W. Priority Effects Alter the Colonization Success of a Host-Associated Parasite and Mutualist. *Ecology* **2022**, *103*, e3720. [CrossRef]
60. Khan, R.A.A.; Najeeb, S.; Chen, J.; Wang, R.; Zhang, J.; Hou, J.; Liu, T. Insights into the Molecular Mechanism of *Trichoderma* Stimulating Plant Growth and Immunity against Phytopathogens. *Physiol. Plant.* **2023**, *175*, e14133. [CrossRef] [PubMed]
61. Rouphael, Y.; Kyriacou, M.C. Enhancing Quality of Fresh Vegetables through Salinity Eustress and Biofortification Applications Facilitated by Soilless Cultivation. *Front. Plant Sci.* **2018**, *9*, 400810. [CrossRef] [PubMed]
62. Halifu, S.; Deng, X.; Song, X.; Song, R. Effects of Two Trichoderma Strains on Plant Growth, Rhizosphere Soil Nutrients, and Fungal Community of *Pinus sylvestris* Var. *Mongolica Annual Seedlings. Forests* **2019**, *10*, 758. [CrossRef]
63. Wang, Y.; Yang, X.; Zhang, X.; Dong, L.; Zhang, J.; Wei, Y.; Feng, Y.; Lu, L. Improved Plant Growth and Zn Accumulation in Grains of Rice (*Oryza sativa* L.) by Inoculation of Endophytic Microbes Isolated from a Zn Hyperaccumulator, Sedum Alfredii H. *J. Agric. Food Chem.* **2014**, *62*, 1783–1791. [CrossRef] [PubMed]

64. Li, R.X.; Cai, F.; Pang, G.; Shen, Q.R.; Li, R.; Chen, W. Solubilisation of Phosphate and Micronutrients by *Trichoderma harzianum* and Its Relationship with the Promotion of Tomato Plant Growth. *PLoS ONE* **2015**, *10*, e0130081. [CrossRef]
65. Abdu, A.O.; De Groote, H.; Joy, E.J.M.; Kumssa, D.B.; Broadley, M.R.; Gashu, D. Zinc Agronomic Biofortification of Staple Crops May Be a Cost-Effective Strategy to Alleviate Zinc Deficiency in Ethiopia. *Front. Nutr.* **2022**, *9*, 1037161. [CrossRef]
66. Galletti, S.; Paris, R.; Cianchetta, S. Selected Isolates of *Trichoderma gamsii* Induce Different Pathways of Systemic Resistance in Maize upon *Fusarium verticillioides* Challenge. *Microbiol. Res.* **2020**, *233*, 126406. [CrossRef] [PubMed]
67. Ning, P.; Fei, P.; Wu, T.; Li, Y.; Qu, C.; Li, Y.; Shi, J.; Tian, X. Combined Foliar Application of Zinc Sulphate and Selenite Affects the Magnitude of Selenium Biofortification in Wheat (*Triticum aestivum* L.). *Food Energy Secur.* **2022**, *11*, e342. [CrossRef]
68. Wu, J.; Zhu, J.; Zhang, D.; Cheng, H.; Hao, B.; Cao, A.; Yan, D.; Wang, Q.; Li, Y. Beneficial Effect on the Soil Microenvironment of *Trichoderma* Applied after Fumigation for Cucumber Production. *PLoS ONE* **2022**, *17*, e0266347. [CrossRef] [PubMed]
69. Tan, B.; Li, Y.; Deng, D.; Pan, H.; Zeng, Y.; Tan, X.; Zhuang, W.; Li, Z. Rhizosphere Inoculation of *Nicotiana benthamiana* with *Trichoderma harzianum* TRA1-16 in Controlled Environment Agriculture: Effects of Varying Light Intensities on the Mutualism-Parasitism Interaction. *Front. Plant Sci.* **2022**, *13*, 989155. [CrossRef] [PubMed]

Disclaimer/Publisher's Note: The statements, opinions and data contained in all publications are solely those of the individual author(s) and contributor(s) and not of MDPI and/or the editor(s). MDPI and/or the editor(s) disclaim responsibility for any injury to people or property resulting from any ideas, methods, instructions or products referred to in the content.

sustainability

MDPI

Article

Climate Change Effects on Dates Productivity in Saudi Arabia: Implications for Food Security

Abda Emam

Department of Agribusiness and Consumer Sciences, College of Agricultural Science and Food, King Faisal University, Al-Ahsa 31982, Saudi Arabia; aaeali@kfu.edu.sa

Abstract: This study aimed to assess the impact of climatic alteration on food security in Saudi Arabia. Date productivity, temperature, and precipitation represent the data which were collected from various sources linked to the study subject and cover the period from 1980 to 2023. The Engle–Granger two-step procedure, the VECM, and forecast analysis were applied to test the long-term relationship, short-term integration, and forecasting, respectively. Moreover, qualitative analysis was used to reveal the influence of climatic change on food security. The results discovered long-term co-integration between date productivity and temperature. Additionally, the results revealed that there has been long-running co-integration between date productivity and the precipitation series. Temperature and precipitation negatively and significantly impacted date productivity during the study period. With reference to forecast results, the graph was validated using various forecast indicators: the Alpha, Gamma, Beta, and Mean Square Error equivalents were 1.0, 0.0, 0.0, and 5.47, respectively. Moreover, the growth rates of date productivity were equal to 0.82 and 0.08 for the periods from 1980 to 2022 and 2023 to 2034 (forecast), respectively, indicating that there is a decrease in the growth rate of date productivity (0.08) during the forecast period. From these results, the conclusion is that climatic change (temperature and precipitation) negatively impacts date productivity. In addition, the growth rate during the forecast period decreased, indicating that climatic change is affecting food security currently and will continue to do so in the future. This study recommended specific policy interventions and innovations in agricultural practices, including developing and implementing a national framework focused on climate-smart agriculture, balancing productivity, adaptation, and mitigation. This could be aligned with Vision 2030 and the Saudi Green Initiative. Additionally, this could include investing in research and development by increasing public–private partnerships to support agricultural R&D in arid regions, with a focus on heat- and drought-resistant crop varieties and water-efficient farming systems. Regarding agricultural innovations, these could include the use of renewable energy, particularly solar energy, the expansion of rainwater harvesting infrastructure, recycling treated wastewater for agriculture, and reducing reliance on groundwater sources.

Keywords: date productivity; temperature; precipitation; Saudi

1. Introduction

Climate change is a chief environmental topic at the global level [1]. International warming primarily arises from carbon dioxide (CO_2) emission levels [2]. Agricultural sectors and ecosystems are presently affected by climate change at national and global levels [3,4]. Climate change is demarcated as any change in the state of the climate which continues over a period of time and is accounted as one of the chief ecological problems troubling humankind in the twenty-first century [5–7]. Climatic factors are altering quickly owing to the rise in greenhouse gas emissions [3]. The rise in the intensity of greenhouse gases has resulted in a rise in average temperatures and altered the size and spread of precipitation universally. The universal temperature is increasing owing to human behaviors such as burning fossil fuels and clearing forests to build buildings [8]. High temperatures disrupt crop growth and development, which causally affects their economic yield [5,6,8,9], and, hence, economic sustainability [5,10,11], and, finally, the food security situation [12].

Food security is demarcated as "existing when all people, at all times, have physical and economic access to sufficient, safe, and nutritious food that meets their dietary needs and food preferences for an active and healthy life" [13,14]. This definition includes the broad concept of food security, which is classified according to four pillars, related to availability, access, utilization, and stability [14–17]. The pillars are connected by agricultural novelty [16,18], social capital [16,19], kitchen appliances [16,20], and global tremors [16,21]. One of the pillars of food security is the availability of food [22]. The availability of food is linked directly to agriculture [12]. The agricultural sector acts in a strategic role to cultivate food readiness and attain food security. In other words to realize food security, the agricultural sector should acts in a strategic role to enhance food availability [23–26].

Many authors have documented the impact of climate change on agricultural production. Agricultural production is threatened by climate change in food-insecure counties. Many extreme climate events, such as droughts, heat waves, erratic and intense rainfall patterns, storms, floods, and emerging insect pests, have negatively impacted farmers' livelihoods [27]. Future climate projections show substantial rises in temperatures and uneven rainfall with greater amounts.

The studies [28,29] examined the potential risks posed by climate change on agriculture—an essential sector of the economy—particularly focusing on its effects on agricultural productivity and farm incomes across nations. The studies emphasize the possible hazards of climate change in relation to agriculture, including its effects on crop harvests, agricultural infrastructure, and water resources, and propose adaptation and extenuation strategies.

The studies [30,31] confirm the significant threat climate change poses to crop yields, water assets, soil health, developing agricultural production, and country-level food security. To meet this challenge, solid global inventive and collaborative actions are immediately necessary.

The study [32] deliberated on the influence of rising temperatures on the productivity of the main cereal crops worldwide and the prominence of the temperature–moisture link in adaptable crop reactions to heat from an international viewpoint. The researchers found that in some areas, lower precipitation and evaporation, combined with greater temperatures, led to extra noticeable production declines in crops, including corn and soybeans, suggesting a combined effect of heat and drought on yields. The study outcomes indicate that climate change and moisture stress factors affected crop production.

State-of-the-art climate and crop models have been used to forecast alterations in harvests of maize, soybean, and rice [28,33]. Some studies have examined the different drivers of climate change, comprising rising temperatures, changes in rainfall forms, drought patterns, and significant carbon dioxide uptake. The outcomes suggest that the influence of climate change on agricultural productivity may be greater than earlier estimated and that relatively high-latitude areas may encounter high-yield harvests whereas low-latitude tropical areas may see lower yields.

The most common crops have been analyzed using different models, taking into account the impact of climate change on their productivity [28,34]. The results specify that climate change considerably disturbs yields, and prospective climate variation predictions suggest that there will be a shrinkage in average crop production worldwide, particularly in hotter areas. Consequently, modifications and alterations in agricultural structures will be necessary in response to future climate change.

Another study aimed to assess the relationship between date production and CO_2 in Saudi Arabia, using econometric analysis tools [35]. The study appeared to show that there has been a long running relationship between these variables.

Climatic change significantly and profoundly impacts soil, which, in turn, adversely affects food security [36]. In addition, hazardous weather caused by climate change threatens crops around the world, making it difficult to sustain stable food systems [37,38].

The date palm (*Phoenix dactylifera*) is one of the main crops produced in the Kingdom of Saudi Arabia [39]. Among other crops, dates are considered a good food source with high nutritional value [40]. Globally, dates are cultured on 1 396.727 ha of land, with an annual production of 9 248.033 tonnes [41]. In Saudi Arabia alone, they are grown on 157,444 ha of land, with a production of 1,642,992.53 tonnes in 2023 [42]. Also, date palms accounted for 56% of total (primary) fruit production in 2022 [42] in Saudi Arabia. Local date consumption amounted, on average, to about 985.94 thousand dates in 2020/2021 [43] The National Agricultural Strategy 2030 includes date production in its plans to achieve food security in the Kingdom of Saudi Arabia [40]. As a result of the position of date palms and their product in Saudi Arabia, they were selected as variables in this study.

Building on prior studies, the importance of this study emerged from its intended aim, which is mainly to examine the influence of climatic change on food security in Saudi Arabia at present and in the future. Accordingly, the study was divided into three parts: part one related to testing the dynamic relationship between the variables under study, part two considered forecast analysis, and part three determined the effect of climatic change on food security.

2. Materials and Methods

2.1. Description of Data

The aim of this study was to determine the relationships between date productivity (kg/ha), temperature (°C), and precipitation (millimeter) and their impacts on food security. The data on such variables were gathered, covering the period from 1980 to 2023, and were examined using the EVIEWS 12 program. Information about the variables is summarized in Table 1.

Table 1. Variables' descriptions.

Variable	Unit	Sources
Date productivity (DP)	(kg/ha)	https://www.fao.org/faostat/ar/#data/QCL (accessed on 10 February 2025)
Annual average temperature	°C	https://climateknowledgeportal.worldbank.org/country/saudi-arabia (accessed on 10 February 2025)
Average precipitation	Millimeter	

2.2. Methods of Analysis

2.2.1. The Graphical Analysis Method

General index curves were drawn for the study's time series to initially shed light on the possibility of a stability link among the variables in the long run. An elementary analytical technique was used to establish this relationship, which also allowed us to proceed with performing a co-integration test.

2.2.2. Co-Integration Tests: Testing the Long-Term Relationship

1. Unit root test: The order of integration of the series was specified using an ADF test [35,44]. Equations (1) and (2) were used:

$$\Delta X_t = C_{t1} + b_1 X_{t-1} + E_{t1} \tag{1}$$

$$\Delta X_t = C_{t2} + B_t + b_2 X_{t-1} + E_{t2} \tag{2}$$

where b_1 and b_2 are the ADF parameters to be evaluated, B is the trend, C is the constant, E is the error term, and t is the period of time. In testing the null hypothesis (H_0), X has a unit root, while in the alternative hypothesis (H_1), X has a constant root (stationary). Variables are stationary if the t-statistics of the ADF coefficient are bigger than the critical t-values. The choice of method for examining the long-term association depends on the order of the series. The Engle–Granger test is applied when the two series' orders are equal to 1 (1) [45,46].

2. Engle–Granger test: This test is used to determine long-term associations.

To perform the co-integration test using the Engle–Granger method, the order of integration of the series should be determined first (this can be achieved using a similar order of integration as above (1)). Next, the Engle–Granger test [45] can be run. This comprises the following steps, and the regression equations are as follows:

$$X_t = a_1 \ + B_3 \ Y_t + z_{t1} \tag{3}$$

$$Y_t = a_2 \ + B_4 X_t + z_{t2} \tag{4}$$

where B_3 and B_4 are the slopes from Equations (3) and (4), respectively.

The ADF test is practiced on the residues (z_{t1} and z_{t2}) to examine whether the series are linked or not. If the ADF statistic is negative and larger than the critical t-value (order 1), then the coefficients B3 and B4 are expected to exist. Therefore, the series is coupled in the order 1, 1.

2.2.3. Error Correction Model (VECM)

The VECM test is implemented after the co-integration test has captured the long-term connection between the series. It is utilized to measure the speediness factor of the short-term link between the chains [47]. The VECM equations are as follows:

$$\Delta(\mathrm{LY}) = \Delta \ B_5 LY_{-1} + \Delta B_6 LX_{-1} + B_7 \ V_{1(t-1)} + U_1 \tag{5}$$

$$\Delta(LX) = \Delta \ B_8 LX_{-1} + \Delta B_9 LY_{-1} + B_{10} V_{2(t-1)} + U_2 \tag{6}$$

where B_5 and B_8 are coefficients of the difference in the lag-dependent variable; B_6 and B_9 are coefficients of the difference in the lag-independent variable; B_7 and B_{10} are used to modulate the speediness (must be negative and significant to prevent model unsteadiness); V_1 and V_2 are error correction terms; U_1 and U_2 are error terms; and L is the logarithm. In addition, VECM feasibility was confirmed by conducting residual diagnostics tests.

2.2.4. Forecast Analysis (Date Productivity)

The VECM was subjected to forecast analysis (using date productivity as the dependent variable). To test the forecast, the ETS (Error, Trend, Seasonal) model was used. Alpha (α), Beta (β), Gamma (γ), and Mean Square Error (MSE) were used as the parameters of the model. The graphical feasibility of the forecast results was evaluated using Alpha (α), Beta (β), and Gamma (γ); these are smoothing parameters that regulate how the model responds to new data. Alpha (α) is responsible for level smoothing and determines how fast the model updates the overall average. Beta (β) is responsible for trend smoothing and controls how quickly the model adjusts to changing trends. Gamma (γ) is responsible for seasonality smoothing and measures how quickly seasonal patterns are updated. These factors range between 0 and 1: a greater value means extra load on new observations, resulting in the construction of a model that is more responsive to changes. Mean Square Error (MSE) is an average of the squared differences between forecasted and actual values. It is used to evaluate model precision and to optimize the smoothing factors. The smaller the RMSE, the better the forecast result. The growth rate of date productivity was also calculated for the period from 1980 to 2022 and for the period from 2023 to 2034 (the projection period) using the following equation [12]:

$$G_t = Y_t - Y_{t-1} \div Y_{t-1} \quad (7)$$

where G_t represent the growth rate between two subsequent years, Y designates date productivity, and t is a year. Finally, the growth rate of date productivity was measured for the periods from 1980 to 2022 and 2023 to 2034 (forecast) as follows:

$$G_{t\ldots n} = (G_t \ + G_{t+1} \ldots + G_{n)})/N \quad (8)$$

where $G_{t\ldots n}$ and N signify the growth rate for a specific period and the number of years, respectively.

2.2.5. The Impact of Climatic Change on Saudi Arabia's Food Security

Based on the study outcomes of integration and forecast analysis, qualitative analysis was performed to shed light on the impact of climate change on food security.

3. Results and Discussion

3.1. Descriptive Statistics Results

The outcomes of the descriptive tests are listed in Table 2. The Jarque–Bera probabilities were lower than the 0.05 levels for date productivity (DP) and precipitation (P), meaning that the variables had a non-normal distribution. Therefore, the data ought to be transformed (into logarithm form) to correct this non-normality condition.

Table 2. Descriptive statistics results.

	DP	T	P
Mean	7380.87	25.37	78.52
Median	6595.90	25.44	74.99
Skewness	1.23	−0.06	3.12
Kurtosis	3.58	2.60	15.61
Jarque–Bera	11.78	0.32	362.49
Probability	0.003	0.85	0.000
Observations	44	44	44

Source: Data collected and analyzed by the author.

3.2. Graphical Analysis

As a pre-test for co-integration, Figure 1 shows that there may have been a long-running relationship between date productivity and annual average temperature during the period from 1980 to 2023 in Saudi Arabia. This result led us to conduct further co-integration analyses.

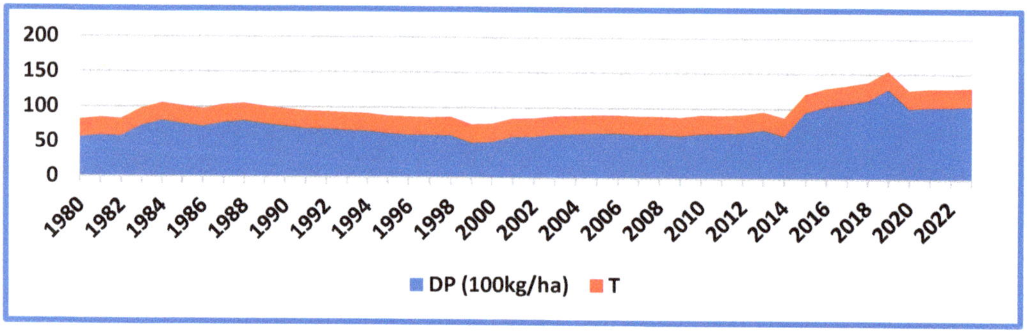

Figure 1. Graph of the performance of the series. DP: date productivity (100 kg/ha). T: temperature in °C. ha: hectare. Source: Drawn by author.

3.3. Co-Integration Analysis Test (Date Productivity and Temperature)

3.3.1. Unit Root Test Results

The stationarity of the series of date productivity and temperature was estimated using the unit root (ADF) test. Table 3 shows that the series remained stable after achieving an initial difference of 1 (1); accordingly, the ADF test statistics were significant at the 1% level. Therefore, the Engle–Granger test was used to estimate the associations between the series.

Table 3. Results of unit root test.

Time Series	Intercept	Intercept and Trend	Stationarity	Intercept	Intercept and Trend	Stationarity
	at Level			at First Difference		
LDP	−1.22	−1.59	Non-stationary	−6.96 *	−6.9 *	Stationary
LT	−0.65	−5.95 *	Stationary	−5.651 *	−5.96 *	Stationary

Source: Data collected and analyzed by the author. * At 1% level of significance.

3.3.2. Engle–Granger Test Results (Date Productivity and Temperature)

Table 4 shows the results of the ADF test on the residuals (z_{t1} and z_{t2} in Equations (3) and (4), respectively). The ADF statistics are negative (−6.81 and −6.81) and statistically significant at the 1% level, leading us to accept the alternative hypothesis of integration which proposes that date productivity and temperature are co-integrated. This result matched with that of an earlier study [12] which assessed date production and CO_2. It found that there was a long-running relationship between climate change (CO_2) and date productivity.

Table 4. Co-integration test. Engle–Granger test results with ADF residuals.

	LT	LDP
LDP	−6.81 *	
LT		−6.81 *

Source: Data collected and analyzed by the author. * Indicates 1% level of significance.

3.4. Results of VECM

To strengthen the finding of the Engle–Granger test (a long-term link between date productivity and temperature), the VECM was applied. To use the VECM, lag must be selected. Lag 1 was chosen, as shown in Table 5. Table 6 shows the results of the VECM. The estimated adjustment coefficient for date productivity (DP), defined as the dependent variable, was negative (−0.08) and statistically insignificant (t-value = −1.22). This means that the model does not demonstrate a significant ability to correct imbalances in previous periods. From the table, the long-term results suggested that temperature influenced the performance of date productivity negatively (coefficient = −9.44 with t-value = −1.76), resulting in lower date productivity. These results (showing a long-term relation between climate and date production) matched with those of previous studies [13–15,19]. Also, the results show that temperature has no effect on date productivity in the short term.

Table 5. Lag selection.

Lag	LogL	LR	FPE	AIC	SC	HQ
0	105.38	NA	2.80×10^{-5}	−4.81	−4.73	−4.78
1	156.25	94.66 *	3.16×10^{-6} *	−6.99 *	−6.74 *	−6.90 *

Source: Data collected and analyzed by the author. * LR—sequential modified likelihood ratio (LR) test statistic (each test at 5% level); logL—log lag variable (lag order selected by the criterion); AIC—Akaike information criterion; HQ—Hannan–Quinn information criterion; FPE—final prediction error; SC—Schwarz information criterion; NA—not available.

Table 6. Results of the VECM.

Long-Term Results: LDP (Dependent Variable)			Short-Term Results: LDP (Dependent Variable)	
Error Correction	**Coefficient**	***t*-Value Statistic**	**Coefficient**	***t*-Value Statistic**
CointEq1	−0.08	−1.22		
LT(−1)	−9.44	−1.76		
D(LDP(−1))			−0.04	−0.27
D(LT(−1))			−0.98	−0.97
C	21.63		0.02	0.94
- VECM residual serial correlation LM tests:			Lags LM-Stat	Prob
	1		7.20	0.13
- VEC residual heteroskedasticity tests:			Chi-sq 14.99	Prob. 0.66
- VEC residual normality tests:			Chi-sq 0.58	Prob. 0.75

Source: Data collected and analyzed by the author.

To verify the adequacy of the VECM, residual diagnostic indicators were used, including the Brosch–Pagan–Godfrey heteroscedasticity test, the LM serial correlation test, and the Jarque–Bera test: Chi-sq. equals 14.99 with Prob. = 0.66, M-statistics (lag 1) equal 7.20 with Prob. = 0.13, and Jarque–Bera test result statistics equal 0.58 with Prob. = 0.75, respectively. Hence, the results for model suitability indicate acceptance of the null hypothesis, a lack of heteroscedasticity, and a lack of serial correlation for the residual distribution and residual normal distribution, respectively.

3.5. The Results of the Engle–Granger Test (Date Productivity and Precipitation)

As a pre-test for co-integration, Figure 2 shows that there might be a long-running relationship between date productivity and precipitation. This result led us to conduct further co-integration analyses.

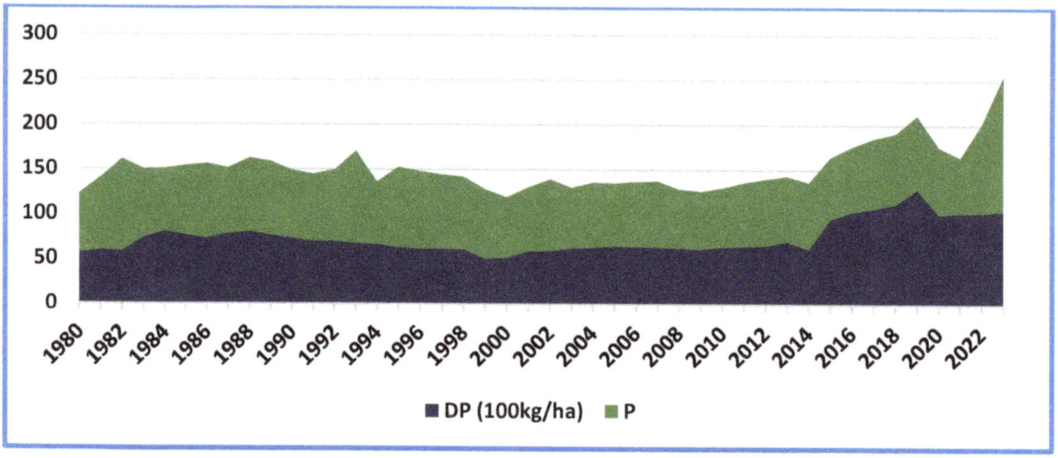

Figure 2. Graphical presentation of the variables under study. DP: date productivity (100 kg/ha). P: precipitation (millimeter). Source: Drawn by author.

3.5.1. The Results of Unit Root Tests

The unit root test outcomes (ADF) documented that date productivity and precipitation were stationary after achieving an initial difference of 1 (1) (Table 7). Consequently, date productivity and precipitation showed a stationarity order of 1 (1); therefore, the Engle–Granger test was selected to assess the associations between the series.

Table 7. Results of the unit root test.

Time Series	Intercept	Intercept and Trend	Stationarity	Intercept	Intercept and Trend	Stationarity
	at Level			at First Difference		
LDP	−1.22	−1.59	Non-stationary	−6.96 *	−6.91 *	Stationary
LP	−2.90	−2.77	Non-stationary	−6.70 *	−6.78 *	Stationary

Source: Data collected and analyzed by the author. * At 1% level of significance.

3.5.2. Engle–Granger Test Results

The results of the ADF test on the residuals (z_{t1} and z_{t2} in Equations (3) and (4), respectively) are presented in Table 8. The ADF statistics are negative (−7.52 and −7.23) and statistically significant at the 1% level. These results led us to accept the alternative hypothesis of integration, suggesting that the series are co-integrated. This result suggests that there is indication of a long-running association between date productivity and precipitation during the study period.

Table 8. Co-integration test. Engle–Granger test results for the ADF residuals.

	LP	LDP
LDP	−7.52 *	
LP		−7.23 *

Source: Data collected and analyzed by the author. * Indicates 1% level of significance.

3.6. Results of the VECM

To strengthen the finding of the Engle–Granger test (a long-term link between date productivity and precipitation), the VECM was used. To apply the VECM, lag must be selected.

Lag 1 was chosen (Table 9). Table 10 shows the results of the VECM. The adjustment coefficient for date productivity (the dependent variable) and precipitation (the independent variable) was negative (−0.01) and insignificant (critical t-value = −0.25), concluding that the model was unable to improve its earlier time imbalance. From the table, the long-term results show that precipitation affected the performance of date productivity negatively and significantly (coefficient = −5.10 with t-value = −3.05). These results (showing the long-term relationship between climate and date production) corroborate the findings of previous studies [28–30,34]. In addition, precipitation has no effect on date productivity in the short term (insignificant). To confirm the adequacy of the VECM, residual diagnostic indicators were used, including the Brosch–Pagan–Godfrey heteroscedasticity test, the LM serial correlation test, and the Jarque–Bera test: Chi-sq. equals 40.96 with Prob. = 0.51, M-statistics (lag 1) equal 8.46 with Prob. = 0.08, and Jarque–Bera test result statistics equal 0.58 with Prob. = 0.75, respectively. The results of model suitability indicated acceptance of the null hypothesis, a lack of heteroscedasticity, and a lack of serial correlation for the residual distribution and residual normal distribution, respectively.

Table 9. Lag selection.

	LogL	LR	FPE	AIC	SC	HQ
0	24.47	NA	0.0012	−1.05	−0.96	−1.02
1	62.47	70.70 *	0.00025 *	−2.63 *	−2.38 *	−2.54 *

Source: Data collected and analyzed by the author. * logL—log lag variable; LR—sequential modified likelihood ratio (LR) test statistic (each test at 5% level) (lag order selected by the criterion); AIC—Akaike information criterion; HQ—Hannan–Quinn information criterion; FPE—final prediction error; SC—Schwarz information criterion; NA—not available.

Table 10. Results of the VECM.

Long-Term Results: LDP (Dependent Variable)			**Short-Term Results: LDP (Dependent Variable)**	
Error Correction	**Coefficient**	***t*-Value Statistic**	**Coefficient**	***t*-Value Statistic**
CointEq1	−0.01	−0.25		
LP(−1)	−5.10	−3.05		
D(LDP(−1))			−0.08	−0.47
D(LP(−1))			−0.003	−0.02
C	13.25		0.02	0.87
- VECM residual serial correlation LM tests:		Lags	LM-Stat	Prob.
		1	8.46	0.08
- VEC residual heteroskedasticity tests:			Chi-sq 40.96	Prob. 0.51
- VEC residual normality tests:			Chi-sq 0.58	Prob. 0.75

Source: Data collected and analyzed by the author.

3.7. Forecast Results (Date Productivity)

Figure 3 displays the forecast graph of date productivity: the blue and red parts indicate date productivity for the periods from 1980 to 2022 and 2023 to 2034 (forecast), respectively. The graph was proven to be valid using various forecast indicators, with Mean Square Error, Alpha, Gamma, and Beta values equal to 5.47, 1.0, 0.0, and 0.0, respectively. Likewise, the growth rates of date productivity are equal to 0.82 and 0.08 for the periods from 1980 to 2023 and 2023 to 2034 (forecast), respectively, concluding that there is an increase in date productivity but at a decreasing rate (0.08) during the forecast period.

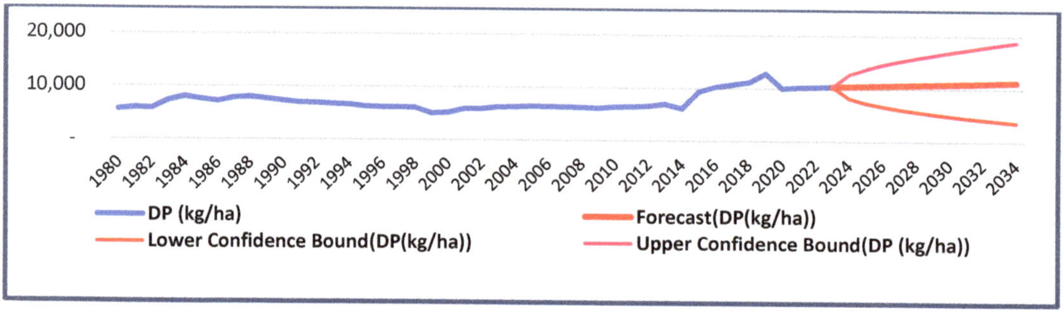

Figure 3. Forecast graph (date productivity (100 kg/ha)). Source: Drawn by author.

3.8. The Impact of Climate Change on Saudi Arabia's Food Security

One of the pillars of food security is the availability of food [22]. The availability of food is determined directly by the agriculture sector [12]. The agricultural sector plays a strategic role in improving the availability of food and realizing food security [23–26]. This study aimed to evaluate the impact of climatic change on Saudi Arabia's food security. From Table 5, the long-term results show that temperature negatively influenced (coefficient = −9.44 with t-value = −1.76) date productivity. In addition, the long-term results show that precipitation affected date productivity negatively and significantly (coefficient = −5.10 with t-value = −3.05) (Table 10). Also, these results (showing a long-term relationship between the climate and date production) align with the results of previous studies [28–30,34]. Referencing the forecast results, the graph was shown to be valid using various forecast indicators: Beta, Alpha, Mean Square Error, and Gamma values were equal to 0.0, 1.0, 5.47, and 0.0, respectively. In addition, date productivity growth rates reached 0.82 and 0.08 for the periods from 1980 to 2022 and from 2023 to 2034 (forecasts), respectively, showing that there was a decrease in the growth rate of date productivity (0.08) during the forecast period. From these results, the conclusion is that climatic change (temperature and precipitation) impacts negatively on date productivity, hindering food security, since food security can be improved through increasing date productivity. This result aligns with that of a previous study [12].

4. Conclusions

This study aimed to evaluate the impact of climatic change on Saudi Arabia's food security (identifying the connection between date productivity and climate (temperature and precipitation) and its impact on food security). The data were collected from various resources linked to study subjects in the period from 1980 to 2023. The Engle–Granger two-step procedure and forecast analysis were applied to test the long-term relationship and enable forecasting, respectively. Also, qualitative analysis was used to determine the effect of climatic change on food security. The results uncovered long-term co-integration between date productivity and temperature. Also, the long-term results demonstrated that temperature negatively influenced date productivity (coefficient = −9.44 with t-value = −1.76). In addition, the coefficient of adjustment for date productivity was shown to be negative (−0.08) and insignificant (critical t-value = −1.22), concluding that the model was powerless to improve its previous time instability. Also, the results uncovered long-term co-integration between date productivity and precipitation series. In addition, the long-term results showed that precipitation negatively and significantly impacted date productivity (coefficient = −5.10 with t-value = −3.05). With reference to the forecast results, the graph was validated using various forecast indices: the Mean Square Error, Alpha, Beta, and Gamma values were equal to 5.47, 1.0, 0.0, and 0.0, respectively. In addition, the growth rates of date productivity were found to be 0.82 and 0.08 for the periods from 1980 to 2022 and from 2023 to 2034 (forecast), respectively, concluding that there was a decrease in the growth rate of date productivity (0.08) during the forecast period. From these results, the conclusion is that both temperature and precipitation negatively affect date productivity in the long run, with a statistically significant impact, resulting in a decline in date productivity directly and an indirect impact on food security. To combat this, this study recommends specific policy interventions and innovations in agricultural date palm practices. With regard to policy interventions, a national framework focused on climate-smart agriculture, balancing date productivity, adaptation, and mitigation, should be developed and implemented. This could be aligned with Vision 2030 and the Saudi Green Initiative. Also, there should be greater investment in research and development with a focus on heat- and drought-resistant varieties and water-efficient farming systems. Regarding agricultural innovations, there should be an increase in the use of renewable energy, particularly solar energy, an expansion of rainwater harvesting infrastructure, the introduction of recycling treatment for wastewater for agriculture, and a reduction in the reliance on groundwater sources.

Funding: This work was supported by the Deanship of Scientific Research, Vice Presidency for Graduate Studies and Scientific Research, King Faisal University, Saudi Arabia [GRANT No. KFU251674].

Institutional Review Board Statement: Not applicable.

Informed Consent Statement: Not applicable.

Data Availability Statement: There are many data that were used in this study to support the reported results: Date productivity in tonnes/ha (https://www.fao.org/faostat/ar/#data/QCL (accessed on 10 February 2025)); annual average temperature in °C and average precipitation in millimeters (https://climateknowledgeportal.worldbank.org/country/saudi-arabia (accessed on 10 February 2025)).

Conflicts of Interest: The author declares no conflicts of interest.

References

1. Tiba, S.; Anis, O. Literature survey on the relationships between energy, environment and economic growth. *Renew. Sustain. Energy Rev.* **2007**, *69*, 1129–1146. [CrossRef]
2. Talbi, B. CO2 emissions reduction in road transport sector in Tunisia. *Renew. Sustain. Energy Rev.* **2017**, *69*, 232–238. [CrossRef]
3. Aroyehun, A.R.; Ugwuja, V.C.; Onoja, A.O. Determinants of melon farmers' adaptation strategies to climate change hazards in south-south Nigeria. *Sci. Rep.* **2024**, *14*, 17395. [CrossRef] [PubMed]
4. Ogunnaike, M.G.; Oyawole, F.P.; Afolabi, O.I.; Olabode, J.O. Determinants of smallholder farmers adaptation strategy to climate change in Nigeria. *NIU J. Soc. Sci.* **2021**, *7*, 243–251.
5. Droulia, F.; Charalampopoulos, I. Future climate change impacts on European viticulture: A review on recent scientific advances. *Atmosphere* **2021**, *12*, 495. [CrossRef]
6. Pachauri, R.K.; Reisinger, A. Climate Change 2007: Synthesis Report. Contribution of Working Groups I, II and III to the Fourth Assessment Report of the Intergovernmental Panel on Climate Change 2007, IPCC. Available online: https://www.ipcc.ch/site/assets/uploads/2018/02/ar4_syr_full_report.pdf (accessed on 20 February 2025).
7. IPCC. *Climate Change. Synthesis Report. Contribution of Working Groups I, II and III to the Fifth Assessment Report of the Intergovernmental Panel on Climate Change*; Core Writing Team, Pachauri, R.K., Meyer, L., Eds.; IPCC: Geneva, Switzerland, 2014; p. 151, ISBN 978-92-9169-143-2. Available online: https://www.ipcc.ch/report/sr15/summary-for-policymakers/ (accessed on 20 February 2025).
8. Fatima, Z.; Ahmed, M.; Hussain, M.; Abbas, G.; Ul-Allah, S.; Ahmad, S.; Ahmed, N.; Ali, M.A.; Sarwar, G.; Haque, E.U.; et al. The fingerprints of climate warming on cereal crops phenology and adaptation options. *Sci. Rep.* **2020**, *10*, 18013. [CrossRef]
9. Rosenzweig, C.; Parry, M.L. Potential impact of climate change on world food supply. *Nature* **1994**, *367*, 133–138. [CrossRef]
10. Lavalle, C.; Micale, F.; Houston, T.D.; Camia, A.; Hiederer, R.; Lazar, C.; Conte, C.; Amatulli, G.; Genovese, G. Climate change in Europe. 3. Impact on agriculture and forestry. A review. *Agron. Sustain. Dev.* **2009**, *29*, 433–446. [CrossRef]

11. Jones, G.V. Climate, grapes, and wine: Structure and suitability in a changing climate . In Proceedings of the XXVIII International Horticultural Congress on Science and Horticulture for People (IHC2010): International Symposium on the Citrus, Bananas and other Tropical Fruits Under Subtropical Conditions, Lisborn, Portugal, 22–27 August 2010; Volume 931, pp. 19–28. Available online: https://www.actahort.org/books/931/931_1.htm (accessed on 20 February 2025).
12. Emam, A.A. Saudi fertilizers and their impact on global food security: Present and future. *Sustainability* **2023**, *15*, 7614. [CrossRef]
13. World Food Summit. *Declaration on World Food Security*; World Food Summit: Rome, Italy, 1996.
14. van Meijl, H.; Shutes, L.; Valin, H.; Stehfest, E.; van Dijk, M.; Kuiper, M.; Tabeau, A.; van Zeist, W.J.; Hasegawa, T.; Havlik, P. Modelling alternative futures of global food security: Insights from FOODSECURE. *Glob. Food Secur.* **2020**, *25*, 100358. [CrossRef]
15. FAO. The State of Food and Agriculture 1996. Available online: http://www.fao.org/3/w1358e/w1358e00.htm (accessed on 14 January 2025).
16. Guiné, R.D.P.F.; Pato, M.L.D.J.; Costa, C.A.D.; Costa, D.D.V.T.A.D.; Silva, P.B.C.D.; Martinho, V.J.P.D. Food Security and Sustainability: Discussing the Four Pillars to Encompass Other Dimensions. *Foods* **2021**, *10*, 2732. [CrossRef]
17. Béné, C. Resilience of local food systems and links to food security–A review of some important concepts in the context of COVID-19 and other shocks. *Food Secur.* **2020**, *12*, 805–822. [CrossRef] [PubMed]
18. Magrini, E.; Vigani, M. Technology adoption and the multiple dimensions of food security: The case of maize in Tanzania. *Food Secur.* **2016**, *8*, 707–726. [CrossRef]
19. Nosratabadi, S.; Khazami, N.; Abdallah, M.B.; Lackner, Z.; Band, S.S.; Mosavi, A.; Mako, C. Social capital contributions to food security: A comprehensive literature review. *Foods* **2020**, *9*, 1650. [CrossRef] [PubMed]
20. Oakley, A.R.; Nikolaus, C.J.; Ellison, B.; Nickols-Richardson, S.M. Food insecurity and food preparation equipment in US households: Exploratory results from a cross-sectional questionnaire. *J. Hum. Nutr. Diet.* **2019**, *32*, 143–151. [CrossRef] [PubMed]
21. Houessou, M.D.; Cassee, A.; Sonneveld, B.G. The effects of the covid-19 pandemic on food security in rural and urban settlements in benin: Do allotment gardens soften the blow? *Sustainability* **2021**, *13*, 7313. [CrossRef]
22. Ahmad, N.; Du, L.; Lu, J.; Wang, J.; Li, H.-Z.; Hashmi, M.Z. Modelling the CO2 emissions and economic growth in Croatia: Is there any environmental Kuznets curve? *Energy* **2017**, *123*, 164–172. [CrossRef]
23. Smutka, L.; Steininger, M.; Miffek, O. World agricultural production and consumption. *Agris Line Pap. Econ. Inform.* **2009**, *1*, 3–12. [CrossRef]
24. Smutka, L.; Steininger, M.; Maitah, M.; Škubna, O. The Czech Agrarian Foreign Trade—Ten Years after the EU Accession. In *Agrarian Perspectives XXIV, Proceedings of the 24th International Scientific Conference, Czech University of Life Sciences Prague, Faculty of Economics and Management, Prague, Czech Republic, 16–18 September 2015*; Smutka, L., Rezbová, H., Eds.; CAB Direct: Glasgow, UK, 2015; pp. 385–392.
25. Otsuka, K. Food insecurity, income inequality, and the changing comparative advantage in world agriculture. *Agric. Econ.* **2013**, *44*, 7–18. [CrossRef]
26. Wegren, S.K.; Elvestad, C. Russia's food self-sufficiency and food security: An assessment. *Post-Communist Econ.* **2018**, *30*, 565–587. [CrossRef]
27. Habib-ur-Rahman, M.; Ahmad, A.; Raza, A.; Hasnain, M.U.; Alharby, H.F.; Alzahrani, Y.M.; Bamagoos, A.A.; Hakeem, K.R.; Ahmad, S.; Nasim, W.; et al. Impact of climate change on agricultural production; Issues, challenges, and opportunities in Asia. *Front. Plant Sci.* **2022**, *13*, 925548. [CrossRef] [PubMed]
28. Yuan, X.; Li, S.; Chen, J.; Yu, H.; Yang, T.; Wang, C.; Huang, S.; Chen, H.; Ao, X. Impacts of global climate change on agricultural production: A comprehensive review. *Agronomy* **2024**, *14*, 1360. [CrossRef]
29. Alam, A. Rukhsana. Climate change impact, agriculture, and society: An overview. In *Climate Change, Agriculture and Society: Approaches Toward Sustainability*; Springer: Berlin/Heidelberg, Germany, 2023; pp. 3–13. Available online: https://link.springer.com/chapter/10.1007/978-3-031-28251-5_1 (accessed on 20 February 2025).
30. Prajapati, H.A.; Yadav, K.; Hanamasagar, Y.; Kumar, M.B.; Khan, T.; Belagalla, N.; Thomas, V.; Jabeen, A.; Gomadhi, G.; Malathi, G. Impact of climate change on global agriculture: Challenges and adaptation. *Int. J. Environ. Clim. Chang.* **2024**, *14*, 372–379. [CrossRef]
31. Eekhout, J.P.; de Vente, J. Global impact of climate change on soil erosion and potential for adaptation through soil conservation. *Earth-Sci. Rev.* **2022**, *226*, 103921. [CrossRef]
32. Lesk, C.; Coffel, E.; Winter, J.; Ray, D.; Zscheischler, J.; Seneviratne, S.I.; Horton, R. Stronger temperature–moisture couplings exacerbate the impact of climate warming on global crop yields. *Nat. Food* **2021**, *2*, 683–691. [CrossRef]
33. Jägermeyr, J.; Müller, C.; Ruane, A.C.; Elliott, J.; Balkovic, J.; Castillo, O.; Faye, B.; Foster, I.; Folberth, C.; Franke, J.A.; et al. Climate impacts on global agriculture emerge earlier in new generation of climate and crop models. *Nat. Food* **2021**, *2*, 873–885. [CrossRef] [PubMed]
34. Zhu, P.; Burney, J.; Chang, J.; Jin, Z.; Mueller, N.D.; Xin, Q.; Xu, J.; Yu, L.; Makowski, D.; Ciais, P. Warming reduces global agricultural production by decreasing cropping frequency and yields. *Nat. Clim. Chang.* **2022**, *12*, 1016–1023. [CrossRef]

35. Emam, A. Present and Future: Does agriculture affect economic growth and environment in the Kingdom of Saudi Arabia? *J. Agric. Econ. Czech Acad. Agric. Sci* **2022**, *68*, 361–370. [CrossRef]
36. Wijerathna-Yapa, A.; Pathirana, R. Sustainable agro-food systems for addressing climate change and food security. *Agriculture* **2022**, *12*, 1554. [CrossRef]
37. Soffiantini, G. Food Insecurity and Political Instability during the Arab Spring. *Glob. Food Secur.* **2020**, *26*, 100400. [CrossRef]
38. Mbow, C.; Rosenzweig, C.; Tubiello, F.; Benton, T.; Herrero, M.; Pradhan, P.; Barioni, L.; Krishnapillai, M.; Liwenga, E.; RiveraFerre, M.; et al. Chapter 5: Food Security. In *IPCC Special Report on Land and Climate Change*; IPCC: Geneva, Switzerland, 2019.
39. Aleid, S.M.; Al-Khayri, J.M.; Al-Bahrany, A.M. Date palm status and perspective in Saudi Arabia. In *Date Palm Genetic Resources and Utilization*; Al-Khayri, J.M., Jain, S.M., Johnson, D.V., Eds.; Springer: Dordrecht, The Netherlands, 2015; pp. 49–95.
40. National Center for Palms and Dates, Report. 2022. Available online: https://ncpd.gov.sa/en/reports (accessed on 28 February 2025).
41. Mohammed, M.; Sallam, A.; Munir, M.; Ali-Dinar, H. Effects of deficit irrigation scheduling on water use, gas exchange, yield, and fruit quality of date palm. *Agronomy* **2021**, *11*, 2256. [CrossRef]
42. FAO. Crops and Livestock Products. [Dataset]. Food and Agriculture Organization of the Nations (FAO). 2023. Available online: https://www.fao.org/faostat/en/#data/QCL (accessed on 14 March 2025).
43. Alhamdan, A.; Alamri, Y.; Aljuhaim, F.; Kotb, A.; Aljohani, E.; Alaagib, S.; Elamshity, M. Economic Analysis of the Impact of Waste on the Production and Consumption of Dates in Saudi Arabia. *Sustainability* **2024**, *16*, 9588. [CrossRef]
44. Dickey, D.A.; Fuller, W.A. Distribution of the estimators for autoregressive time series with a unit root. *J. Am. Stat. Assoc.* **1979**, *74*, 427–431. [CrossRef]
45. Chen, L.; Chang, J.; Wang, Y.; Guo, A.; Liu, Y.; Wang, Q.; Xie, Z. Disclosing the future food security risk of China based on crop production and water scarcity under diverse socioeconomic and climate scenarios. *Sci. Total Environ.* **2021**, *790*, 148110. [CrossRef]
46. Peseran, M.H.; Peseran, B. *Working with Microfit 4.0: Interactive Econometric Analysis*; Oxford University Press: Oxford, UK, 1997.
47. Venujayakanth, B.; Swaminathan Dudhat, A.; Swaminathan, B.; Ardeshana, N.J. Price integration analysis of major groundnut domestic markets in India. *Econ. Aff.* **2017**, *62*, 233–241. [CrossRef]

Disclaimer/Publisher's Note: The statements, opinions and data contained in all publications are solely those of the individual author(s) and contributor(s) and not of MDPI and/or the editor(s). MDPI and/or the editor(s) disclaim responsibility for any injury to people or property resulting from any ideas, methods, instructions or products referred to in the content.

sustainability

MDPI

Article

Gendered Analysis of Agro-Based Climate Adaptation in the Santchou Landscape of Cameroon

Tosam Hycinth Ngong [1], Banseka JaneFrances Yenlajai [2], Ngwa Kester Azibo [3], Constantine Nwune Alusoh [4] and Jude Ndzifon Kimengsi [5,6,*]

1 Department of Maritime Transport, Higher Institute of Transport and Logistics, The University of Bamenda, North West Region, Bambili P.O. Box 39, Cameroon; hycinton@gmail.com
2 Department of Geography and Planning, The University of Bamenda, North West Region, Bambili P.O. Box 39, Cameroon; jbanseka@yahoo.com
3 Higher Institute of Agriculture and Rural Development, Bamenda University of Science and Technology, Bamenda P.O. Box 277, Cameroon; ngwakester@gmail.com
4 Department of Economics, The University of Bamenda, Bamenda P.O. Box 39, Cameroon; alusoh.consty@gmail.com
5 Forest Institutions and International Development (FIID) Research Group, Faculty of Environmental Sciences, Technische Universität Dresden, 01737 Tharandt, Germany
6 Department of Geography, Higher Teacher Training College, The University of Bamenda, North West Region, Bambili P.O. Box 39, Cameroon
* Correspondence: jude_ndzifon.kimengsi@tu-dresden.de or jude.kimengsi@catuc.org

Abstract: Agriculture remains the backbone and major source of livelihood for men and women in most parts of sub-Saharan Africa. However, the gender-differentiated roles in agricultural transformation as a coping strategy to climate change in this context still beg for empirical substantiation. Using the Santchou Landscape of Cameroon as a case, this study sought to (a) examine the effects of climate change on agricultural practices, (b) characterize gender-differentiation in agro-based climate adaptation interventions, and (c) explore the gender-based challenges to agro-based climate adaptation planning. A representative sample of 159 households was conducted in five communities in the study area, complemented by key informant interviews (N = 5). The data collected were analyzed descriptively. The findings of this study revealed the following conclusions: Firstly, climate change significantly affects agricultural practices in the Santchou Landscape as mirrored in faming season fluctuation as well as the alteration of the farming calendar. Secondly, men and women play differentiated roles in agro-based climate adaptation, especially through farming practices such as the introduction of drought-resistant crops, the the practice of intercropping and agroforestry. Thirdly, gender-based challenges to agro-based climate adaptation include unequal access to land between men and women and unequal access to farm inputs, agricultural training, and technology. This study provides empirical evidence to substantiate the theoretical position on gender-differentiated roles in agro-based climate adaptation. Further studies are required to establish the incidence of gender variations in agro-based climate adaptation on livelihoods.

Keywords: men; women; farming season; adaptation; Santchou; Cameroon

1. Introduction

The attendant effects of climate change on agricultural systems are no longer new in the geographical lexicon around the world. Many scholars have theorized about climate change and its impacts on agricultural productivity and livelihoods around the world. For instance, between 1991 and 2017, climate change led to losses in agriculture in the

USA to the tune of 27.0 billion USD [1]. Furthermore, crop production losses across West Africa between 2000 and 2009 were associated with historical climate change effects, which accounted for losses to the tune of 2.33–4.02 billion USD for millet and 0.73–2.17 billion USD for sorghum [2]. In terms of the gender-differentiated effects, women are known to be highly vulnerable to climate change impacts compared to men. This is partly due to differences in the ownership of economic resources such as land, labor, and capital and also entrenched socio-cultural norms and beliefs that disfavor women [3]. In addition to this, the poverty levels among women are high, and this renders them vulnerable to severe climate impacts. These challenges, which stall progress towards several interlinked sustainable development goals, are rooted, among other things, in poor farming practices and farmers' perceptions of and adaptation and resilience approaches to climate change [3–5].

Faced with these realities, climate adaptation has become one of the key strategies to deal with the issue of climate change. While adaptation is sector-specific, it seems to have gained prominence in the agricultural sector, considering the fact that most of the economies in sub-Saharan Africa (SSA) depend significantly on agriculture [6–10]. Scholars have theorized about the varied systems of adaptation in different parts of SSA [11–13]. Adaptation to climate change as a strategy has been reported to be defined by socio-demographic determinants, such as sex, age, education, household size, ethnicity, and factors like agro-ecology, livestock holding, off-farm income, access to credit, and extension services [14–20]. Studies conducted by [18] revealed that family size affected adaptation strategies toward climate change in Indonesia. According to them, large families can divert part of the family labor to non-farming activities to earn more income and to ease the consumption pressure imposed by large families. However, [16] stated that it is better to have a large household with excellent education rather than a small family with inadequate knowledge, as the probability of adaptation is higher in large families with a high level of education than in smaller families with low levels of education. As explained by [21,22], people with a higher level of education have a greater impact on farmers' decisions in comparison to those with a lower level of education.

Adaptation depicts variations in terms of the practices and systems employed. Agro-based climate adaptation has a long history in different parts of SSA [5]. For instance, some of the agro-based adaptation systems and practices include monoculture, mixed cropping, irrigation, crop rotation, agroforestry, the planting of drought-resistant crops, and the alteration of the planting calendar. Scholars [5,23,24] have examined the evolution of farm-based climate adaptation practices in the vegetable subsector of the Western Highlands of Cameroon and found out that agro-based practices were significantly reflected in crop performance, mirrored through an increase in vegetable quantity, promoted by both state and non-state actors. While both state and non-state actors are engaged in agro-based climate adaptation practices, there are variations in the roles played by men and women in implementing agro-based climate adaptation. These variations are linked to cultural, socio-economic, and political factors. Substantial empirical evidence indicates that gender disparity exists in access to resources, information, and agricultural inputs [25,26]. For example, in rural Kenya, a significant proportion of women were found to adopt food crop-related strategies, whereas men employed livestock and agroforestry-related strategies. The authors of [26,27] demonstrated that in India, there are geographically determined gendered preferences and adoption of strategies regarding adaptation options, which are influenced by the socio-ecological context and institutional dynamics. Scholars recognize that agro-based climate adaptation practices are gender specific and mediated by social, cultural, institutional, and economic structures and processes [28,29]. This explains the fragmented nature of climate change adaptation as well as the suboptimal outcomes attached to such practices.

While success has been registered in some parts of SSA in terms of agro-based climate adaptation, failures have been registered in other parts. This therefore calls for adaptation planning that takes into consideration gender disparities. There is, however, limited evidence on gender roles in several parts of sub-Saharan Africa. This hinders progress towards a comprehensive policy approach to address climate adaptation. In other words, context-specific knowledge gaps exist in many parts of sub-Saharan Africa in terms of the differentiated roles of men and women in agro-based climate adaptation, with Cameroon being a good example. Described as Africa in miniature, Cameroon embodies almost all of the agro-ecological zones that can be found across Africa, making it the example par excellence to study the differentiated roles of agro-based climate adaptation practices, systems, and planning approaches. Furthermore, Cameroon is rated as one of the most ethnically diverse countries in sub-Saharan Africa, with an ethnic fractionalization score of 0.73 [30]. This ethnic diversity is rooted in a culture which tends to shape the differentiated roles of men and women within the society [30]. While there is evidence in the northern and coastal regions of Cameroon of some of the adaptation practices, there is limited evidence in the transitional zone between the coastal areas and the Western Highlands [3]. Santchou, being a transitional zone between the coastal areas and the Western Highlands, provides a good example. This highly dynamic transition zone is characterized by significant agricultural activities [3,5], making it a useful case to illustrate the differentiated roles of men and women in agro-based climate adaptations. It is in this light that this study seeks to (a) examine the impacts of climate on agricultural practices in Santchou, (b) categorize the differentiated roles of men and women, and (c) explore the gender-based challenges to agro-climatic adaptation in Santchou.

2. Analytical Framework

State and non-state actors, such as NGOs, migrants, and natives, are engaged in agro-based climate adaptation at different levels. These are viewed in the spectrum of farming practices, systems, and adaptation planning approaches (Figure 1). Gender variations in agro-based climate adaptation in Santchou depend on a host of determinants, such as income, culture, access to information, capital, and land. These shape the outcomes and necessitate adaptation planning at different levels by state and non-state actors [5]. Adaptation planning includes the abolition of culture, the restriction of land redistribution, ensured equal rights, land inheritance for both men and women, equal opportunities for assistance, subventions to female farmers, and subsidized input costs. At the level of other non-state actors such as NGOs, adaptation planning strategies include access to free agricultural training programs, the sensitization of women on the importance of training programs, agricultural technology and innovation, capacity building on the adoption of new technologies, and the promotion of gender-balanced and peasant-friendly technological innovations. Adaptation planning approaches also vary depending on legal, socioeconomic, sociopolitical and natural contexts.

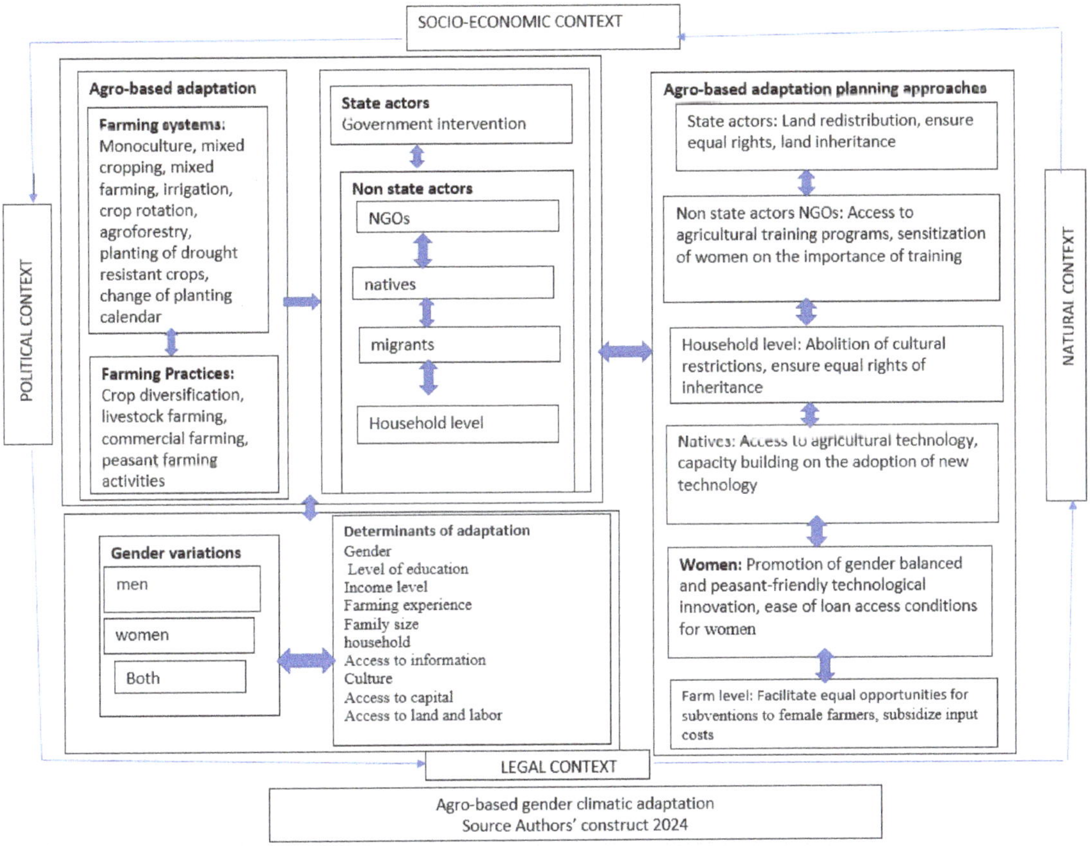

Figure 1. Gender variations in agro-based climate adaptation.

3. Study Area and Methods

The Santchou landscape is found in the Menoua Division of Cameroon (latitudes 5°18′0″ north and longitudes 9°54′0″ east of the Greenwich Meridian). It constitutes part of the Mbo Plain (Figure 2). It is bounded to the north by Mount Manengouba, to the south by the Dschang Cliff, and to the east by Kekem. Santchou has a population of 46,249 inhabitants who are mostly migrants. It has a surface area of 335 km^2 and a population density of 137 person/km^2. It is made up of 60 villages and (Santchou Rural) 4 semi-urban settlements [31].

The main instrument used for data collection was a questionnaire, structured into socio-demographic characteristics of respondents, farming practices, gender differences in agro-based climate adaptations, and the implications of agro-based climate adaptation practices. Two enumerators were recruited and trained for data collection. With the assistance of two community members who have lived in the study area for over 10 years, our research group was able to identify the five communities that were targeted for this study. Data were randomly collected from 159 households in the five communities to include Santchou Ville (18.9%), Madagascar (33.3%), Bessouck (18.2%), Behind Albatros (5%), and Kassala Farm (24.4%) (Table 1). These communities were identified on the basis of the engagement of the inhabitants (men and women) in agricultural innovation as a coping strategy against climate change. Data were collected from May to June 2024. We used a random sampling approach in which an ad hoc household numbering process was employed to determine the total number of households. From this, we proceeded to carry out raffle draws to randomly select numbers corresponding to 159 households. The population size of 250 inhabitants representing the population of the five targeted communities was obtained from [31]. The selected numbers guided us to select the precise households for data collection. A 5 to 8% sampling fraction was assured for all five communities under study, with the unit of analysis being the households. Household representatives were used as proxies for the households.

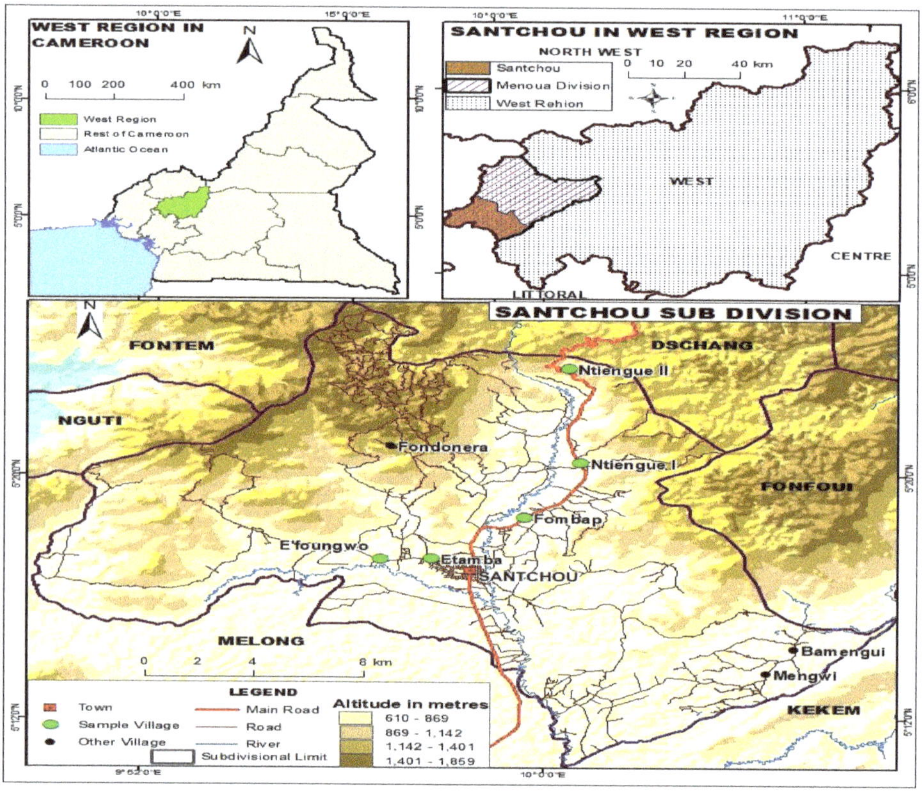

Figure 2. Location of Santchou.

Table 1. Distribution of questionnaires.

Communities	Households Sampled	% Sampled	Sampling Fraction %
Santchou Ville	30	18.9	6
Madagascar	53	33.3	8
Bessouck	29	18.2	5
Behind Albatros	8	5.0	5
Kasala Farm	39	24.5	6

The questionnaires were entered into Kobo Collect (Version 2023). Subsequently, the spreadsheets were downloaded and exported to the Statistical Package for Social Sciences (SPSS Version 25.0), where frequencies and percentages were calculated. The results have been presented using tables, charts, and graphs. Information collected using the questionnaire copies was complemented with other sources such as field observations and interviews with key informants to obtain insights on agro-based climate adaptation practices and gender.

4. Results

4.1. Socio-Demographic Characteristics

In this study (Table 2), five communities were sampled to include Santchou Ville (18.9%), Madagascar (33.3%), Bessouck (18.2%), Behind Albatros (8%), and Kassala Farm (24.4%). With respect to the level of education, 6.9% of the respondents had no formal education, 22.6% had primary education, 49.1% had secondary education, 15.1% had vocational education, and 6.3% had at least a university degree. In terms of occupation, 52.2% were farmers, 8.2% were bike riders, 10.2% were petty traders, 11.9% were civil servants, 6% worked with NGOs, 6% were builders, 1.9% were sand miners, and 7.5% were in other occupations.

Table 2. Demographic characteristics of the respondents.

Variable	Categories	Percentage (%)
Santchou Community	Ville	18.9
	Madagascar	33.3
	Bessouck	18.2
	Behind Albatros	5.0
	Kasala Farm	24.5
	Female	24.5
Level of Education	No formal	6.9
	Primary	22.6
	Secondary	49.1
	Vocational	15.1
	BSc and above	6.3
Main Occupation	Farmer	52.2
	Bike/Taxi Rider	8.2
	Petty Trader	10.7
	Salary State	11.9
	Salary Private	6.3
	NGO	0.6
	Builder	0.6
	Sand Miner	1.9
	Others	7.5

Furthermore, 47.8% of the respondents were household heads, while 52.2% were household members (Figure 3). It is worth mentioning that this study comprised 64.8% respondents who are migrants and only 35.2% who are indigenous. In terms of gender distribution, 44.7% were males while 55.3% were females. Moreover, male household heads constitute 75.5% of the sample, while female household heads account for the remaining 24.5%. Therefore, while there is a great representation of the female population in this study (55.3%), only 24.5% were heads of their households.

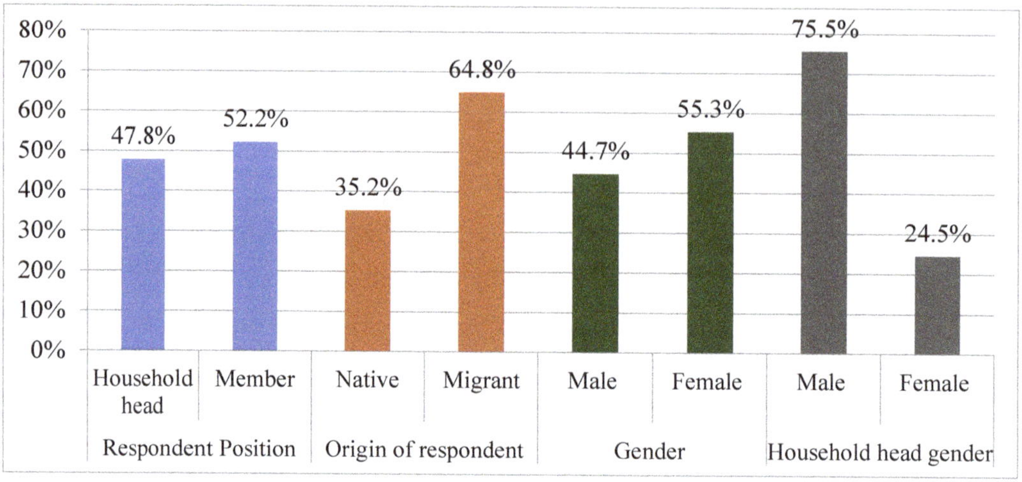

Figure 3. Socio-demographic attributes of respondents.

4.2. Farming Practices and Interventions by Both State and Non-State Actors

In the Santchou landscape, the population practices both large- and small-scale farming Figure 4. The key agricultural practices include large-scale plantation farming (58.5%), small-scale plantation farming (33%), and livestock rearing (reported by just over 6%). The results also indicate that there are female-inclined farming practices such as mulching, mixed cropping, vegetable cultivation, cassava cultivation, and farming close to the house.

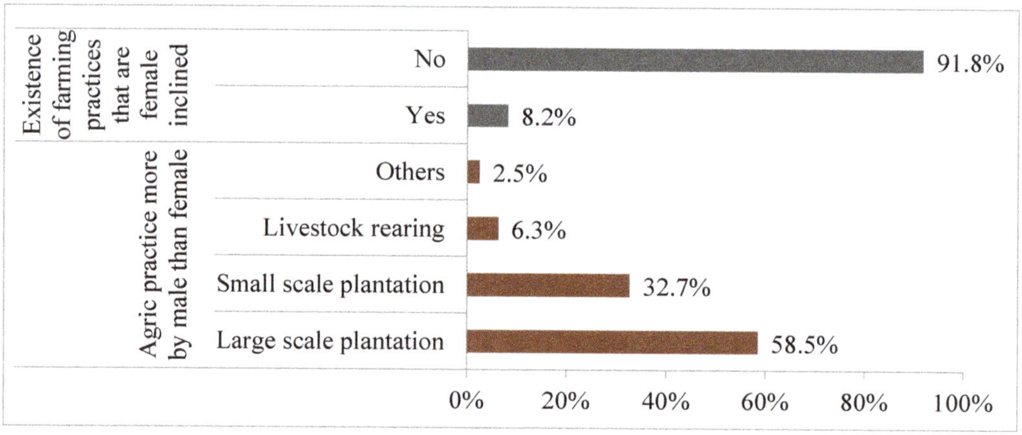

Figure 4. Variations in farming practices in the Santchou landscape.

Given the above-mentioned farming practices and the advent of climate change, farmers (Figure 5) reported that their farming practices have been affected by climate change, especially with respect to fluctuations in the farming seasons (75.5%), reductions in soil fertility (8%), and the complete alteration of the farming calendar (6%).

To promote adaptation to climate change, several interventions from different stakeholders in both the government and private sectors have been undertaken (Table 3). The results indicate that the promotion of mixed farming was the most significant (32%) intervention by the state. This was followed by interventions to promote cash crop farming

(26%). Interventions were also undertaken by the state (12%) and NGOs (17%). This was carried out for men (7.5%) and women (7%).

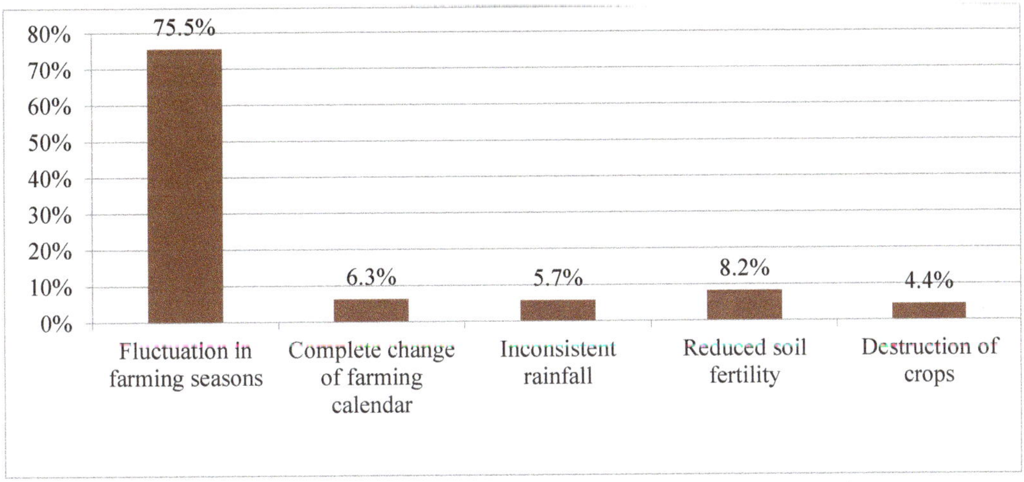

Figure 5. Perceived effects of climate change on farming practices.

Table 3. Actor interventions in farming in the Santchou landscape of Cameroon.

	State	NGOs	Household	Men	Women	Natives	Migrants
Monocropping	22%	21.4%	00	6.9%	1.9%	7.5%	7.5%
Mixed cropping	00	3.8%	5.7%	3.8%	6.9%	6.9%	6.9%
Mixed farming	32.1%	17.0%	8.8%	48.4%	61%	61%	61%
Agroforestry system	10.1%	15.7%	61.6%	8.8%	3.1%	4.4%	4.4%
Market gardening	2.5%	1.3%	5%	1.3%	8.2%	3.8%	3.8%
Cash crop farming	25.8%	11.9%	8.2%	17.0%	7.5%	6.9%	6.9%
Apiculture	0.6%	1.9%	2.5%	1.9%	1.9%	2.5%	2.5%
Snail farming	00	3.8%	00	1.9%	1.9%	0.6%	0.6%
Add perennial crops	6.9%	23.3%	8.2%	10.1%	7.5%	6.3%	6.3%

It is, however, worth mentioning that the promotion of agroforestry systems was mainly practiced at the household level (62%), with some interventions from NGOs (18%) and the state (10%).

4.3. Gender Variations in AgroBased Adaptation

The gender-differentiated dimension of agro-based climate adaptation in the Santchou landscape of Cameroon is presented in Table 4. The general trend is that both the men and women are highly involved in climate adaptation. For instance, change in planting season is practiced by almost all the men and women (99.4%). Similarly, a majority of the farmers (98.7%) reported that the introduction of drought-resistant species is practiced by men (1.3%) and women (98%). Moreover, 96.9% of women practice crop rotation versus 3.1% of men, and the practice of organic farming by men (3.1%) and women (96.9%) shows variations. However, land rotation and zero tillage were practiced more by women (90.6%) compared to men (9.4%). On the other hand, men (84.4%) were more involved in intercropping than women (11.4%). In the case of mulching, women (88.7%) were more engaged than men (8.2%).

Table 4. Gender variations in agro-based climate adaptation in the Santchou landscape of Cameroon.

Strategy	Male	Female
Change of planting season/periods, dates	0.6	99.4%
Introduce drought-tolerant varieties	1.3	98.7%
Crop rotation	3.1	96.9%
Land oration	10.9	89.9%
Intercropping	88.4	11.4
Repeated sowing	3.8	96.2%
Zero tillage	9.4	90.6%
Cover cropping	4.4	95.6%
Mulching	8.2%	88.7%
Organic farming	3.1	96.9%
Green manuring	6.9	93.1%
Irrigation	3.8	96.2%

4.4. Intensity of Each of the Following Farming Adaptation Strategies

Table 5 shows the intensity of the various adaptation strategies; the results show that the most intense agro-based climatic adaptation practice in the Santchou Landscape was the alteration of the planting season (39%). This is followed by crop rotation (30.8%). On the other hand, the introduction of drought-tolerant varieties and the practice of organic farming were less significantly reported, accounting for 47.2% and 21.4%, respectively. Similarly, land rotation (49.1%), zero tillage (45.9%), green manuring (44%), irrigation (42.8%), mulching (42.1%), repeated sewing (36.5%), and cover cropping (34%) were moderately intense in their application.

Table 5. Intensity of farming practices in the Santchou landscape of Cameroon.

	Not Intense	Less Intense	Moderately Intense	Intense	Very Intense
Change planting season/dates	5%	8.2%	35.8%	11.9%	39%
Introduction of drought-tolerant varieties	6.9%	21.4%	47.2%	22.6%	1.9%
Crop rotation	4.4%	9.4%	35.8%	19.5%	30.8%
Land oration	3.8%	20.1%	49.1%	15.1%	11.9%
Intercropping	3.8%	17%	32.7%	16.3%	30.2%
Repeated sowing	1.9%	12.6%	36.5%	22%	27%
Zero tillage	4.4%	25.2%	45.9%	16.3%	8.2%
Cover cropping	8.8%	19.4%	34%	20.8%	17%
Mulching	6.3%	24.5%	42.2%	15.7%	11.3%
Organic farming	6.3%	23.3%	35.2%	19.5%	15.7%
Green manuring	5%	27%	44%	12.7%	11.3%
Irrigation	13.2%	23.3%	42.8%	10.7%	10%

4.5. Implications for Agro-Based Climate Adaptation Planning

The strategies adopted by farmers in the Santchou landscape have implications for present and future agricultural adaptation planning. Women are more involved in the cultivation of short-cycle crops such as maize, beans, and vegetables, and this has influenced their adaptation practices as a coping strategy to climate change. On the other hand, though men are equally involved in the practice of short-cycle crops, they are more involved in the cultivation of long-term crops such as cocoa farming and the cultivation of palms. This is explained by the fact that the long-term crops require more land for cultivation, and in terms of access to land, existing social and cultural norms favor the men in terms of land access and ownership over women. The involvement of women in the cultivation of short-cycle crops is driven by the motive that such crops are most often cultivated on small farm spaces. The acquisition of land for the cultivation of short-cycle crops is easier through land rental arrangements at prices that are quite affordable to the women.

Women do not only face the challenge of access to land ownership, which limits adaption; they are further hindered by the lack of farm inputs such as fertilizers, improved seeds, access to new technology, and agricultural training. On the challenges faced by farmers in adopting strategies in the agricultural sector to fight against the negative consequences of climate change, the key challenges faced by men and women include the lack of improved seeds: 45% for women and 38% for the men. This was followed by inadequate access to modern and sustainable technologies (31% of the women and 32% of the men) and inadequate access to new knowledge and skills in farming (24.5% of the women and 30% of the men) (Figure 6).

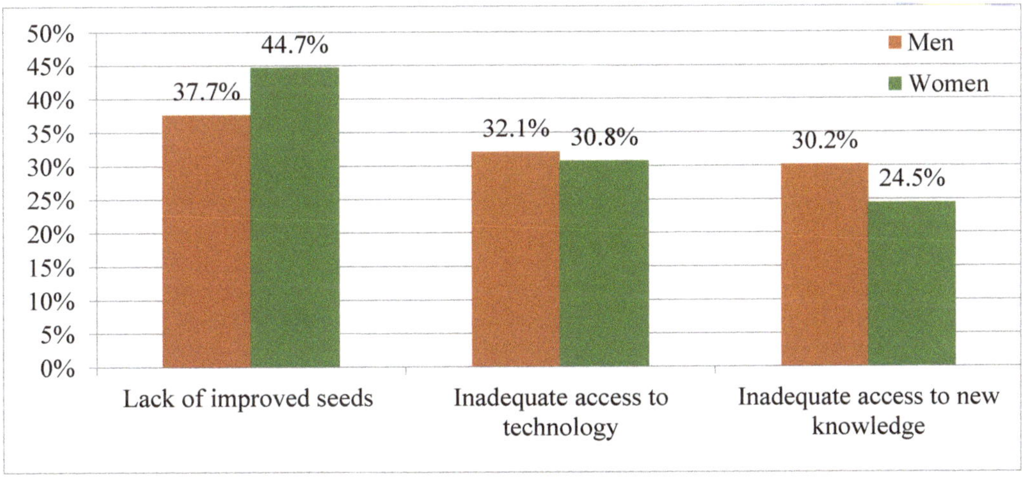

Figure 6. Constraints on agricultural adaptation to climate change.

4.6. Strategies to Bridge Agro-Based Climate Adaptation Gaps Between Genders

As can be inferred from Table 6, improving the participation of women in the adoption of sustainable farming practices in the advent of climate change can be achieved through increasing access to land, farm inputs, agricultural training, and new technologies, as well as access to financial capital. Based on field evidence, the sensitization of women on the importance of agricultural training was reported as close to 99%, followed by the promotion of farmers (75.5%), capacity building through the adoption of new technology (74%), and the provision of equal opportunities to both men and women (69%). The abolition of cultures that restrict access to land for women was reported at 63.5%. Moreover, the promotion of gender balance was reported at 32.6%, land redistribution at 18.2%, and ensuring equal rights to inheritance for both men and women at 17.6%, among other factors reported (Table 6).

Table 6. Strategies to bridge agro-based climate adaptation gaps between genders.

Strategy	Category	Response (%)
Access to land	Abolition of cultural restrictions	63.5
	Land redistribution	18.2
	Ensured equal rights of land inheritance	17.6
	Others	0.6
Access to farm inputs	Equal opportunities for assistance	69.2
	Subventions to female farmers	13.8
	Subsidized input cost	17.0
Access to agricultural training	Sensitization of women on the importance of training	98.7
	Others	1.3
Access to new technologies	Capacity building on the adaptation of new technology	73.6
	Promotion of gender balance	13.2
	Peasant-friendly technological innovations	13.2
Access to financial capital	Promotion of farmers	75.5
	Provision of interest-free/low-interest loans to farmers	13.8
	Preferential treatment of women when it comes to loan conditions	10.7

5. Discussion

This study sought to analyze the gender-differentiated roles in agro-based climate adaptation. The results on the differentiated roles of men and women in terms of agro-based climate adaption are that men and women play different roles depending on the agricultural practices that they are engaged in. For instance, men are more engaged in the cultivation of long-cycle crops such as cocoa and palms, while women are involved in the cultivation of short-cycle crops such as vegetables, maize, and beans. This explains the variations in their adaptation strategies. Adaptation strategies are influenced by socio-cultural factors which influence land access as well as access to farm inputs and technology. These findings are consistent with the works of [24,25] who argued that men and women play different roles in agro-based, climate-based adaptation. The findings also revealed that even though men and women are involved in agro-based climatic adaptation, women are more inclined to specific adaptation practices than men. For example, land rotation and zero tillage were practiced more by women compared to the men. On the other hand, men were more involved in intercropping and mulching (9.4% and 8.2%, respectively) as compared to the women (1.9% and 3.1%, respectively). This finding conforms with studies carried out by [29] who argued that men and women carry out varied adaptation practices to fight against climate change.

The results also revealed that while men are involved in long-term agro-based climate adaptation practices, women are more involved in short-term farming activities, as this provides returns, income, and sustenance for their livelihoods. This is also consistent with the findings of [3] in the Western Highlands of Cameroon, where women carried out more of the short-term farming activities. The findings are also in line with other studies carried out across Africa and other parts of the Global South [21,29,32–35], which indicate major gender differences in agro-based adaptation by farmers and livestock keepers. In terms of agro-based climate interventions, the study revealed that state and non-state actors play varied roles in agro-based climate adaptation in the Santchou landscape. The promotion of mixed farming receives the highest intervention (32%) from the state. This was followed by interventions promoting cash crop farming (26%). While both state and non-state actors are engaged in agro-based climate adaptation practices, there are variations in the roles played by men and women in implementing adaptation. This finding is consistent with initial studies conducted by [25,35]. The variations are linked to cultural, socio-economic, and political factors. Substantial empirical evidence indicates that gender disparity exists in access to resources, information, and agricultural inputs. For example, in rural Kenya, a higher percentage of women were found to adopt food crop-related strategies, whereas men employed livestock- and agroforestry-related strategies [26,27]. Studies also reveal that in India, there are geographically determined gendered preferences and adoption of strategies regarding adaptation; these are influenced by the socio-ecological context and institutional dynamics.

6. Conclusions

Climate change has significant effects on agricultural practices in the Santchou landscape of Cameroon. Agricultural innovations have been introduced through gender-based adaptation as a coping strategy. As the gender-differentiated roles in agricultural transformation still beg for empirical substantiation, this paper used the Santchou Landscape of Cameroon as a case to (a) examine the effects of climate change on agricultural practices in Santchou, (b) characterize gender differentiation in agro-based climate adaptation interventions, and (c) explore the gender-based challenges to agro-based climate adaptation planning. The analysis led to the following conclusion: while men and women are both involved in the agro-based adaptation strategies, variations exist based on the type of agricultural practice. The study also concludes that socio-cultural factors act in various dimensions to influence agro-based climate adaptation processes, while state and non-state actors have played varied roles in influencing gender-differentiated agro-based climate adaptation. For example, the broad-based approach to agro-based climate adaptation has influenced men and women to adopt similar strategies, such as the introduction of new crop varieties and the adjustment of the planting calendar. Gender-based challenges to agro-based climate adaptation include unequal access to land between men and women and unequal access to farm inputs, agricultural training, and technology. The provision of farm input, agricultural training, and innovation has reduced the gender gap between men and women in adapting to climate change. This study provides empirical evidence to substantiate the theoretical positions on gender-differentiated roles in agro-based climate adaptation. Further studies are required to establish the incidence of gender variations in agro-based climate adaptation on livelihoods.

Author Contributions: Conceptualization, T.H.N., B.J.Y. and J.N.K.; Methodology, T.H.N., B.J.Y., C.N.A. and J.N.K.; Formal analysis, N.K.A.; Writing—original draft, T.H.N., B.J.Y., N.K.A., C.N.A. and J.N.K.; Writing—review & editing, J.N.K.; Supervision, J.N.K. All authors have read and agreed to the published version of the manuscript.

Funding: This research received no external funding.

Institutional Review Board Statement: The Board of the Higher Institute of Transport and Logistics of the University of Bamenda hereby confirms that the study on the Gendered Analysis of Agro-based Climate Adaptations in the Santchou Landscape of Cameroon focused on analyzing perceptions of Agro-based Climate Adaptation. The data collection process was conducted with strict respect for the integrity and privacy of all respondents. As the research does not involve data collection on human specimens, it is therefore recommended for an ethics review waiver.

Informed Consent Statement: Informed consent was obtained from all subjects involved in the study.

Data Availability Statement: The original contributions presented in this study are included in the article. Further inquiries can be directed to the corresponding author.

Conflicts of Interest: The authors declare no conflict of interest.

References

1. Noah, S.D.; Davenport, F.V.; Burke, M. Historical warming has increased U.S. crop insurance losses. *Environ. Res. Lett.* **2021**, *16*, 084025.
2. Sultan, B.; Defrance, D.; Iizumi, T. Evidence of crop production losses in West Africa due to historical global warming in two crop models. *Sci. Rep.* **2019**, *9*, 12834. [CrossRef] [PubMed]
3. Awazi, N.P.; Kimengsi, J.N.; Forje, G.D. Gender-differentiated capture of agro-based climate adaptation interventions: Implications for agricultural systems and practices in Cameroon's Western highlands. *SN Soc. Sci.* **2024**, *4*, 116. [CrossRef]
4. Campbell, B.M.; Vermeulen, S.J.; Aggarwal, P.K.; Corner-Dolloff, C.; Girvetz, E.; Loboguerrero, A.M.; Ramirez-Villegas, J.; Rosenstock, T.; Sebastian, L.; Thornton, P.K. Reducing risks to food security from climate change. *Glob. Food Secur.* **2016**, *11*, 34–43. [CrossRef]
5. Kimengsi, J.N.; Silberberger, M. How Endogenous Cultural Institutions May (Not) Shape Farmers' Climate Adaptation Practices: Learning from Rural Cameroon. *Soc. Nat. Resour.* **2023**, *36*, 460–478. [CrossRef]
6. Allen, M.R.; Barros, V.R.; Broome, J.; Cramer, W.; Christ, R.; Church, J.A.; Clarke, L.; Dahe, Q.; Dasgupta, P.; Dubash, N.K.; et al. *IPCC Fifth Assessment Synthesis Report-Climate Change 2014 Synthesis Report*; Intergovernmental Panel on Climate Change (IPCC): Geneva, Switzerland, 2014.
7. Ambagna, J.; Kane, G.; Oyekale, A. Subsistence Farming and Food Security in Cameroon: A Macroeconomic Approach. *Life Sci. J.* **2012**, *9*, 3949–3954.
8. Calzadilla, A.; Zhu, T.; Rehdanz, K. Climate change and agriculture: Impacts and adaptation options in South Africa. *Water Resour. Econ.* **2014**, *5*, 24–48. [CrossRef]
9. Nhamo, L.; Matchaya, G.; Mabhaudhi, T. Cereal production trends under climate change impacts and adaptation strategies in Southern Africa. *Agriculture* **2019**, *9*, 30. [CrossRef]
10. Emeiegwu, L.E.; Ada, W.; Alastair, H. The impacts of climate change on agriculture in sub-Saharan Africa: A spatial panel data approach. *World Dev.* **2022**, *158*, 105967. [CrossRef]
11. Abass, R.; Mensah, A.; Fosu-Mensah, B. The role of formal and informal institutions in smallholder agricultural adaptation: The case of Lawra and Nandom Districts, Ghana. *West Afr. J. Appl. Ecol.* **2018**, *26*, 56–72.
12. Tume, S.J.P.; Kimengsi, J.N. Indigenous and modern agro-based climate adaptation practices in rural Cameroon. *Int. J. Environ. Stud.* **2021**, *78*, 1057–1066. [CrossRef]
13. World Bank. Agriculture, Forestry, and Fishing, Value Added (% of GDP)—Sub-Saharan Africa. 2021. Available online: https://data.worldbank.org/indicator/NV.AGR.TOTL.ZS?locations=ZG (accessed on 17 December 2021).
14. Lawson, E.T.; Alare, R.S.; Salifu, A.R.Z. Dealing with climate change in semi-arid Ghana: Understanding intersectional perceptions and adaptation strategies of women farmers. *Geo J.* **2020**, *85*, 439–452. [CrossRef]
15. Opiyo, F.; Wasonga, O.V.; Nyangito, M.M.; Mureithi, S.M.; Obando, J.; Munang, R. Determinants of perceptions of climate change and adaptation among Turkana pastoralists in northwestern Kenya. *Clim. Dev.* **2016**, *8*, 179–189. [CrossRef]
16. Arunrat, N.; Wang, C.; Pumijumnong, N.; Sereenonchai, S.; Cai, W. Farmers' intention and decision to adapt to climate change: A case study in the Yom and Nan basins, Phichit province of Thailand. *J. Clean. Prod.* **2017**, *143*, 672–685. [CrossRef]
17. Phuong, L.T.H.; Biesbroek, G.R.; Sen, L.T.H.; Wals, A.E.J. Understanding smallholder farmers' capacity to respond to climate change in a coastal community in Central Vietnam. *Clim. Dev.* **2017**, *10*, 701–716. [CrossRef]
18. Fachrista, I.A.; Irham, M.; Suryantini, A. Livelihood resilience of vegetables farmers: Efficacy of organic farming in dealing with climate change in Java, Indonesia. *Appl. Ecol. Environ. Res.* **2019**, *17*, 11209–11232. [CrossRef]
19. Ngo, C.C.; Poortvliet, P.M.; Feindt, P.H. Drivers of flood and climate change risk perceptions and intention to adapt: An explorative survey in coastal and delta Vietnam. *J. Risk Res.* **2019**, *23*, 424–446. [CrossRef]
20. Gomez, N. Climate change and adaptation on selected crops in Southern Philippines. *Int. J. Clim. Change Strat. Manag.* **2015**, *7*, 290–305. [CrossRef]
21. Ngigi, M.W.; Mueller, U.; Birner, R. Gender differences in climate change adaptation strategies and participation in group-based approaches: An intra-household analysis from Rural Kenya. *Ecol. Econ.* **2017**, *138*, 99–108. [CrossRef]

22. Evengelista, P.; Young, N.; Burnett, J. How will climate change spatially affect agriculture production in Ethiopia? *Int. J. Devoted Descr. Causes Implic. Clim. Change* **2013**, *119*, 855–873.
23. Balgah, R.A.; Kimengsi, J.N. Building an indigenous agro-pastoral framework to climate change in Sub-Saharan Africa: Experiences from North West Region of Cameroon. *Procedia Environ. Sci.* **2015**, *29*, 126–127. [CrossRef]
24. Amawa, S.G.; Kimengsi, J.N. The implications of climate on market gardening in Santa Sub-Division, North West Region of Cameroon. *Environ. Nat. Resour. Res.* **2015**, *5*, 14–23. [CrossRef]
25. FAO. *The State of Food and Agriculture 2010–2011: Women in Agriculture, Closing the Gender Gap for Development*, 2011th ed.; Food and Agriculture Organization of the United Nations: Rome, Italy, 2011.
26. Peterman, A.; Behrman, J.A.; Quisumbing, A.R. A review of empirical evidence on gender differences in nonland agricultural inputs, technology, and services in developing countries. In *Gender in Agriculture. Closing the Knowledge Gap*; Quisumbing, A.R., Dick, R.M., Raney, T.L., Croppenstedt, A., Behrman, J.A., Peterman, A., Eds.; FAO: Rome, Italy; Springer Science: Washington, DC, USA, 2014.
27. Ravera, F.; Martín-López, B.; Pascual, U.; Drucker, A. The diversity of gendered adaptation strategies to climate change of Indian farmers: A feminist intersectional approach. *Ambio* **2016**, *45*, 335–351. [CrossRef] [PubMed]
28. Kaijser, A.; Kronsell, A. Climate change through the lens of intersectionality. *Environ. Politics* **2014**, *23*, 417–433. [CrossRef]
29. Sugden, F.; Maskey, N.; Clement, F.; Ramesh, V.; Philip, A.; Rai, A. Agrarian stress and climate change in the Eastern Gangetic plains: Gendered vulnerability in a stratified social formation. *Glob. Environ. Change* **2014**, *29*, 258–269. [CrossRef]
30. Fearon, J.D. Ethnic and Cultural Diversity by Country. *J. Econ. Growth* **2023**, *8*, 195–222. [CrossRef]
31. Santchou Council. *Santchou Council Development Plan*; PNDP: Santchou, Cameroun, 2015; p. 152.
32. FAO: Food and Agricultural Organization. *Crop Prospects and Food Situation—Quarterly Global Report No. 1*; FAO: Rome, Italy, 2022. [CrossRef]
33. Mckune, S.L.; Borresen, E.C.; Young, A.G.; Auria, T.D.; Russo, S.L.; Diao, A. Climate change through a gendered lens: Examining livestock holder food security. *Glob. Food Secur.* **2015**, *6*, 1–8. [CrossRef]
34. Mersha, A.A.; Van Laerhoven, F.A. Gender approach to understanding the differentiated impact of barriers to adaptation: Responses to climate change in rural Ethiopia. *Reg. Environ. Change* **2016**, *16*, 1701–1713. [CrossRef]
35. Stål, H.I.; Bonnedahl, K.J. Provision of climate advice as a mechanism for environmental governance in Swedish agriculture. *Environ. Policy Gov.* **2015**, *25*, 356–371. [CrossRef]

Disclaimer/Publisher's Note: The statements, opinions and data contained in all publications are solely those of the individual author(s) and contributor(s) and not of MDPI and/or the editor(s). MDPI and/or the editor(s) disclaim responsibility for any injury to people or property resulting from any ideas, methods, instructions or products referred to in the content.

MDPI AG
Grosspeteranlage 5
4052 Basel
Switzerland
Tel.: +41 61 683 77 34

Sustainability Editorial Office
E-mail: sustainability@mdpi.com
www.mdpi.com/journal/sustainability

Disclaimer/Publisher's Note: The title and front matter of this reprint are at the discretion of the Guest Editors. The publisher is not responsible for their content or any associated concerns. The statements, opinions and data contained in all individual articles are solely those of the individual Editors and contributors and not of MDPI. MDPI disclaims responsibility for any injury to people or property resulting from any ideas, methods, instructions or products referred to in the content.

www.ingramcontent.com/pod-product-compliance
Lightning Source LLC
Chambersburg PA
CBHW040751250926
51850CB00007BB/1466
* 9 7 8 3 7 2 5 8 7 5 9 3 1 *